Herbert A. Simon

Herbert A. Simon

The Bounds of Reason in Modern America

Hunter Crowther-Heyck

The Johns Hopkins University Press

Baltimore and London

© 2005 The Johns Hopkins University Press
All rights reserved. Published 2005
Printed in the United States of America on acid-free paper
9 8 7 6 5 4 3 2 1

The Johns Hopkins University Press
2715 North Charles Street
Baltimore, Maryland 21218-4363
www.press.jhu.edu

Library of Congress Cataloging-in-Publication Data
Crowther-Heyck, Hunter, 1968–
  Herbert A. Simon : the bounds of reason in modern America/
  Hunter Crowther-Heyck.
      p. cm.
  Includes bibliographical references and index.
  ISBN 0-8018-8025-4 (hardcover : alk. paper)
    1. Simon, Herbert Alexander, 1916–2001. Economists—United States—
Biography. 3. Social scientists—United States—Biography. 4. System
theory—History. 5. Information theory in the social sciences—History.
I. Title.
√ HB119.S47C76 2005
    330′.092—dc22   2004012459

A catalog record for this book is available from the British Library.

Frontispiece: Herbert Simon at chess. Carnegie Mellon University
Libraries.

For Mom, Dad, and Kathleen; with love, all things are possible.

Contents

Acknowledgments

Novelists and poets supposedly write best in the exquisite torment of solitude; whether that is true I cannot say. Fortunately, it is certainly not true of historians. At every step in the long process that brought this book into the world, I have been guided, supported, critiqued, challenged, encouraged (and sometimes funded) by warm hearts and sharp minds.

This project began, years ago, as a dissertation at the Johns Hopkins University in the Department of History of Science, Medicine, and Technology. There I had the great benefit of working with many outstanding scholars. Among the faculty, I owe the greatest debts to my adviser, Bill Leslie, whose patience with me, confidence in me, and sound advice for me all were crucial. He instructed me in much more than history. Dorothy Ross taught me almost everything I know about the history of the social sciences, and her seminar on history and social theory showed me how to read and think at another level. The third of my major professors at Johns Hopkins was Lou Galambos, who provided me with the invaluable lesson that a critical reading of a text should reveal not only what is wrong with it but also what is right. Constructive, creative synthesis was his goal, and it has become mine as well.

Other teachers left their marks on me also. The first seed of this project was sown in a seminar on metaphor in science taught by Dan Todes back in the fall of 1992, my first semester in graduate school. Dan's twin enthusiasms for the history of science and for good writing about it were infectious; I hope I have passed these benign contagions on to others. Harry Marks challenged me to meet ever higher standards, and Sharon Kingsland, Mary Fissell, Bob Kargon, and Gert Brieger all provided valuable comments, criticism, and advice on this project and others.

The graduate student community at Johns Hopkins was a big part of my education and a vital part of keeping me sane. Lloyd Ackert, Keith Barbera, Jesse Bump, Greg Downey, Trudy Eden, Sue Ferry, Carl-Henry Geschwind, Sandy Gliboff, Melody Herr, Christine Keiner, Scott Knowles, Tom Lassman,

David Munns, Buhm Soon Park, Shahana Sarkar, Karen Stupski, and Harry York all were (and are) good critics and even better friends.

Mike Sappol, Patricia Tuohy, and the exhibition staff at the National Library of Medicine provided comradeship and support, not to mention a paycheck, once I departed Johns Hopkins. My colleagues in the History of Science at the University of Oklahoma—Peter Barker, Kathleen Crowther-Heyck, Steven Livesey, Marilyn Ogilvie, Katherine Pandora, Jamil Ragep, Ken Taylor, and Stephen Weldon—have made OU a warm and welcoming place. The anonymous reviewers for the Johns Hopkins University Press provided invaluable advice on the manuscript as it evolved from dissertation to book, and Melody Herr, Amy Zezula, and Robert J. Brugger have given expert editorial advice.

Books do not get written without financial as well as intellectual support. In this regard, I thank the National Science Foundation for a four-year graduate study fellowship, the Department of History of Science, Medicine, and Technology at Johns Hopkins for stipends for both teaching and research, the U.S. Department of Education for student loans (and American taxpayers for supporting the loan program), the Rockefeller Foundation for a research travel grant, and the University of Oklahoma's Junior Faculty Summer Research Grant program for the grant that enabled me to complete this manuscript.

Archivists and librarians at the Carnegie Mellon University Archives, the Rockefeller Foundation Archive Center, the University of Chicago Archives, the Johns Hopkins University Special Collections, the Library of Congress, and the Harvard University Archives all provided much needed assistance in conducting my research. In particular, Gabrielle Michalek and Jennie Benford of the CMU Archives have gone out of their way to be helpful. Their assistance is greatly appreciated.

I also owe a deep debt to Herb Simon. Not only was he was generous in allowing me access to his papers, he was a thoughtful, helpful critic of my work. One simply has to admire the energy, commitment, and enthusiasm he brought to his work. I hope I have learned from him as well as about him.

My deepest debt of all is to my family. My sister, Shannon Heyck-Williams, is smart, fun, and committed to putting her talents to work in the world. It's a better, greener place because of her. The warmth, wisdom, and love of my parents, Bill and Deni Heyck, have guided me and sustained me from the beginning. The best thanks I can give them is to say that I hope that I can be for my child what they have been for me.

Words simply cannot convey how much my wife Kathleen has meant to me. She is my best friend, my best critic, and my only love. On top of that, she also happens to be a fine historian, combining deep knowledge of the history of science and medicine with careful reasoning and creative insight. Her touch is everywhere among these pages. Our son Max, who arrived during the final stages of this project, has brought more joy than I thought possible. I am grateful to them both.

(Un)bounded Rationality

On a cold day in early January 1956, Herbert Simon opened his class at Carnegie Tech's Graduate School of Industrial Administration with a startling announcement: "Over the Christmas holiday, Al Newell and I invented a thinking machine."[1] A machine that thinks! This was a stunning claim, one bordering on the alchemical. But perhaps even more remarkable than the statement were the speaker—a man with a doctorate in political science, not engineering or mathematics—and the context—a class in a school of industrial administration.

Simon's "thinking machine" was a computer program called the Logic Theorist. Devised in late 1955 and first run on a computer in August 1956, the Logic Theorist was taken by many to be an "existence proof" of artificial intelligence. It had been designed to prove the theorems of Bertrand Russell and Alfred North Whitehead's *Principia Mathematica,* using only their basic axioms and any theorems it could prove along the way. It was remarkably successful at doing so, even proving one theorem more elegantly than had Russell and Whitehead. (When informed of this by the irrepressible Simon, Lord Russell good-naturedly replied, "I am delighted to know that *Principia*

Mathematica can now be done by machinery. I [only] wish Whitehead and I had known of this possibility before we both wasted ten years doing it by hand.")[2]

The Logic Theorist was a remarkable accomplishment—and quite a revealing one as well. The story behind its creation tells us a great deal about the emergence of artificial intelligence as a field of research. It also illuminates an important set of changes in how behavioral scientists understood both science and mind, changes that made mind and machine nearly synonymous.

This book explains what Simon believed he had accomplished, why he thought it mattered, and how a political scientist interested in efficient public administration came to be a leader of the cognitive revolution in psychology and a founder of one of the world's top schools of computer science. In so doing, it aims not only to illuminate the career of one of the most influential scientists of the twentieth century but also to shed light on the world in which he lived and worked, a world he did much to shape.

Herbert Simon

Herbert Alexander Simon was born in Milwaukee, Wisconsin, on June 15, 1916. He died on February 9, 2001, at the age of eighty-four, and he was an active researcher and a vital part of Carnegie Mellon University (CMU) to the end. His was a varied and fascinating career.

Simon began as a political scientist, receiving his Ph.D. from the University of Chicago in 1943; in 1947, he published his dissertation on public administration as *Administrative Behavior.* By the late 1950s this book, along with his later work *Organizations,* had become staples in courses on business education, public administration, and organizational sociology.[3] *Administrative Behavior* is now in its fourth edition and is arguably one of the twentieth century's ten most influential works in political science, public administration, and management. *Organizations* likewise remains an essential text in sociology and management training.

He did not remain a political scientist for long, however. In the late 1940s, he began the work in econometrics for which he would receive the 1978 Nobel Prize in Economics. In the mid-1950s, he started researching the psychology of problem-solving; later, this work would earn him the American Psychological Association's highest award for lifetime achievement. Also in the mid-1950s, he wrote his first computer programs, starting down a path that would lead him and his colleague Allen Newell to receive the Association for Computing Machinery's Turing Award, that discipline's highest honor.

His career path straightened out after the mid-1960s, as he eased into an endowed chair at CMU in computer science and psychology. He still found time, however, to deepen his explorations into the philosophy of science, the theory of design, and sociobiology. His list of publications runs to more than eight hundred items, and if they had to be categorized by discipline, the fields would include (at the least): political science, public administration, and management; operations research, systems theory, organization theory, decision theory, and economics (including the theory of the firm, game theory, economic history, and econometrics); sociology, sociobiology, social psychology, and cognitive psychology; pure mathematics, philosophy, linguistics, and computer science. "Diverse" does not even begin to describe his intellectual interests or achievements.

Simon created institutions as well as ideas. He helped build Carnegie Tech's Graduate School of Industrial Administration (GSIA), which pioneered the introduction of concepts and methods from the behavioral sciences into management education. The ideas and techniques Simon and his colleagues advocated soon found their way into almost every business school, helping to create the modern master of business administration (MBA) degree.

He also led the transformation of CMU's Psychology Department from a second-tier program into one of the most influential in the nation, and he was one of the founders of its renowned Department (now School) of Computer Science. Before he came to Carnegie Tech (now CMU), the Psychology Department was largely unknown outside of Pittsburgh. Thanks in large part to Simon, however, it became a leader in the cognitive revolution of the 1960s and 1970s and in recent years has ranked first in citation impact in the discipline. Similarly, the CMU Department of Computer Science consistently has ranked in the top three nationally since its inception in the mid-1960s. The team that developed Deep Blue, the computer that defeated chess grandmaster Gary Kasparov in 1997, for example, contained several members trained at CMU.

On another level, Simon was active and influential in the world of science policy. He was a key adviser to Bernard Berelson of the Ford Foundation regarding its program in the behavioral sciences, an effective lobbyist for the behavioral sciences in the National Academy of Sciences, and an important member of its Committee on Science and Public Policy. He played an influential role as a member of the Social Sciences Research Council, serving on its Policy and Planning Committee and board of directors, including a three-year stint as chairman of the board. In addition, he was the first behavioral scientist to serve on the President's Science Advisory Committee. He used these positions to encourage both the behavioral revolution in social science and the inclusion of behavioral scientists among the nation's science policy elite.

As this litany of accomplishments indicates, Simon's influence extended across many fields and in many venues, from political science to computer science, from department meetings to executive councils. Indeed, when I first began to study Simon, it seemed as if the only constant in his career was success: so many fields, so many projects, so many papers. It seemed that the key questions to ask about him were why so many changes, why so many sudden movements? Those questions seemed too personal to have a larger meaning. He was simply a gifted mathematician with the soul of a pioneer. That explained everything, it seemed. His mathematical skills allowed him to contribute to a variety of fields, all desperately seeking mathematical rigor, and his pioneering urge meant that he was ready to move on to the next frontier as soon as he saw the smoke of a neighbor's fire. Simple enough.

No, too simple (fortunately). Simon did have the classic second child's need to be bold and different, to produce something bright and shiny and new, but that was not his deepest passion. His mathematical skills were remarkable, but his greatest contributions were concepts, not equations. As to the sudden changes in his course, the seemingly scattershot nature of his work? I will let him speak to that: "I am often complimented, sincerely I think, on the range of my dilettantism . . . the compliments were largely undeserved [however] . . . what appeared to be scatteration was really closer to monomania."[4]

Surely this was some *post hoc* rationalization, a reconstruction of his past to suit his future? Yes, it was that, but it was more than that as well. Despite the variety in his work that his drive for novelty produced, it is possible to find a pattern beneath the surface complexity. This pattern can be difficult to see because it does not map onto accepted disciplinary structures. When one stops trying to fit Simon into a mold defined by present disciplinary boundaries, however, one can see that Simon truly was remarkably consistent in his pursuit of a single set of tightly linked goals. He did redefine these goals over the course of his life, so the trajectory of his career is a curve, not a straight line, but it is a continuous curve, unbroken.

What was the shape of that curve? Where did it begin? Where did it end? The place to begin to answer these questions is at the end—not Simon's final destination, for that was unanticipated, but his goal, which was constant. This basic goal, the one that animated all Simon's work, was simple but broadly ambitious. It was nothing less than to bring the complex and chaotic world of human thought and action within the ambit of rational, empirical science.

The Sciences of Choice and Control

To Simon, the place to begin this quest was with the "atomic phenomena of human behavior in a social environment."[5] The most atomic of such phenomena, he believed, was the essential act of choice. If one could discover how and why the human social atom chose to do one thing rather than another— how the forces of heredity and personality and education and environment conspired to lead an individual to "take the road to the right or to the left"— if one could understand the choices that individuals made, then perhaps one could assemble a true science of human behavior.

When Simon was forming this goal in the 1930s and 1940s, there were two powerful, widespread, and apparently inimical ways of looking at the problem of choice. Proponents of each saw theirs as the true path to a grand synthesis and sought to incorporate elements of the other into it, usually without success. One way emphasized the freedom of choice of the individual social atom, seeing individuals as rational choosers intent on maximizing their own values. This approach to understanding human action was embraced by the rising vanguard of mathematical economics, especially by those at the Cowles Commission for Research in Economics in Chicago. This outlook drew heavily on physics and engineering as sources of ideas and inspiration, and efficiency was one of its central concepts. It was epitomized by game theory, neoclassical utility theory, and statistical decision theory—the *sciences of choice.*

The other basic approach to human behavior assumed that individuals were plastic creatures, shaped and molded by their social environment. The majority of the social sciences embraced this view, including behaviorist experimental psychology, sociology, social psychology, anthropology, and political science—the *sciences of control.* Biology, especially experimental physiology, was the model science for the sciences of control, and adaptation was a vital concept. This approach was epitomized by Harold Lasswell's and Charles Merriam's psychological studies of propaganda and power, by John B. Watson's radical behaviorism, and by Talcott Parsons's structural-functional sociology.[6]

Although the proponents of each saw their path as the high road to a unified behavioral science, Simon thought both lacked something essential that only the other could provide. Parsons, for example, called his social theory a "voluntaristic" theory of social action, but there seemed precious little room for individual choice in structural-functionalism. This did not sit well with a man who believed that "the only real certainty" in life was the necessity of accepting "the burden of personal ethical choice."[7]

At the same time, in the sciences of choice it appeared that the players chose every move in the game of life as if they had full knowledge and perfect reason. Simon thought these, quite simply, were "fantastic" assumptions.[8] As he wrote to a colleague, "we need a less God-like and more rat-like chooser."[9] The world was too complex for people even to approximate such supremely rational behavior.

How to create a unified science out of these two disparate models of human behavior? It was not an easy task, nor was the road straight. In the end, however, Simon's constant drive to harmonize all aspects of his work led him to develop an integrated worldview that brought together the sciences of choice and control in a nested set of models of science, humanity, and nature.

Simon did not start his career seeking to create this integrated, multilayered schema, however. Rather, he began with a few basic assumptions that he sought to elaborate, specify, and formalize. First, he believed that there was an order to nature, even human nature. Second, he assumed that this order was universal, meaning that the complex and the local was always a manifestation of the simple and the global. Third, he held that this order was accessible to humans through observation and reason, not revelation. Fourth, he never doubted that the human capacity for reason was both limited *and* meaningful.

Within the broad frame formed by these assumptions, Simon worked to develop focused theories about specific aspects of human behavior, particularly those aspects having to do with decision-making in organizations. These investigations led him to reexamine, refine, and formalize his assumptions. These more fully elaborated assumptions then formed the basis for connecting his research in a range of fields. Thus, while Simon's early projects were linked to each other mainly by their relation to the problem of choice and by a gut feeling that it all had to be connected somehow, his later works were related to each other intimately and explicitly. All his many projects became parts of an integrated program, and after the mid-1950s, he took each new step with the development of this comprehensive schema in mind.

This comprehensive schema, this nested set of models, represented Simon's *bureaucratic worldview*. I give it this name because Simon came to define mind and machine, organism and organization, individual and institution, all as highly specialized yet tightly integrated hierarchical systems, each locked in a continual struggle to adapt to its environment as best it could, given its limited powers. To him the mind and the computer were model bureaucracies, and a bureaucracy was a model mind.

Fittingly, this bureaucratization of Simon's world picture took place on multiple levels. It involved a redefinition of the world as a complex, hierarchical system; of the various sciences as the study of the subunits of that system; and of human science as the study of that class of complex systems characterized by purposeful, adaptive behavior. In keeping with this model of science, Simon redefined the sciences of choice and control as components of this new, higher-level, science of adaptive systems. The goal of this new science was the construction of formal models of human behavior, and its method was to develop programs that would enable one complex system—such as the computer—to simulate the behavior of another, such as the human mind. Thus, the Logic Theorist was not merely a novel computer program but a theory of how both science and mind worked.

The ideas of *system* and of *bounded rationality* were the pillars of this new outlook. To Simon, all the world was a system. The economy, the family, the individual organism, the cell, the atom all were complex, hierarchically structured systems. That they were systems meant that their component elements were strongly interdependent. That they were hierarchical meant that they had a tree-like structure and so were decomposable into subsystems, sub-subsystems and so on. That they were complex meant that the behavior of the system at one level of the hierarchy was difficult to predict from knowledge of the properties of the elements at lower levels.

This prefiguration of the world as system had important implications for the questions Simon asked and the methods he used to answer them. First, seeing the world as system focused Simon's attention on systemic properties, such as the organization of a system's components, the ways they communicated with each other, how that the system maintained equilibrium, and the means by which the overall system adapted to its environment.

Second, seeing the world as system encouraged a *behavioral-functional* mode of analysis. In Simon's view, individuals could be known only by their behaviors, and those behaviors could be known and identified only by their effects on the other elements of the system to which the individual belonged. This was as true for objects as it was for humans: to Simon, even such a seemingly natural and individual quality as mass actually was a property of an object in a particular system, not of the object itself.[10]

Behavioralism and functionalism also were products of a systems-based perspective in that both enabled the radical simplification of the analysis of phenomena by concentrating on the relational rather than the intrinsic, individual properties of system components. An individual performing a function is far easier to understand than an individual with a unique history and nature, and individuals can be analyzed in terms of their functions only if they are parts of systems.

Third, seeing the world as a system encouraged mathematical formalization. It did not lead inevitably to mathematical analysis (witness Talcott Parsons), but it did make the development of mathematical social science seem a natural next step. To Simon, there was no question that a reformed behavioral science would be mathematical, for mathematics was the essential "language of discovery" of science. He was fond of quoting Fourier's ode to mathematics: "Mathematics is as extensive as nature itself; it defines all perceptible relations, measures time, spaces, forces . . . Its chief attribute is clearness; it has no marks to express confused notions. It brings together phenomena the most diverse, and discovers the hidden analogies which unite them. It seems to be a faculty of the human mind destined to supplement the shortness of life and the imperfections of the senses."[11]

Simon believed that the kind of mathematical formalism appropriate to the modeling of complex adaptive systems, such as humans and the worlds they make, would *not* look like the mathematical formalism that had characterized past science, however. The new formalism would not be the system of differential equations that characterized classical physics or the system of stochastic equations that characterized quantum mechanics. Rather, the formalism peculiar to the description of the behavior of complex adaptive systems would have to mirror the hierarchical structure of those systems while still being able to describe coherent sequences of behavior. The formalism that did so was the *program*. Simon did not invent the concept of the program, but he was the leading exponent of the idea that the program was the essential formalism for the sciences that studied adaptive systems.

The second conceptual keystone of this bureaucratic worldview was Simon's trademark principle of bounded rationality. To him, human reason was bounded. These bounds are not set by the passions or the unconscious but by the inherent limits of the human organism as an information processor. Simply put, "the capacity of the human mind for formulating and solving complex problems is very small compared with the size of the problems whose solution is required for objectively rational behavior in the real world." As a result, the human actor must "construct a simplified model of the real situation in order to deal with it."[12] Humans behave rationally with regard to these simplified models, but such behavior does not even approximate objective rationality. Rational choice exists and is meaningful, but it is severely bounded.

We just cannot know enough—or process it fast enough—to be rational all by ourselves. We need help, and that help is provided by the organizations to which we belong: our families, our places of work, our political institutions. Indeed, in Simon's view, simplifying the problem-solving process is the main reason why we create organizations in the first place. Quite simply, "The behavior patterns which we call organizations are fundamental ... to the achievement of human rationality in any broad sense. The rational individual is, and must be, an organized and institutionalized individual."[13] The constraints that organizations impose on rationality thus are not Weberian "iron cages" of bureaucratic domination. Rather, they are what make rationality possible.

This principle of bounded rationality was Simon's basic building block in everything from public administration to economics to artificial intelligence. Though a seemingly simple concept, it had revolutionary implications in every one of these fields. In economics, for instance, it undercut the reigning assumption of profit-maximizing rationality embodied in the neoclassical economist's *homo economicus*. Similarly, the concept of bounded rationality played an important role in the "cognitive revolution" against strict behaviorism in experimental psychology, for it presumed that the mind was an active agent in constructing models of the world.

Bounded rationality also had deep implications for Simon's philosophy of science, for it suggested that the construction and testing of such simplified models was the essence of all thought, even scientific thought. An idea of the importance Simon attached to models and modeling can be gleaned from the titles of some of his books: *Models of Man, Models of Thought, Models of Discovery,* and *Models of Bounded Rationality.* He even titled his autobiography *Models of My Life.*

Bounded rationality also had implications for how Simon thought the scientific enterprise should be organized. It led him to advocate interdisciplinary research, for example, because disciplines could limit reason in unhealthy ways. While he did support the expansion of specialized research—social organizations, such as disciplines, were essential to rational problem-solving—he believed that such research had to be coordinated and synthesized else it would become sterile. Therefore, Simon worked to make his institutional home, Carnegie Tech's GSIA, into an interdisciplinary research center, and he consciously supported people and projects that crossed disciplinary boundaries.

Change and Continuity

Although it was a result of his lifelong struggle to unite the sciences of choice and control, Simon's embrace of this new bureaucratic worldview led him to redefine his most basic questions. Where he first had sought to understand how *social* organizations affect the choices made by their individual members, he later tried to discover how our *internal* mental organization affects the ways in which we, as individuals, solve problems.

With this redefinition of the problem of choice came a number of other shifts, each subtle but significant. These included a movement from organization theory to theories of complex information processing as the focus of his work; a corresponding shift from a more social to a more individual psychology; and a change from a focus on "decision-making" to an interest in "problem-solving." Simon also moved from a science whose models were embodied in texts and tested by interviews, surveys, and the observation of group behavior, preferably in field experiments, to a science whose models were designed and tested for internal consistency on computers and tested for external consistency through controlled laboratory experiments. Even the group Simon considered his peers underwent a dramatic alteration, from an interdisciplinary community of mathematical social scientists to a more firmly bounded community of cognitive scientists and AI researchers.

Despite these transformations, there was as much continuity as change in Simon's work. It always was characterized by his relentless pursuit of synthesis, by his steadfast faith in the power of reason, especially organized reason, and by his lasting concern that abstraction be wedded to application. He was ever an evangelist for the gospel of reason and always a prophet of the secular creed: humanity is sacred, but not divine.

In these qualities, as in so many things, Simon was both unique and representative. He was unique in the degree to which his science took as its subject the very transformations in science and society that shaped his outlook and in the degree of consistency he demanded from his philosophical, theoretical, institutional, and practical agendas. He was unusual in the breadth of his ambition and astonishing in the reach of his mind; he was remarkable in his self-confidence and extraordinary in his zeal. The salient characteristics of his work, however—his focus on system properties, his ideas regarding the hierarchical structure of complex systems, his drive for a synthesis of choice and control, his behavioral-functional mode of analysis, his emphasis on interdisciplinary work, his adoption of computer modeling and simulation, and his fascination with strategies, programs, and plans—did come to characterize work across a wide range of fields in the 1950s, 1960s, and 1970s, and their legacy continues today.

If we want to understand the postwar transformation of the sciences, especially the behavioral sciences, we need to understand Herbert Simon. And if we want to understand Herbert Simon, we must start with his life in the community that first shaped him and his values: the Simon family, of German heritage but firmly rooted in the new world of Milwaukee, Wisconsin.

CHAPTER ONE

The Garden of Forking Paths

In Herbert Simon's autobiography there is a chapter titled "Mazes Without Minotaurs." In this chapter, Simon describes a conversation he had in 1970 with Jorge Luis Borges and presents Simon's own short story entitled "The Apple: A Story of a Maze."[1] The story, written in 1956, was inspired by Simon's paper of that year titled "Rational Choice and the Structure of the Environment."[2] Both the conversation and the story are revealing. They are linked by a common theme: the labyrinth or maze as a metaphor for life. It is a metaphor that has multiple meanings for the understanding of Simon's life.

Simon had requested to meet with the Argentine writer because of his fascination with Borges's stories "The Library of Babel" and "The Garden of Forking Paths," both of which depict worlds of ceaselessly branching ways.[3] In the course of their conversation, Simon asked if Borges had begun that story with an abstraction, an idea he instantiated in the tale. Borges replied:

> Not true! I can tell you how this story spewed out. I worked in a small public library on the west side of Buenos Aires. I worked nine years in this library with a miserable wage, and the people who worked there were very disagreeable. They were ignorant people, stupid really.

And this gave me nightmares.

One day I said to myself that my entire life was buried in this library. Why not invent a universe represented by an interminable library? A library where one can find all the books that have been written ... The concept of this library evokes in me my deepest, most intrinsic pleasures ... One *feels* this kind of bliss.[4]

From the horrible experience of the mundane library in which he worked, Borges created a library of infinite wonder. Yet the Library of Babel did reveal its roots: there is a sense of confinement and exhaustion in this labyrinth despite all the wonders it holds. The walls, heavy with musty books, close in on the reader; his steps falter as he contemplates the unending turnings that lie ahead, the choices he must make without a compass, without knowledge of the final destination. Even the library's books are "formless and chaotic," and "for every sensible line of straightforward statement, there are leagues of senseless cacophonies, verbal jumbles and incoherencies."[5] (As a result, a great many of the librarians have committed suicide.)[6] Borges is aware that what makes the library wonderful also makes it terrible; irony is an old friend in his journey through the maze.

Simon's fascination with mazes ran in rather the opposite direction. He turned inward from the infinite complexities of the world, seeking the rules that governed the generation of the complex from the simple. Where Borges began with a real library in Buenos Aires, Simon began with a theory. Where Borges reveled in the experience of journeying through the labyrinth, of seeking heaven, though "my place be in hell," Simon wanted to know what rule governs how we decide to take one path instead of another.

Indeed, the world of "The Apple" and "The Library of Babel" are almost perfect inversions of each other: Borges's library—his world—is full of unintelligible books. It is a catalog of all the possible random permutations of letters and spaces and punctuation marks. There is no order underlying this world, save the order of chance. Simon's world is far more orderly. Although we find our way through it as much by trial and error as by plan, there is an order for us to discover. Thus, while Borges's is an ordered search through a chaotic world, Simon's is a heuristic search through an ordered, if complicated, world.

In "The Apple," Hugo (an "ordinary man") lives in a "castle with innumerable rooms. Since the rooms were windowless, and since he had lived there since his birth, the castle was the only world he knew." Hugo lives alone in this vast castle, but he "had become so accustomed to his solitary life that he was not bothered by loneliness."[7] Hugo spends his days wandering from room to room, looking at the murals on the wall, daydreaming in comfortable chairs, and eating the food that he finds laid out for him. Significantly, he finds food only in some rooms, not in all; when hungry he often has to search for some hours until he finds a room with food prepared for him.

The rooms have glass doors that prevent him from going back the way he had come, indicating that time has an arrow, even in this infinite castle. Hugo does not waste much time trying to pry the doors open, nor does he linger at them, casting wistful glances back at the past. Indeed, whereas the labyrinth in "The Garden of Forking Paths" *is* time, Hugo's castle is strangely timeless; the only evidence of time's passage is that eventually Hugo discovers he prefers some foods and some murals to others. He then conducts his travels through the castle with a sense of purpose, seeking his favorite pleasures. He attempts to find patterns in the castle's arrangements, trying to see if certain kinds of murals presage the kinds of foods he desires, and he keeps a notebook to record the results of his tests of these hypotheses. (One can imagine Hugo crying "Hypothesis!" whenever he discerned a correlation between a mural and a meal, as Simon was wont to do.) This search proves exhausting, however, and he cannot always find what he wants before he must eat again.

> Only this distinguished his new life from that of his boyhood: then he had never been pressed for time, and his leisure had never been interrupted by thoughts of uncompleted tasks. What should he do from moment to moment had presented no problem. The periodic feelings of hunger and fatigue, and the sight of a distant dining room had been his only guides to purposeful activity.
>
> Now he felt the burden of choice . . . he realized that he would never again be free from care.[8]

Hugo had in his possession a Bible, and Simon ends the story with Hugo reading from it: "and when the woman saw that the tree was good for food, and that it was pleasant to the eyes . . ." The meaning of the story's title, "The Apple," is thus made clear: knowledge and choice are the twin sources of both our humanity and our fall.

This is a curious tale to tell of one's life: a solitary journey through a world in which all needs, but not all wants, are provided for and in which the only purpose is the pursuit of pleasures—a pursuit that is as random as it is guided by knowledge of the world. The story is an allegory to be sure, one intended to illustrate Simon's theory that the human actor is a simple creature motivated by a few basic drives that attempts to achieve its goals largely by trial and error, guided only by rough heuristic lessons drawn from experience. Borges's library is an allegory as well, but whereas Simon depicts a strangely ordered world and a stark, simple person, Borges portrays a chaotic world and a character rich with wonder—and fear.

Was "The Apple" how Simon understood his own life? At the end of the story, he claims that it is: "My own conjecture is that Hugo found a meaning [in those words from the Bible] not very different from the one I have arrived at, journeying through the maze of my own life. If it were not so, my experience would have falsified my theory, the model from which 'The Apple,' was drawn."[9] Indeed, Simon consistently characterized himself as a simple man, one driven by an intellectual "monomania" despite his multiple disciplinary affiliations, one who would not mind if his whole career was described as a gloss on one paragraph from his dissertation, one who saw the choices he made in his journey through the maze as obvious responses to circumstances.[10] But was he truly such a simple man? Was he but a small bundle of drives and heuristics not terribly different from the other rats in the maze?

The answer is both yes and no. There was a simple, basic core to Simon that remained constant through the years: he was an independent-minded seeker of patterns in human behavior who had faith that such patterns could be found and understood. But he was a complex, changing person as well, and there were many ironies and tensions in him. He was in many ways a kind of evangelist, but he preached a secular, relativist truth. He was an extremely rational man who studied the processes of rational decision-making, but he discovered early on that "reason" does not always (or even often) prevail. He was an independent spirit who worked well on teams, an outsider who was an avid organizational politician, a believer in individual choice and in learning for oneself who emphasized the importance of forces outside the individual in shaping thoughts and actions.

Borges said that he perceived life as "a continual amazement; a continual bifurcation of the labyrinth."[11] Simon understood this wonder, this amazement, when contemplating the simple man exploring the infinite turnings of the world. Borges's saying also holds true when one explores the infinite branchings of the self, even as "simple" a self as Herbert Simon.

Family Values, German Style

Herbert Alexander Simon was born in Milwaukee, Wisconsin, on June 15, 1916, to Edna and Arthur Simon. Both Edna and Arthur were of German descent. Edna Simon (formerly Merkel) was the granddaughter of German immigrants who had fled the turmoil and reaction that followed the abortive revolutions of 1848. Arthur Simon had come to the United States in 1903, shortly after graduating from the Technische Hochschule in Darmstadt with a degree in electrical engineering. Edna Simon's family was admirably ecumenical in its background: of the three great-grandparents Simon lists in his autobiography, one was Jewish, one was Lutheran, and one was Catholic. Arthur Simon was Jewish; Herbert Simon speculates that he may have emigrated because of the rising anti-Semitism in Germany at the turn of the century, noting that his father had challenged a classmate to a duel in 1899, presumably in the course of an argument over the Dreyfus affair.[12]

Though thoroughly German in his professional demeanor and personal habits, Arthur Simon was given neither to nostalgia nor to hostility toward Germany. Edna Simon's views on her heritage are not documented, but one suspects she shared a similarly benign regard, but not reverence, for her ancestral home. She did marry a man with a thick German accent, after all, and they did choose to live in Milwaukee, perhaps the most German city west of the Rhine.

In addition to his mother and father, Simon's close kin included his older brother Clarence ("usually protective," later a lawyer, not mentioned more than three times in Simon's autobiography), his uncle Harold Merkel (studied economics with John Commons at Wisconsin, died young at age 30, left his books and Phi Beta Kappa key to young Herbert), and his Grandmother Merkel. Grandmother Merkel was a source of some friction in the Simon household, especially after she came to live with her daughter's family after her husband's passing.[13] (Apparently she provoked numerous arguments between Edna and Arthur.)

Simon describes his mother as warm and caring but "more than a little neurotic."[14] She plays a small role in his autobiography, one that dwindles to nothing as Simon leaves his childhood years behind. It is hard to know whether this small role is fact or artifact of his recollection, but it is clear that the adult Herbert Simon avidly sought to find connections between his own life and work and that of his father, not his mother. Indeed, Simon's pride in his father and his desire to establish an intellectual linkage to him are almost palpable throughout *Models of My Life*. At one point Simon writes: "It gradually dawned on me that the path I was following through the professional maze was returning me to the paternal calling, and not only because I had chosen to teach at an engineering school. As a designer of control gear, my father had been a significant contributor to the development of feedback devices. Now [in 1948] I was beginning to think of feedback theory as a tool for modeling the dynamic behavior of economic systems and organizations."[15] Simon emphasizes that he was "moved deeply" and filled with "great pleasure" at this discovery. Although such statements may tell us more about Simon's desires late in his life than about the extent of his father's actual intellectual influence, they are in keeping with much other evidence: Simon always was a "closet engineer" interested in the principles underlying the workings of machines, especially that most marvelous machine, the human mind.[16]

Arthur Simon worked at Cutler-Hammer Manufacturing Company in Milwaukee for most of his professional career. Though he later added a license to practice patent law, he thought of himself primarily as an engineer. He was skilled at his craft, and he pursued it with the intellectual zeal appropriate to a true professional, seeking to master the theory (Charles Steinmetz, Oliver Heaviside) behind the practice of electrical engineering. He does not seem to have followed the engineer's typical path of promotion, however, never leaving the engineering of machines for the management of men.[17] As a result, he never made himself a fortune. The Simons thus were solid members of the professional middle class, but they were far from wealthy.

Arthur Simon adhered to the German professional ideal socially and intellectually, which is to say that he was broadly cultured and thus no mere specialist, that he was active in community affairs but nonpartisan and nonideological, that he had a strongly secular outlook, and that he moved in a masculine world. He was active in his professional capacity as a member of the Milwaukee Professional Men's Society, for example, but he had no interest in ethnically or religiously based social groups, nor was he invited to join the elite society of the Social Register. His friends largely were other professional men, generally of German heritage, with whom he discussed business, politics, and matters intellectual. At the dinner table (when the boys were old enough) or with company, there was the men's conversation and the women's conversation, usually taking place in entirely separate spheres.[18]

Herbert Simon adopted many of his father's attitudes and values. He was extremely active in professional societies, for instance, and devoted much time to public service. Similarly, Simon always was leery of ethnically, religiously, or ideologically based groups. In fact, one could describe much of his career as a kind of crusade against the parochial, a crusade oriented around discovering why people join such groups despite the irrational behaviors they induce. In a revealing aside, when describing his paternal grandmother, Simon writes that she was a "kind and genial woman, not at all religious," unlike his troublesome grandmother Merkel.[19]

Also like his father, Simon sought to be broadly cultured as well as highly trained, and he was ever ready to display his prowess in history or philosophy or foreign languages. He claimed, for example, to have reasonable facility in reading twenty languages. (I, for one, did not have the capacity to test him, though I have seen evidence for at least seven and have no reason to doubt the others.) He believed that every member of a university faculty should be able to teach an introductory course in any discipline and certainly was capable of doing so himself. In a similar vein, when the topic of the role of the expert in society came up in one of our conversations, Simon was quick to demonstrate that he was well versed in the long history of debates about the subject, from Plato to the present.[20] Simon also was proud of his mathematical sophistication: when in the course of an interview I inadvertently implied that some of his mathematical papers from the 1950s were less original in the math they employed than in the areas to which they applied it, he made sure to let me know that "this [the math] was cutting edge stuff. I was making a lot of it up as I went along."[21] Finally, though Simon always described his work as "great fun," he approached it with the serious commitment of one who viewed a profession as a secular calling: eighty- to one-hundred-hour weeks were more the rule than the exception for Simon in his prime.

One area of change from father to son had to do with the assumption that professional life was necessarily masculine. Simon did work in fields that were almost exclusively male until the 1970s, and the household division of labor in the Simon family was quite traditional, but he does not appear to have doubted that women were capable of high intellectual endeavor. For example, Simon's wife, Dorothea, was a fellow graduate student in political science at the University of Chicago in the 1930s (she received a master's degree), and he clearly respected her mind as much as he liked her smile, co-authoring several papers with her. In his autobiography Simon notes—with more than a hint of sadness—that his father probably never thought of a wife as a potential source of intellectual companionship, an aspect of Simon's own marriage he clearly enjoyed. Similarly, Simon took women on as graduate students once they began to enter the cognitive sciences, and the ones I have spoken with describe him as a wonderful teacher and mentor, someone who clearly was not prejudiced against them.

Similarly, while his science certainly was gendered, it was not sexist. Indeed, from an early twenty-first-century perspective, one of the striking things about Simon is how gender-neutral his language usually was, even in private correspondence. This neutrality is particularly noticeable when one compares his language with that of many of his contemporaries in psychology (let alone psychiatry).[22] If he took "man" to be his subject, it was because he assumed that the minds of men and women operated according to the same universal mechanisms. To him, the differences between men and women were only interesting scientifically if they pointed the way to larger commonalities. In short, he was typical of a generation of liberal men who were taken rather by surprise by the need for a women's movement in the late 1960s but who were friendly to it so long as it focused on equality of opportunity rather than equality of result.

Perhaps the most important aspects of Arthur Simon's legacy for his son were his attitudes toward personal identity and his "unmovable honesty." For Arthur Simon, a person was defined by what he did for a living, not by his birth or religion. One's identity was a product of one's membership in a professional group and one's relationships with family and close friends, not one's ethnic, religious, or political affiliation. Though Jewish, he allowed his sons to go to the local Congregational Sunday school that their friends attended; though German, he spoke English at home and disparaged ultranationalist "two hundred per centers" of all nationalities, a sentiment that grew in strength during the world wars.[23] Similarly, in the Simon home, "Blacks were Negroes (in the 1920s and 30s), not 'niggers,' the Italians were Italians, not 'wops,' and the Poles were Poles, not 'Polacks.'"[24]

Simon shared these views of his father, writing in 1994 that "I acquired [from him] a deep distrust for the claims of superiority that any ethnic group, mine or another, might make ... There is no place in my imagined future world for a 'chosen people.'"[25] Also like his father, Simon came to associate nationalism and ethnocentrism with a particularly bloodthirsty brand of irrationality (Nazism), and he adopted a strongly secularized view of professional life. Even more, he made nationalisms, large and small, and professional life the focus of much of his research. His first books, from *Administrative Behavior* to *Public Administration* to *Organizations,* all explored the power of organizational identification and professional training in constructing identities and influencing perceptions.

Profession

Herbert Simon both studied and exemplified the rise and transformation of the professions in the twentieth century. The professions had changed greatly in the decades before Simon's birth, as a host of new professions emerged to join the classical set of law, medicine, the military, and the clergy. The new professionals created roles for themselves in the large-scale organizations that loomed so large over the fin de siécle social landscape. They could be found in the modern business enterprise, working for the railroads, the great industrial corporations, the investment banking houses, and the consulting industries that the corporations spawned. They staffed the new regulatory agencies and independent commissions that sprang up at every level of government, from city planning commissions to state boards of health to federal agencies such as the Interstate Commerce Commission and Food and Drug Administration. They held forth in the halls of the new research universities and the land grant agricultural and technical schools, and they took over the direction of the elementary and secondary schools as well. They wrote the columns and the editorials of a news industry newly interested in "truth" and "objectivity," and they formed the officer corps of a reorganized military. Accountants, engineers, schedulers, planners, market researchers, economists, journalists, chemists, physicians, psychologists: everywhere seeking rationality, efficiency, and objectivity; everywhere attempting to build "the one best system."[26]

The new professionals, like the old, staked their claims to a place of importance on the basis of their mastery of a body of knowledge and their ability to apply that knowledge to particular cases in the real world. (Hence the ubiquity of the "problem set" in the training of professionals.) Though influenced by the traditions of the classical professions, their education differed in several ways. First, the "best" kind of knowledge now was the abstract knowledge of the natural sciences, as opposed to knowledge gained through practical experience, revelation, or the study of ancient texts. Second, the new professionals were trained in specialized professional schools, usually attached to universities, rather than through apprenticeship. These two factors combined to produce an emphasis on original research as an important part of professional life.

Even the classical professions experienced a like shift from a "shop" to a "school" professional culture in the late nineteenth and early twentieth centuries: witness the turn to laboratory science in medicine, the movement toward the standardization and academization of legal training, and the transformation of military education and organization.[27] The rise of a new school culture can be seen as well in that most ancient of professions, the clergy, as educational requirements for clerics were stiffened, an accommodation with evolutionary science was sought, and scientific Biblical scholarship was pursued. Of course, it long had been customary for many members of the clergy to be school-trained, but in nineteenth-century America educational attainments of clergy varied widely both within and across denominations. Although Congregationalist, Unitarian, and Episcopalian ministers, for example, tended to be quite highly educated, even their training was not systematized before the turn of the century, and fervent personal faith rather than book learning was the chief credential for many Southern Baptist preachers well into the twentieth century. There was much variation in the training of priests before the turn of the century in even the primal large-scale organization, the Catholic Church.[28]

Perhaps most strikingly, new professionals adopted a strongly secular vision of professional life, a vision that Simon and his father shared. This "secular trend," as it were, was central to the changing relationship of the professional to society and the changing motivations of individual professionals, such as Herbert Simon. As Dorothy Ross has shown in her biography of G. Stanley Hall, *The Psychologist as Prophet,* the newly secular professionals of the late nineteenth and early twentieth centuries possessed the same world-shaping zeal of their (truly) evangelical predecessors; only, in them this drive to remake man and society was based on a faith in science, not scripture; reason, not revelation.[29]

This secular gospel of science and reason inspired many with the kind of passion one associates with a deep religious faith. The scientific sublime could be as powerful a dream as the Puritan city on a hill. For Herbert Simon, solving the puzzle of human behavior was not only a joy but also a calling worthy of his every effort.

In keeping with this sublimation of religious fervor into scientific ardor, by the early twentieth century, professional culture was increasingly Christian in a social rather than a spiritual sense. That is, it was characterized by a common set of ethical and social values rather than by a shared set of religious beliefs. This set of values is difficult to specify precisely, but secular professionals could be identified by the high value they placed on objectivity, order, reason, and public service, and by the importance they ascribed to the pursuit of knowledge, to progress, and to self-control. It generally was assumed (without needing to be said) that those who shared these values would be middle and upper-class Protestant males. Jews, atheists, Catholics, and women were not unknown among the professions, but they were rare—and still more rarely welcomed.

As David Hollinger has written, the influx of refugee intellectuals in the 1930s, the avid pursuit of higher education by second- and third-generation American Jews, and the general reaction in professional circles against the horrors of Nazi anti-Semitism combined in the middle third of the twentieth century to make the professions increasingly tolerant of those who shared their values but not their pedigree.[30] Herbert Simon was part of this first generation of those of Jewish heritage who were able to win their way into the professions and the academy. He, like many others of that generation, had an aggressively secular value-system that meshed well with that of the patrician leadership of the professions.

Though Simon was an extraordinarily calculating man, one who often decided in advance whether to have a fit of anger, his adherence to this set of values was not a matter of "passing." His secular, scientific values came well before he was old enough to make such calculating career decisions. For example, while still in middle school, Simon wrote a letter to the editor of the *Milwaukee Journal* defending the civil liberties of atheists, and by high school he was "certain" that he was "religiously an atheist," a conviction that never wavered.[31] Simon later joined a Unitarian church (in 1949) but only after he had "made it"—and at a time when his religious affiliation meant less and less to his professional colleagues. (Postwar Unitarianism would not have scored many points with traditional Christians in any event.)

Perhaps the best exposition of Simon's views on religion can be seen in his response to a letter from an old friend who had become a Dominican monk. Though written well after his boyhood, it is one of the earliest documents in Simon's collection, and it represents an enduring aspect of Simon's personality. His friend wrote that he believed that science cannot answer the important questions; only faith can do so. Simon replied:

> The possibility that we will come to an understanding on these matters seems very remote to me—unless, of course, your reason finally destroys the faith that you have embraced. My diagnosis of theology in general and Catholicism in particular is very similar to your diagnosis of "modernism"—that it is a symptom of a deep-seated neurosis that warps the reason to the demands of subconscious drives ... Moreover, I have grave doubts that submission to the irrationalism of Catholic doctrine, or to the undemocratic authority of the Roman church is a socially desirable remedy for wounds in the modern mind, if such wounds exist. Perhaps what the modern world really needs are human beings sufficiently unneurotic so as not to require the spurious props of fallacious Aristotelian "proofs," and an external divine and human authority to relieve them of the task of ethical choice—a task that seems to me implicit in the very definition of human dignity. It seems to me more than coincidence that individuals who, like yourself, have fled to Catholicism, are frequently the same ones who previously showed need of the same kind of external reassurance by embracing Stalinism, Trotskyism, or what not, and who rationalized these other isms with the same energy that you now apply to Catholicism.[32]

This letter reveals the keystones of Simon's secular convictions: faith and reason are opposed, and reason is to be preferred to faith; faith and ideology are both irrational products of an unwillingness to ask hard questions and to make tough choices; and individual ethical choice is central to human dignity. These are the articles of belief of a secular evangelist, a preacher to the "tough-minded."

Though strongly held, these beliefs were not uncontested or unmixed. While Simon believed in the importance of personal choice, he also believed in the power of external influences on the individual's thoughts and actions. At the same time that he subscribed to the power of reason, he also understood that there were severe limits to human rationality.

These tensions between individual will and outside forces and between reason and its limits surfaced time and again in Simon's work. His persistent interest in both decision-making and causality, for instance, stemmed in large part from his concern with this ancient problem, for Simon often wondered how his actions could be free if they have been *caused* by something outside himself:

This is the form in which I conceive free will: It resides in the fact that I am that which acts when I take a given action. And the fact that something has caused this behavior in no manner makes me (the I who acts) unfree.

So, when we reach a bifurcation in the road, of the labyrinth, "something" chooses which branch to take. And the reason for my researches, and the reason why labyrinths have fascinated me, has been my desire to observe people as they encounter bifurcations and try to understand why they take the road to the right or to the left.[33]

There were tensions as well in Simon's responses to his ethnic—as opposed to religious—Jewish heritage. He devoted but a page of his autobiography to a section on being Jewish—roughly the same amount of space he devoted to his discovery that he was color-blind. (Both are put forward as examples of "Being Different.") Yet Simon was a fierce hawk regarding U.S. involvement in World War II, and he did not "feel comfortable visiting Germany" until twenty years after the war.[34]

Even more revealing is a short piece he wrote late in life titled, "What It Means to Me to Be Jewish." In this article, written for a book titled *Jewish: Does It Make a Difference?*, Simon notes the ambiguities—and the perils—of ethnic identity. He observes that because the Jewish tradition is matrilineal, by Jewish custom he is not Jewish. By the standards of the anti-Semitic, however, he is more than Jew enough to qualify. He also states that while he believes himself to be a member of but one race, the human race, he nevertheless felt a surge of pride when "'we' whipped the Arabs at the time of partition."[35] He acknowledges this pride to be another example of the bounded rationality of all humans, a part of our nature that must be understood in order for it to be productive rather than deadly.

Arthur Simon's last legacy to his son is one that all parents would like to leave their children: an admiration for their integrity. Simon greatly admired his father in many ways, but of all the things he could hope to inherit, he chose his father's "unmovable honesty" as the most important.[36] It appears that he did: the impression one receives on reading Simon's letters is truly that of an honest man—one who was sometimes fiercely so. Though he could be charming and diplomatic—or polemical—at need, Simon did not say one thing and do another, nor did he take credit that was not his due. As in the case of his partnership with Allen Newell, Simon often took great pains to emphasize the contributions of his co-workers. He insisted, for example, that the papers he and Newell co-authored have their names listed alphabetically so that he would not be mistaken for being the primary author of a collaborative work.[37] Similarly, though Simon was no stranger to conflict or controversy—in fact, he often sought them out—the vast majority of his blows hit above the belt. His opponents frequently felt that he hit harder and more often than was strictly necessary—Simon could get into "wonderful rages," which he rather enjoyed—but he fought fairly, and he usually liked it when people fought back.[38] In sports parlance, he would be described as a "tough" and "physical" player but not a "dirty" one.

Schooling, Learning, and Independence

Herbert Simon was a proud product of Milwaukee's public schools, and an even more proud product of his own self-education. This independent attitude toward education is important to understanding Simon. He liked to do well by the public standards of tests and grades (and Nobel Prizes), but he preferred to achieve such goals on his own, in his own way. As a youth, he always enjoyed doing his own reading in the public library on school subjects (or other things), and he had great confidence in his ability to master any subject by himself. This appealed to his curiosity and his vanity—he was quite proud of being "smarter than his classmates"—and it prepared him well for the unstructured education he later would receive as both undergraduate and graduate at the University of Chicago under the "old New Plan" of the mid-1930s.[39]

This admiration of independence became a strong part of Simon's value system and stayed with him throughout his career. One of the reasons he was such a strong advocate for interdisciplinary work, for example, was his belief that too strong an identification with any one discipline could limit a person's intellectual independence. "If you see any one of these doctrines dominating you join the opposition and you fight it for a while . . . Most people want more structure in their lives than is good for innovation in the field."[40] In addition, Simon's high valuation of independence fostered his desire to be an iconoclast. Few things prove one's independence of mind better than smashing the idols of the "received view," after all.

The importance Simon attached to independence of mind could conflict at times with his evangelical drive. He wanted others to see the light of mathematical social science, but he wanted them to travel the road to Damascus (largely) on their own. Typically, he resolved this potential conflict through the belief that an unfettered mind naturally would come to accept the proper view (i.e., his own) and that persistent opposition was the result of an unwillingness to think clearly. Such opposition, as a result, tended to evoke a steamroller-like response from Simon that was quite unlike his collegial response to those who shared his basic assumptions but disagreed on technicalities. As the political scientist Dwight Waldo wrote after an exchange with Simon, "Professor Simon seems to me that rare individual in our secular age, a man of deep faith. His convictions are monolithic and massive. His toleration of heresy and sin is nil. The Road to Salvation is straight, narrow, one-way, and privately owned."[41]

This tension would emerge most frequently in Simon's attempts to build a research program at Carnegie Tech's Graduate School of Industrial Administration (GSIA). Simon and his ally, G. L. Bach, the dean of the GSIA, had a clear vision for the school's research program. Its economists would play an important role in realizing this vision, but for them to do so, they needed to adopt Simon's understanding of the psychology of economic behavior. As he later wrote, he was "prepared to preach the heresies of bounded rationality to economists, from the gospel of *Administrative Behavior,* Chapter 5, in season and out."[42] The economists did not enjoy being told how to do economics, and, possessing their own expansive creed, they were intent on winning converts as well. For several years, Simon strove to get the economists in line without dictating to them, but he had only limited success in restraining his evangelistic impulses.

The value Simon placed on independence was closely related to his genuine relish at being an outsider and iconoclast. Like a prophet or missionary, he saw himself as standing apart from the crowd, and his letters are full of evangelical metaphors. For instance, he often referred to his "missionary work" in bringing mathematics to the "primitive" regions of political science, and he spoke of his "hopes for your conversion" [to positivism] to Dwight Waldo, a fellow political scientist.[43] Similarly, he writes in his autobiography that he felt like "the young Jesus in the Temple" when addressing a group at the Maxwell School of Public Affairs.[44]

It may seem odd to characterize such a skilled organizational politician and institution builder as an outsider, and there was indeed a tension between Simon's outsider and insider selves. This was not an irresolvable tension, however. One good example of how Simon blended these two sides was his decision to stay at Carnegie Mellon despite lucrative offers from the University of Chicago and MIT in the 1950s and Harvard in the 1960s. At CMU, Simon could be the insider, the campus power broker, while still playing the role of the outsider struggling to disturb the placid equilibrium of academia's powers that be. As he put it in an interview with Pamela McCorduck, "there's a little extra fun of sort of building up a place and just doing it and being able to thumb your noses at the prestigious schools and say who the hell needs you?"[45] There also was a little extra fun, undoubtedly, in being both faculty member and trustee for life at Carnegie Mellon.

His independent spirit and outsider values combined to produce a taste for insurrection. Simon thoroughly enjoyed the role of the destroyer of ancient shibboleths, of the innocent who observes that the emperor has no clothes. Examples of this penchant for provocation are legion, from his high school debates where he "always took the unpopular side" to his controversial attack on the "Proverbs of Administration," to his (in)famous 1957 speech in which he predicted that within ten years a computer would be world chess champion, make scientific discoveries, and even compose aesthetically pleasing music.[46] It is only fitting that the young Simon's church youth group in Milwaukee should have called themselves (proudly), "The Heretics."[47]

In a similar fashion, Simon thoroughly enjoyed the role of the "Young Turk" in a disciplinary revolt. From the "behavioral revolution" in political science to the "cognitive revolution" in psychology to the inherently revolutionary enterprise of artificial intelligence, Simon always took up the mantle of the challenger. Indeed, he often appeared to lose some of his interest in a field once his position in it was no longer controversial.

As a result of this taste for insurrection, Simon was no stranger to disputes and engaged in many polemical exchanges both in private and in public. Simon rarely let a disparaging comment on his work go unchallenged, and it quickly became known that one had better come well armed to any debate with Simon. It is not for nothing that he titled a chapter of his autobiography "On Being Argumentative."[48]

To be a revolutionary instead of merely an iconoclast, however, one must do more than topple idols. One must offer something *new,* and Simon strongly desired to produce something novel. One can see this desire in his youthful disappointment at not discovering any new insect species despite his many hours of work at the local natural history museum.[49] One can see it as well in his adult decision not to pursue physics despite excelling in it in high school—

to Simon it was clear that "physics was a science that was finished."[50] The social sciences, on the other hand, seemed to him to be uncharted territory. The one to map it could hardly help but produce something new and would doubtless make a name for himself as well. While it would be nothing special to be a physicist capable of "doing the math," it certainly would be novel to be the social scientist who could.

Iconoclast, outsider, independent thinker, pioneer—these are the ways in which Simon liked to think of himself. Because he tried to fashion himself in keeping with those ideals, the terms do describe him in important ways. They do not describe all of him, however. He was as much an insider as an outsider: he wanted to institutionalize his revolutions.

These same tensions—this more complicated Simon—also can be seen in the truths that this pragmatic prophet preached. What was the gospel according to Simon?

The key to finding truth was to identify the patterns hidden in nature, for pattern was the product of law, of rule, of mechanism. Though a master of the details, Simon always sought the rule, not the exception, to take the instance and find the law, to take the complex and the chaotic and find the simplicity and order that must lie beneath them. Quite simply, "the rule is the important thing ... I want some lackey to take care of the exceptions."[51] Indeed, for Simon, the entire purpose of science was to reduce the complex to the simple, the phenomena to the mechanisms that generated them. Thus, he had from an early age an intense desire—a "drive"—to "see pattern in things."[52] He called this trait his inherent "Platonism."[53]

It was an ambitious desire, this drive to find the pattern, especially since the patterns he sought were the most elusive of all: those underlying human thought and action. He was sustained in his search by an unblinking faith that such patterns existed, that they were universal, and that they were reducible to a set of mechanisms simple enough to be accessible to human understanding. He was confident that he was capable of finding them, and he believed that others would listen when he spoke of what he found. In these beliefs he was not disappointed.

Not everyone listened, however, and in addition to the evangelist's unblinking confidence, Simon also had the prophet's difficulty in understanding why some did not see the truth when it was revealed. The logic was difficult to escape: if truth required no unique revelation (once discovered, at least), anyone who did not accept it was not merely ignorant but either willful or irrational. Though Simon was well aware that different people could have different truths, he had great difficulty in communicating with those who believed that a truth could be mysterious and ineffable, uniquely personal, or purely abstract: he wrote many a reply to his critics within political science, economics, and psychology, but he never directly addressed humanist critics of artificial intelligence, such as Hubert Dreyfus and Joseph Weizenbaum, because "You don't get very far arguing with a man about his religion, and these are essentially religious issues to the Dreyfuses and Weizenbaums of the world."[54]

The irony of Simon's particular quest is that the truth he found was that human access to truth is inevitably partial. A strange creed for a prophet, this principle of bounded rationality; yet it was universal, had observable consequences, and was accessible to human understanding, however limited that understanding might be. It served the evangelist in him well.

Leaving Home

It is September 1933. A young man, his bags stuffed to overflowing, stands at a train station, waiting nervously for the train that would take him to the University of Chicago and a new stage in his life. He has won a full scholarship to the university in the competitive examinations it sponsors, much to the relief of his parents, who are feeling the bite of the Great Depression. What baggage does this eager freshman take with him? What parts of his home remain within him as he departs? Simon himself could not be sure: in his autobiography, he shifts from third to first person when describing the events he associates with his transition to adulthood. The youth he once was had become "the boy" to him.[55]

Yet it is possible to recognize Simon amid the crowd at the station, possible to see that the "Child is the Father of the Man." At heart, he was and always would be a preacher of the gospel of reason, always a proud heretic, always a believer in the lawfulness of all nature, even human nature. But his were not simple, unchanging faiths: Simon believed in reason—and its limits; he believed in the importance of choice—and the influence of external forces on those choices; he valued independence of mind—and the structures imposed by organizational memberships and specialized training. He was a lifelong democrat—and an advocate of expert-led social planning.

As Simon soon would discover, he was not alone in these beliefs. The tension between independence, free will, and democratic politics on the one hand, and external causation, deterministic science, and expert authority on the other lay at the very heart of social science, particularly as practiced at the University of Chicago. At Chicago, Simon would find both teachers and kindred spirits, fellow outsiders learning to move comfortably in the inner circles of American intellectual and political life.

CHAPTER TWO

The Chicago School and the Sciences of Control

As William Cronon so ably describes in *Nature's Metropolis*, in the late nineteenth century Chicago grew to be the rail hub of the Midwest and thus the gateway to the East.[1] Chicago became "hog butcher to the world," a "teeming tough among cities," and the great western frontier became Chicago's hinterland.[2] When Herbert Simon boarded the North Shore train for Chicago in September 1933, he, like so many other raw materials, came to the new metropolis and was transformed by his entry into a larger world.

The railroad brought all things to the city, changing both the city and that which it brought from the hinterland. It enabled a new scale of life, a new level of interconnection and interdependence, and a new kind of diversity. The numbers are simply staggering: in 1840, before the Michigan Southern and Michigan Central railroads connected Chicago to New York, the town at the southwest end of Lake Michigan had 4,470 inhabitants; in 1880, it had over 503,000; in 1920, over 2.7 million.[3] In 1833, the newly incorporated city of Chicago was little more than a small fort stuck in a sea of mud; in 1933, more than three million Chicagoans celebrated their "Century of Progress" amid the skyscrapers they had introduced to the world.[4] Simon's hometown of

Milwaukee was no small village, but in Chicago he, like Carl Sandburg, found a "tall bold slugger set vivid against the little soft cities."[5]

Chicago's tremendously rapid growth from frontier town to metropolis meant that all the characteristics of urban life, all the wrenching changes from *gemeinschaft* to *gesellschaft,* were thrown into sharp relief.[6] Unlike other cities, Chicago did not appear to have a facade of history or tradition to obscure its essential nature. The structures and functions of this social organism were not hidden, or so it seemed. Indeed, to Max Weber, who visited the city in 1904, Chicago was like a man whose skin had been peeled off and whose intestines could be seen at work.[7] Chicago was the raw material of urban life, and to the social scientists at the University of Chicago—like the city, a new creation of boundless ambition—their city was the perfect laboratory for the study of modern man and the society he had built.

The social scientists of the various "Chicago Schools" were not alone in see-ing the city as the prime exemplar of modern life and thus as the proper focus for social scientific research. Robert Park, W. I. Thomas, Charles Merriam, and the other social scientists of the University of Chicago, however, took the metaphor of the city as laboratory very seriously. As a result, they, more than any other group of their time, sought to apply modern civilization's most characteristic intellectual product, science, to its most characteristic social form, the city.[8]

There was a great need to apply science to the city, for the "City of Big Shoulders" also was a city of big problems. Chicago's boastful motto was "I Will," and for a century it had. But the corruption of the 1920s and the Great Depression of the 1930s took some of the swagger out of the city's stride. When Herbert Simon arrived in the fall of 1933, for example, Chicago's new mayor, Edward Kelly, had just taken the reins from his predecessor, Anton Cermak, who had been slain by a bullet meant for President Franklin D. Roosevelt. Kelly inherited a city on the verge of bankruptcy, a government in disrepute due to Samuel Insull's scandalously obvious buying of influence, and a populace desperate for work. For Kelly, these proved to be all the right disadvantages; though he was not as colorful as the near-legendary "Big Bill" Thompson, "Boss Kelly" quickly proved himself even more the master of machine politics, Chicago style.[9] The rise of Boss Kelly, combined with the social disorganization that accompanied the Depression, meant to Simon and his fellow students and teachers at the University of Chicago that their city was not only thrillingly alive but also "first in violence, deepest in dirt; loud, lawless, unlovely."[10]

Simon and his fellow students of politics soon found that Chicago politics was politics in its rawest form. This meant excitement for the scholar, frustration for the reformer. The encounter with the promise and the squalor of the city played a major role in shaping Chicago social science in the 1930s. In that decade many of Simon's teachers turned from the city in disgust, pinning their hopes for rational reform on the federal government instead. As such researcher-reformers shifted from a local to a national orientation, they sought a harder, purer, more tough-minded social science. This movement toward sharper scientism was *not* a rejection of social science's traditional association with reform, however. It was not a withdrawal from engagement with the world but rather a redirection of efforts to reform it. This new path to the union of research and reform would lead social science—and Herbert Simon—to an unexpected destination.[11]

The University and the City: Research and Reform

The founding of the Johns Hopkins University in 1876 is often taken to be *the* great turning point in the history of the American university.[12] The institutionalization of the Germanic research ethos, and the increased emphasis on science that Hopkins pioneered, marked a change in how universities conceived their mission. That Daniel Coit Gilman invited T. H. Huxley, "Darwin's Bulldog," to give an address at Hopkins's inaugural ceremonies indicated just how revolutionary he intended his university to be. Another signal year in American higher education was 1892, during which the University of Chicago opened its doors.

The University of Chicago was a hybrid of the Hopkins-style research university and the traditional American community-based denominational college. It was founded under the aegis of the American Baptist Educational Society and funded (largely) by the living symbol of corporate greed, John D. Rockefeller. It had strong ties to local midwestern (especially Chicago-based) elites, but it also looked for approbation and financial support from a more nationally oriented urban upper class—the same class that flocked to Chautauqua every summer.[13] Rockefeller was an extreme example of such national-level patronage: unlike other, earlier, founding patrons of American universities, he had no particular connection to the city in which his university was built, did not want to have it named after him, and left its administration to the president and faculty.

In a manner befitting this novel combination of local and national sources of support, the University of Chicago sought to combine teaching and research, local civic involvement and national intellectual leadership. The result was a new kind of university: a comprehensive institution in which national ambitions found local expression in both teaching and research. As historian Martin Bulmer writes: "The University of Chicago set out, from the beginning, to be a university of national and international stature. This was a necessary part of its research orientation. So long as universities confined themselves to teaching, they were visible only locally and to those whom they had taught ... Achievements in research, however, were public property, and individuals, departments, and institutions could be assessed vis-à-vis competitors nationally and internationally."[14]

In keeping with its national mission as a research institution, it was only fitting that the new university should be located near the site of the 1893 World's Columbian Exposition, the event at which Chicago announced itself to the world.[15] It is important to remember, however, that researchers at the University of Chicago, particularly in the social sciences, expressed their ambitions for national recognition through research on their local community and through their work as civic leaders and as educators of the sons (and daughters) of Chicago's elite.[16] To the political scientist Charles Merriam, for example, Chicago was the perfect case study from which to generalize, just as involvement in local civic life was the avenue to national political influence.[17]

University president William Rainey Harper's leadership, Rockefeller's millions, and the kind of good fortune that favors enterprises that suit their times enabled the University of Chicago to make this hybrid form work. Rockefeller gave the university a generous endowment, and up through 1910 he continued to contribute large sums in response to Harper's pleas ($35 million in all). Rockefeller's monies, however, were not sufficient—or reliable enough—to satisfy Harper's ambitions or his desire to escape from dependence on a single patron.[18] As a result, Harper and many members of his faculty cultivated links to Chicago's reform-minded elite: the members of the City Club, the Commonwealth Club, and the Chicago Civic Federation, as well as the philanthropist Julius Rosenwald and the Progressive elements of both the

Republican and the Democratic Parties.[19] The result was a close-knit fabric of interactions between the university and the city.

The combination of Rockefeller's generous endowment, strong local support, and Clark University's near-collapse in spring 1892 allowed Harper to assemble a first-rate faculty almost overnight.[20] Clark University had been the brainchild of G. Stanley Hall, a professor of experimental psychology at Johns Hopkins. Hall was a true believer in the Germanic research ideal, and he was not inclined to compromise that ideal as he felt Gilman had been forced to do at Hopkins.[21] When the possibility arose in 1889 of establishing a new university wholly devoted to research and to graduate level instruction (with money provided by Jonas G. Clark), Hall jumped at the opportunity. Within three years, however, the new university in Worcester, Massachusetts, was in dire straits, thanks to the weakness of the university's connections to the local community, Hall's abrasive personality, and Clark's persistent meddling in university affairs. A majority of the members of the Clark faculty resigned in January 1892, and the alert Harper snapped up no fewer than seventeen by the end of 1893.[22]

The failure of Clark's program of exclusive focus on research, particularly "pure" research, and its consequent isolation from local support taught Harper that if his university was to be a viable institution, it must be both college and university.[23] That is, it must provide something to satisfy both local and national interests in its teaching and research. Harper, his successor, Horace Judson, and the faculty they assembled were remarkably successful at doing just that—at least until the 1940s, when the advent of federal patronage changed the equation for success for American universities.

Despite this early triumph, the provision of intellectual leadership to the city and the advancement of knowledge through research remained potentially divergent goals since different constituencies defined success in each area. So long as local civic leaders and academic professionals shared similar values and were linked by close personal connections, the two goals could be seen as reinforcing one another. Indeed, up through the 1920s, the perception of a disharmony between research and civic engagement typically was taken to indicate a misunderstanding of both. If the values of the researchers and the reformers were to diverge, however, or if the personal links between individual members of each group were to fail, then research and reform might part ways—or even come to blows.

Two examples, both of particular relevance to Simon, illustrate this fusion of research and reform in the early twentieth century at the University of Chicago. The first was the pragmatic, democratic educational philosophy of John Dewey. To Dewey, education was a process of inquiry, of constant dialogue between ideas and experience.[24] Truths were provisional, to be accepted only so long as they were useful in explaining and ordering the world. The best education, exemplified by the Laboratory School of the University of Chicago, would train the student for a life of perpetual, independent inquiry, the result of which (on a larger scale) could only be the rational reform of society.[25] The intellectual and moral transformation of the individual through education was vital to all reform, and the freedom to explore, experiment, and inquire was the key to social progress.

Though Dewey had left Chicago long before Simon arrived, the Chicago curriculum under the "New Plan" of the early 1930s still reflected these ideals.[26] The university allowed its students great latitude in designing their own educational programs, encouraged independent inquiry, and fostered an atmosphere where "nothing was too new, too arcane, or too absurd," and in which "everything had to be explored, tested, before it could be accepted or rejected."[27]

Similarly, in political science as practiced by Charles Merriam, chair of the Chicago department from 1923 to 1940, science and civic involvement were seen as natural allies. Merriam was a Chicago alderman from 1912 to 1917, a candidate for mayor in 1915, and an important member of the Progressive wing of the Illinois Republican Party until the 1930s. In the 1930s, Merriam became closely allied with Franklin D. Roosevelt and the New Deal, and he was appointed to the vice-chairmanship of the National Resources Planning Board, thanks in large part to his close friendships with Harold Ickes and city planner Frederic Delano. Merriam's ties to the Delano family were so strong, in fact, that Frederic Delano's nephew (President Roosevelt) was fond of calling Merriam "Uncle Charlie" in private.[28]

Merriam saw no conflict between his career as a political scientist and his career as a politician; rather, he saw each as informing the other. According to his biographer, "Merriam believed that participation in politics was the only base for an experimental approach to the study of politics."[29] Taking reform as one's goal and assuming democracy as an ultimate good did not imply a lack of scientific objectivity to Merriam. Rather, the advance of the science of politics necessarily would advance the progressive reform of democracy. Thus,

Merriam's persistent inveighing against "do-gooders" and his efforts to exclude them from social science were not rejections of reform as the goal of political science but of amateurism in its pursuit.

By the 1920s, however, this union of research and reform was beginning to show some cracks. Merriam's major research project of the early 1920s, for example, was a study of civic education, a traditional theme of the political researcher-reformer. The disillusionment that followed World War I, however, combined with Merriam's personal political defeats of the 1910s and 1920s to give this project a new, sour flavor.[30] The international team of scholars Merriam organized studied the methods used by several nations to indoctrinate their young with a fervent patriotism, not techniques used to produce rational citizens. Similarly, although the dedication of the university's new social science building in 1929 was in many ways a triumph of the fusion of research and reform, of national ambitions given local expression, it also marked the beginning of a movement toward seeing the university, rather than the city, as the true social laboratory.

On an intellectual level, this growing split between research and reform was associated with an increasingly instrumental view of the purposes of knowledge. Men like Harper and the professors he recruited to establish philosophy and the social sciences at Chicago had held what might be called an "idealist" view of the purposes of knowledge, one in keeping with their religious concerns and their disciplines' roots in moral philosophy.[31] To them, knowledge and its uses were not so distinct as we see them today, and right knowledge largely was synonymous with right action. The utility of social knowledge inhered in the knowing itself, just as religious truth's utility inheres in the knowing: knowledge necessarily transforms the knower, and that transformation is its use. For example, as David Roberts has shown in his study of mathematical education in the late nineteenth century, mathematics was valued less as a tool than as a provider of a healthy mental discipline. That is, it was valued for the way in which it transformed the individual, contributing to his *Bildung*.[32]

In this older, idealist view, to educate the public was necessarily to reform it. Knowledge was not something that could be misused. Errant action necessarily implied errant belief; if one sinned, one did not really know the Truth.

The historian Fritz Ringer captured the idealist purposes of education best when he wrote that for the German academic "mandarins," "specialized knowledge [of an instrumental character] lacked precisely that dimension

that had connected Idealist *Wissenschaft* with Bildung and with Weltan-
schauung. The specialist was a mere expert; his knowledge was narrowly technical and dealt only with means. This revulsion against specialization was
in effect a demand for wisdom, for reflection about ends, for the knowledge
of the sage, the prophet, or the harmoniously cultivated man."[33] Although
the aristocratic aspects of this idealist view often repelled American social scientists of the 1880s–1900s, they found the linkage of research to reform via
individual enlightenment very appealing, perhaps because of their own evangelical roots.[34]

The professional academics who came of age intellectually in the 1910s and
1920s, however, increasingly came to reject this idealist view of the purposes of
knowledge in favor of an instrumentalist one. In this new view, knowledge was
seen as a tool, not as a state of being, and "facts" and "values" were thought to
govern separate spheres. Instrumentalist scholars consciously concerned
themselves with means rather than ends, though the instrumentalists' ends
often lay hidden within their understanding of "natural" laws.[35] To them, technical knowledge was "objective" while moral knowledge was "subjective"—
and therefore suspect. The shift toward an instrumental view of knowledge
was particularly marked in the social sciences at Chicago, where scholars such
as the sociologist William Ogburn strove mightily to separate the "is" from the
"ought."[36]

Such instrumentalist scholars continued to have a strong desire for their
work to make a difference in the world, but they saw their role as advisers
rather than educators. Instead of pursuing their research in order to reveal
truths whose knowledge would effect personal (and therefore social) reform,
the instrumentalists' goal was to develop value-neutral tools to place in the
hands of the nation's leaders. To put it bluntly, idealist educators worked to
foster enlightenment, while instrumentalist advisers sought to facilitate government.[37]

The older, idealist understanding of the purposes of knowledge did not die
out overnight, nor was the shift to instrumentalism complete when Herbert
Simon arrived at the University of Chicago in 1933. The undergraduate curriculum at Chicago under the "Old New Plan" of the early 1930s, for example,
still reflected the earlier faith in the inherent harmony of research and
reform.[38] It still was based on a Deweyan valuation of independent, problem-
oriented inquiry as vital to both individual and social development. The 1930s
were a time of change in the undergraduate experience, however, for the uni-

versity's new president, Robert Maynard Hutchins, and his chief ally, the philosopher Mortimer Adler, sought to move the university toward their own Neo-Thomistic synthesis and away from Deweyan perpetual inquiry and the pragmatic, instrumental approach to values it fostered.[39]

During Simon's years at Chicago, the virtues of the university's untraditional tradition and Hutchins's conservative program of change were the subject of vigorous debate.[40] Both approaches still were viable at the university, and the continual debate between them forced students not only to take sides but also to defend their positions against fierce attacks. Adler, for instance, would not rest until every student in his class agreed with Aquinas's proof of the existence of God, even telling his classes that he was "not sure that I would hesitate to say that the church was right in burning heretics."[41] Hutchins and Adler saw their eventual victory as a triumph for the basic, core values necessary for the preservation of democracy against the totalitarian threat (against which the "relativist defense of democracy" offered by the Deweyans was no defense at all).[42] But to those, such as Simon, who studied at Chicago in the days before Hutchins's triumph, this was not victory but betrayal.

Simon counted himself one of the fortunate ones who experienced the old University of Chicago in its final flowering. To Simon, it was a place full of new ideas, new friends, and new freedoms, a place where "the exposure to everything new and modern was massive."[43] Traditional truths, whether in classical physics, classical economics, or classical political science, all were open to challenge. Although his teacher Charles Merriam had come to see education as indoctrination, Simon found independence the doctrine at Chicago, and he took full advantage of a system based on it.[44]

The emphasis the University of Chicago put on independence and novelty meshed perfectly with Simon's personal values. He attended few classes, preferring to read on his own for the common exams: as he later wrote, "early in my second year, I terminated my formal education in mathematics when a calculus professor insisted that I attend class."[45] He did well enough by this method to graduate in three years. He was introduced early (and often) to the university's research ethos, particularly through the graduate-level courses in economics and political science he took in his second year. In his third year he took almost entirely graduate courses, even attending some of them, though boxing remained the only course on his graduate transcript. (As happened so often with Simon, even this course, chosen half at random, turned out to be quite useful for him in advancing his career: his instructor was the economist

William Cooper, who became a good friend of Simon's. Cooper later introduced him to the Cowles Commission for Research in Economics and played a major role in recruiting Simon to Carnegie Tech in 1949.)

In keeping with the blending of research and social engagement that characterized Chicago, Simon and his college friends were very much concerned with the issues of the day. The Depression, the darkening shadow over Europe, the transformation of the Soviet Union under Stalin, and the machinations of Boss Kelly and his cronies fueled many a late-night argument and many a debate in class (especially Adler's).[46] Several of Simon's friends responded to the crisis of the West and the seeming promise of the Soviet Union by developing a curious "Aristotelian-Thomistic-Catholic-Trotskyism," taking Hutchins and Adler in a direction they doubtless never expected.[47] Similarly, Simon and his fellow students of political science served as election watchers in the 1935 Chicago mayoral elections and were astounded by the blatant, open fraud.[48]

This was a bookish sort of engagement, to be sure, one expressed through speeches and readings and political clubs that largely were debating societies. An effort to canvas for voters for an independent aldermanic candidate netted only fifteen votes, and the Progressive Club Simon co-founded would have to go through many changes before its descendant organization would join the Americans for Democratic Action in the 1960s. Nevertheless, the events of the larger world were more than mere background noise for Simon and his friends. They were the facts of experience against which the novel concepts of the intellectual had to be tested.[49]

One result of Simon's engagement with these issues was that he came to invest in science the fierce emotional attachment that some of his friends invested in political or religious ideologies and that his professors once had invested in democracy. He describes his fascination with science and its role in society as "quasi-religious."[50] In this, he and his fellow Chicagoans reflected the larger trends of the Western world, as the prewar faith in the progressive nature of liberal, democratic society was replaced by faith in "blood and history" on the one hand and faith in science and management on the other. In particular, the rise of the Nazi threat and the example of the Soviet Union as a planned society committed to a "scientific" ideology brought the question of the relationship between science and politics, facts and values, to the fore, just as the chaos of World War I had brought the question of the limits to human rationality to the center of European intellectual life.[51] Could objective science prove the Nazis wrong? Was communism the "one best system" social engineers had been seeking? Was Nazi science "perverted" because it took rationality to extremes or because it warped reason in a vice of twisted values?

Simon never forgot the lessons he learned from the Deweyan aspects of his education: knowledge always must be tested through experience; research and reform must be linked; and independent, interdisciplinary inquiry is best. He, like most social scientists of his generation, would find that the gap between research and reform continually threatened to grow into a chasm, but in the mid-1930s the bridge still was strong. One did not have to choose between graduate study and civic engagement, which made Simon's decision to continue on to graduate school so easy it seemed no decision at all.

In the spring of 1936 Simon wrote a paper in one of his seminars on how to account for marginal costs in municipal budgeting. This paper impressed his instructor, Clarence Ridley of the International City Managers' Association (ICMA), leading to a graduate assistantship that fall. The assistantship paid enough to live on, which was no small thing during the Depression, and it offered Simon a chance to pursue research in a way that was immediately connected to active reform. The city still was his laboratory.

The Chicago School of Political Science

The department Simon entered in 1936 was *the* leader in its field. Harvard and Columbia had dominated political science during the first quarter of the twentieth century, but after Charles Merriam took the departmental helm in 1923, Chicago quickly rose to equal and then to surpass its eastern rivals. Harvard did retain a lead in the total number of Ph.D.s granted through the 1930s and 1940s, but Chicago closed that margin to virtual equality, with Columbia (the former leader in the field) falling to a distant third.[52] More importantly, the political scientists based in or trained at Chicago in the 1920s and 1930s were easily the most innovative and productive group in the discipline, with a faculty that included Charles Merriam, Harold Lasswell, Harold Gosnell, and Leonard White, and a graduate student corps that included Quincy Wright, Herbert Beyle, Gabriel Almond, David Truman, Herman Pritchett, Avery Leiserson, V. O. Key, Don K. Price, and, of course, Herbert Simon. Even this abbreviated list gives an indication of the remarkable quality of the "Chicago School": five of the above (Almond, Key, Lasswell, Simon, and Truman) were named in a 1963 survey as being among the ten most important contributors to the discipline in the years since 1945—and that does not include Merriam, White, and Gosnell, who would have been at the top of any similar list for the 1930s![53]

A large part of what made the Chicago School of Political Science so productive and so influential was that its members shared a common outlook, as had the members of the equally famous Chicago School of Sociology that flourished in the 1920s.[54] This outlook was not a rigid, formal doctrine but rather a linked set of beliefs about the science of politics. The most important of these common beliefs was the basic assumption that there could be such a thing as a science of politics.[55] As Simon later wrote: "It was Lasswell's psychologizing and Gosnell's quantitative and empirical methods that most specifically symbolized the Chicago School. But what characterized it even more fundamentally for me, and I think for a number of other graduate students, was its commitment to the proposition that political science is science. Along with that commitment went a dissolving of departmental boundaries that made the whole university, and all of its methodologies, available to the students of political science."[56]

Perhaps this science of politics might never be as precise in its predictions as, say, physics; nevertheless, it could be a proper science characterized by both empirical truth to nature and theoretical rigor. In this view, empiricism demanded that political science take political behavior, not political philosophy, to be its subject. Theoretical sophistication, meanwhile, required that political science look to psychology, especially social psychology, in order to understand the mechanisms underlying that behavior.

Despite the strength of the Department of Economics at Chicago in the 1930s, which numbered Frank Knight, Jacob Viner, Henry Schultz, Paul Douglas, and Henry Simons among its members, the political scientists primarily looked to psychology and to sociology, their fellow sciences of control, for ideas and kindred spirits. Simon was quite unusual among his cohort in finding in economics a powerful set of tools for understanding human behavior. Even so, one of the things for which Simon was best known in economics was his insistence that economists needed a better understanding of social psychology.

This "behavioral" approach to political science still was in its infancy at Chicago during the 1930s. It was largely innocent of the mathematics that would accompany it into the mainstream of political science after World War II, and its scientism was far less pronounced than that of its heirs.[57]

Nevertheless, the political science of Merriam, Lasswell, Gosnell, and their colleagues was noticeably different from that of their teachers and their rivals at other universities. The Chicago School of Political Science constituted a challenge to traditional political science at every level: its members advocated a new understanding of the overall aims of the field, advanced a new set of answers to the core intellectual problems around which the discipline was oriented, and developed a new set of institutional arrangements to support their work. It provided Simon with his first model of what social science should aspire to, intellectually and institutionally, and thereby had a lasting influence on him and his work.

The Chicago School had a new vision of the proper aims of political science, a vision linked to its strongly instrumental view of the purposes of knowledge. As mentioned above, to Merriam and his colleagues the political scientist no longer was an educator of rational public opinion but an adviser to the elites who managed a complex, highly organized social-political system. In Merriam's case, this instrumental view of political science marked a shift from his earlier, more optimistic view of the public's rationality, a shift that was connected to the decline of his own political fortunes.[58] For his younger colleagues, particularly Harold Lasswell, the trials and tribulations of war, corruption, and Depression did not cause so sharp a movement, but that was only because they began where Merriam left off.

This first generation of the Chicago School retained enough of its heritage and traditional ideals to keep its members from marching to technocratic extremes (though Lasswell sometimes veered across the line). Instead of technocracy, Merriam, Gosnell, Lasswell, White, and their colleagues articulated a liberal managerial political philosophy in which they sought to harmonize democracy and expert authority. I call their philosophy "liberal managerialism" in order to emphasize the faith in management that was characteristic of these elite reformers, especially during the interwar period and the first twenty-five years after World War II. In addition, the term emphasizes the persistent tension underlying their ideals, for their goals were liberal at the same time their methods were managerial. They sought the traditional liberal goals of liberty and equality; only, to them, large-scale formal organizations, even giant corporations and government bureaucracies, were potential sources of equality, not threats to it.

This philosophy was liberal in that it emerged from a sincere concern for individual liberty and a deep opposition to special privileges based on birth or wealth. It was managerial in that it also grew out of a belief that freedom could be its own worst enemy if individuals did not recognize their responsibility to the larger community. Because individuals typically do not look beyond their own interests unless they are guided, the "visible hand" of rational, expert management was a necessary complement to individual freedom. This liberal managerial view thus reflected the widespread beliefs that organization was the key to progress and that the expert manager was vital to the smooth functioning of the social machinery.[59]

The central tenet of this liberal managerial view was that conscious, rational coordination ("organized intelligence" in Dewey's phrase) was essential for democracy as well as for efficiency. To Merriam and his colleagues, there was greater liberty under rational democratic government than in any state of nature. Modern society was too complex, its members too interdependent, for order to emerge spontaneously. Societies needed to be led and economies needed to be regulated for there to be progress. As a result, the members of the Chicago School agreed that the decentralized, egalitarian *consumption* of goods, economic or political, required the efficient, centralized *production* of such goods by large-scale organizations, public and private.

Such centralization, however, could be carried too far: to them, socialism was a cure worse than the disease. In their politics, therefore, as in all things, the Chicago School sought to define a rational middle ground. Its members never argued that political scientists (or other experts) should rule, and the final authority always was the public will as embodied in elected leaders. Once the public chose the destination, then the political scientist could step in to help steer the ship of state through the shoals of self-interest.

Perhaps the best examples of this liberal managerial philosophy in action were the city manager movement of the 1920s–30s and the various efforts at federal, state, and local planning of the 1930s, both of which had strong Chicago connections.[60] The city manager movement was fueled by urban businessmen and professionals who believed that politics was interfering with aspects of government—particularly public works—that should be handled in a more efficient, businesslike manner.[61] In their view, the city was a corporation, its citizens were its stockholders, and the city council was its board of directors. The chief executive of a city, by this reasoning, therefore should be a CEO chosen by the board, not a politician chosen by the masses. This CEO would take care of "nonpolitical" issues such as building streets and sewers, while the citizens, through the city council, would set overall priorities. The best man for such a job, of course, was an engineer or businessman.[62] The political scientist's job was to advise this city CEO as to the best means of administrating his corporation—and to remind him and his stockholders where the boundaries lay between administrative and democratic authority.

Many members of the Chicago School were involved intimately in the city manager movement, so much so that it is fair to call the University of Chicago the movement's intellectual and institutional hub. Merriam, for instance, brought the International City Manager Association and the Public Administration Clearing House (PACH) to Chicago, along with Clarence Ridley and Louis Brownlow to manage them. Funded by the Rockefeller Foundation, the ICMA was the quintessential example of 1930s-era attempts to revitalize local government through research sponsored by national—and even international—organizations.[63] The PACH, which also drew support from the Rockefeller Foundation, similarly played a vital role in the promotion of the city manager form of government, as well as in advocating other "rational reforms" of municipal administration. In addition, department faculty and graduate students, such as Leonard White and Don K. Price, wrote a number of studies of city managers in action.[64]

In the hands of the Chicago political scientists, the surface naiveté of the city manager movement was informed by a more astute political sensibility: while it was naive to think that politics could be eliminated from public administration, Merriam and his allies knew full well that the control of patronage in basic public services and public works was vital to the maintenance of political machines.[65] Boss Kelly in Chicago was a case in point, having ridden control over patronage in the Chicago Sanitary District and South Park District to power.[66] If such patronage could be brought under rational control, not only would municipal services be improved, politics itself would be reformed.

Although the city manager movement never won significant support in Chicago, in ironic fashion the career of Boss Kelly proved the wisdom as much as the naiveté of the city manager ideal. Boss Kelly was everything that the city manager was supposed to be: he was an engineer who knew the city's public works intimately and expanded them greatly; he instituted various administrative reforms aimed at making government more efficient and more cost-effective (and at increasing his own executive authority); and he even made the Chicago Planning Commission an official—and powerful—agency.[67] The only problem was that his administration was "maximally corrupt."[68]

The great appeal of the city manager form of government was that it would enable the rational allocation of public resources toward the achievement of public goals. The same dream of rational planning animated efforts in the 1930s to bring about a more reasonable allocation of regional and national resources. These efforts were products of the Depression, as many social and political leaders were struck by the existing economic system's inability to operate rationally. They observed millions going hungry while hundreds of thousands of hogs were being burned and tons of fruit left to rot on the trees because they could not be sold. Order was not emerging spontaneously in this complex system. Rational management—rational planning—was needed.

The first institutional embodiment of this desire for rational planning at a national level was the National Recovery Administration (NRA). While the NRA soon was deemed unconstitutional, the National Resources Council (later named the National Resources Planning Board) that it sired lived on, at first under Harold Ickes (a close friend of Merriam's) and then directly under FDR. Merriam was vice-chairman of the NRPB, and his friend the city planner Frederic Delano (FDR's uncle) was the chairman.

The NRPB never had the authority to carry its plans into action, nor did it really want such power. Merriam and his allies at the NRPB did not seek to develop or implement national five-year plans; rather, they wanted to develop resources and establish mechanisms that would enable communities to plan for themselves.[69] In these aims, they followed John Dewey's admonition that "an immense difference divides the planned society from a continuously planning society. The former requires fixed blueprints imposed from above . . . the latter means the release of intelligence through the widest form of cooperative give-and-take."[70] For Merriam, as for Dewey, planning was not merely reconcilable with democracy, it was essential to it.

The NRPB produced a number of valuable studies of national resources, and a great many social scientists got their first taste of applied social research while working on its studies, many of them going on to develop quite an appetite for it.[71] The organization also acquired a number of powerful enemies in Congress, which did not like to have its pork questioned, and in the Army Corps of Engineers, which viewed large-scale public works as its turf and wanted no intruders.[72] So long as FDR found the NRPB useful enough to defend, it survived, but by 1943 the political costs outweighed the benefits, and he let it go under.

Although Merriam and his allies saw them as great defeats, the demise of the NRPB and the inability of the city manager movement to make headway in the largest of cities should not be taken as a wholesale rejection of the "third way" of liberal managerialism that the Chicago School espoused.[73] As Daniel Rodgers has shown in his study of the transatlantic traffic in municipal reform programs, the cities became laboratories for a host of experiments in planning in the early twentieth century, a role they continue to play today. In addition, as Kenneth Boulding, John Kenneth Galbraith, and Alfred Chandler were to point out in the 1950s and 1960s, the "visible hand" of management, public and private, played a central role in creating the prosperity of the postwar era.[74] It would not be until the 1980s and the rise of another Chicago School that this liberal managerial approach to solving the nation's problems would be forced into a full-scale retreat.

A New Framework, Intellectual and Institutional

In keeping with their new, instrumental vision of the purpose of social knowledge, the Chicago School built a new intellectual and institutional framework for political science. In this new framework, the characteristic challenges of modern society—and the central questions for social science—were reinterpreted as problems of control or governance. The most prominent of these problems were the problems of change, interdependence, and subjectivity. The Chicago School of Political Science represented a distinctive (if not wholly unified) set of answers to these problems—and to the problem of reconciling expert authority with democratic politics that linked them all.

These questions, and his teachers' answers to them, served as the starting point for Simon's intellectual journey, leading him to explore the constraints group memberships impose on our ability to make rational decisions. Unlike his mentors, however, Simon found human subjectivity to have a rational structure, and he came to see the complex webs of interdependence that distinguished modern from premodern society as essential supports for reason, not amplifiers of unreason. Though he thus stood Merriam and Lasswell on their heads, in order to understand Simon one must first understand how they and their colleagues addressed these key problems.

Change

Modern American social science emerged in the last quarter of the nineteenth century as a direct response to the dramatic changes in society wrought by the Second Industrial Revolution.[75] As historian Dorothy Ross writes, "The discovery of modernity is the fundamental context of the social sciences."[76] From the 1880s onward, social science took as its central question the nature of the differences between modern and traditional society, and its primary goal was to find ways of bringing order to this new world. As the 1936 course catalog of the University of Chicago stated, the social science core course was "primarily concerned with the impact of the complex of forces generally described as the industrial revolution on economic, social, and political institutions ... The economic, social, and political order that preceded the industrial revolution is then contrasted with contemporary society."[77]

By the early twentieth century, social scientists had come to believe that modern society was fundamentally different from traditional societies. They saw an epochal break between gemeinschaft and gesellschaft, between societies ruled by status and those governed by contract, between preindustrial America and one dominated by "the Trusts."[78] Frederick Jackson Turner's "frontier thesis" stands as a perfect example of his generation's belief that America had crossed the Rubicon of industrialization and could not turn back.[79] In his famous speech, Turner argued that the frontier had given traditional America its unique character—and that the frontier was now closed. That which made traditional America what it was existed no more. The city was the new frontier, and it was a frontier wholly unlike the one of old. It was, of course, only fitting that Turner's address was given in Chicago at the Columbian Exposition of 1893.

Social science, it followed, needed to employ new methods and new concepts in order to understand this new world. It needed to explain the processes of change, and, perhaps most importantly, it had to try to understand how order (usually understood as equilibrium) could be produced and maintained amid the constant change that all social observers saw as a defining characteristic of modernity.[80]

One result of these mandates to understand change and to bring order was that social scientists were preoccupied with finding the changeless laws underlying change.[81] For the predecessors of the Chicago School, the central concept in that effort had been evolution. Indeed, evolution was so vital a concept to turn-of-the-century social science that its employment was virtually synonymous with taking a scientific approach to the study of society. For example, when Thorstein Veblen, a professor at the University of Chicago in the years before the Chicago School, sought to define the status of economics, he asked whether economics was an "evolutionary science," not whether it was a science.[82]

The study of change could lead in another, nonevolutionary, direction, however. Instead of the historicism inherent in evolutionary theories, one could emphasize the radical separation of present and past that industrialization had brought and so dissolve the study of change into the study of process in the present. As historian Dorothy Ross observes: "The interaction of . . . the movement towards modernist historical consciousness, the growing power of professional specialization, and the sharpening conception of scientific method . . . produced a slow paradigm shift in the social sciences. The result was a broad move away from historico-evolutionary models of social science to specialized sciences focused on short-term processes rather than long-term change over time."[83]

In the 1920s and 1930s at Chicago, this ahistorical study of social process grew in importance, exemplified by Merriam's and Lasswell's works on political power and Gosnell's studies of voting behavior.[84] The leaders of the Chicago School, however, did not move away from history as sharply as did their colleagues in economics—or as their students would in their work.[85] Leonard White's massive administrative histories stand as perfect examples of his cohort's mixed attitude toward historical change: change in administrative structures had occurred over time, but this change was driven by functional adaptation to increasing social complexity, not by historical contingency.[86]

Though it was not complete, the turn from history and evolution to process and equilibrium nonetheless was a significant departure from political science's past. Political science long had been linked to history, so much so that they really were part of a joint tradition of "historico-politics," as Ross terms it.[87] The turn-of-the-century effort to professionalize political science as a science had involved a conscious separation from history as a discipline, but the need to establish a new disciplinary identity had been softened by a desire to retain what was valuable from that joint heritage. For many years after their split, the APSA and AHA held joint annual meetings. In the 1920s, however, they stopped doing so.

One factor in the break with historical approaches was that historicism in social science was associated with social and political radicalism because of its strong connections to the socialist German Historical School of political economy. Such radicalism had been acceptable in the 1880s, but the crises of the 1890s pushed such critiques beyond the pale.[88] This shift was most pronounced in economics, which had been the most radical of the social sciences in the 1880s. Political science had never been so radical, so its movement away from history could be slower and less abrupt.

A sharper break with history came after World War I. For many social scientists, the Great War was a cataclysm signaling the end of the old world and the beginning of a new one, one for which history was no guide. One economist, for example, wrote that "it would perhaps be an exaggeration to say that the European War ... has rendered every text in social science thus far published out of date ... but it would not be a very great exaggeration."[89] Although the reaction to the war was less dramatic in America than in Europe, the members of the Chicago School nevertheless accepted without question that the modern world was radically new and that one of the things that made the modern world modern was that its members were in constant motion, physical and social.

There were four aspects of Simon's work that were directly connected to this problem of change. First, Simon was impressed by what he termed "the time-binding" nature of decision—that decisions of the present moment can determine not only present actions but also future choices. Because individuals (and societies) can construct their future environments of decision through their present choices, decisions are integrated with each other in a hierarchy of values and in a sequence of decisions—in "courses of action," "strategies," "algorithms," "heuristics," and "programs," to use the terms that became popular with the postwar rise of computer science and game theory.

Second, Simon was fascinated by the ways in which future conditions played a role in present decisions through people's expectations about the future. For example, institutions enable reliable expectations about the behavior of others to be formed, allowing people to plan their own courses of action. Similarly, Simon was much impressed by Carl Friedrich's "rule of anticipated reactions," a term used to describe the situation where authority does not need to be exercised visibly because the prospective challenger anticipates being overridden and so does not make a challenge at all.[90] Simon's interest in the phenomena surrounding expectation also carried into his philosophy of science, where he was interested in the nature of prediction and the possible effects of the experimenter's prediction-making on the systems being studied.[91]

Third, Simon was deeply interested in creating a dynamic social theory. Like his contemporary Paul Samuelson, he was fascinated by the relationship between "comparative statics" and "dynamics," and he hoped to build a theory that would encompass both.[92] Simon's concern with comparative statics and dynamics in his early work largely was limited to the question of how people learn to bound their rationality.[93] He was not satisfied with this answer, however, and he soon became convinced that a theory of learning—of the evolution of the individual mind—would be central to any truly dynamic theory of human behavior. *Administrative Behavior* was far from his last word on this matter, and the psychology of the learning process became a crucial part of his postwar research program.

Fourth, and finally, Simon's own perceptions of change in the world around him informed his understanding of administrative science and its place in the modern world. He, like so many who lived in the first third of the twentieth century, was deeply conscious of the end of the old and the birth of the new. A vast gulf appeared to separate the modern world from that of even a generation before. A new system was in place, and a new class of people— the administrative class whose decisions Simon wished to influence—was needed to lead this new urban society.

Interdependence

From the 1880s onward, social scientists also shared a belief that modern society was characterized by specialization, interdependence, and complex organization. These traits could be seen most clearly in the giant industrial enterprises that loomed large over the turn-of-the-century economic landscape, enterprises characterized by an extraordinary division of labor. Social analyst after social analyst wrote in fascination of extremes of specialization far beyond Adam Smith's dreams, dwelled upon the ingenious methods by which these specialized labors were coordinated to serve a common end, and expressed an anxious wonder at how strongly dependent on the actions of myriad unknown others was modern man.[94]

From their studies of such large-scale social organization, many social scientists drew the conclusion that they no longer could explain social events in terms of individual agency. Causes were to be found not in individual will but in the interests of groups, the effects of social institutions, the actions of masses, or in the operation of universal natural laws. As Thomas Haskell argues in *The Emergence of Professional Social Science,* causation "receded" as the environment moved into the foreground of explanation. Perhaps the best example of this tendency was Charles Beard's radical *Economic Interpretation of the*

Constitution, in which the founding document of American democracy was shown to be a product of the struggles between economic interest groups, not the inspired wisdom of select men.[95]

Causation had receded for the leaders of the Chicago School as well, but they differed from their predecessors in their understanding of the external forces that shaped the behavior of the individual. Merriam and Lasswell, for instance, looked to social psychology rather than economic interest as their explanatory agent. This turn to psychology was in keeping with the more general rise of psychology in the 1910s–1930s. Psychology at that time was an imperial discipline with ambitions of becoming the fundamental social science.[96] In the dreams of psychological prophets like G. Stanley Hall and John B. Watson, psychology would provide the foundational data and theoretical structure for all analysis of human behavior. It would supply the tools necessary for the control of behavior and so would lead to the most fundamental social reform of all.

This turn toward psychology as explanatory agent, however, posed serious problems for anyone committed to individual rights, free will, and democracy. After all, how can my decisions be mine, be free, be meaningful, if even my thoughts are caused by something beyond my control? How can the ideal of free choice be reconciled with the reality of social control? Merriam and Lasswell both were troubled by this problem, and neither was able to find satisfying answers. Indeed, for social scientists who came of age in the 1930s, perhaps the most vital task was to work out a way in which free will and individual rights could be reconciled with social influences and collective needs.[97] As one political scientist wrote:

> A central conceptual controversy, probably inescapable for political scientists because of their disciplinary heritage, is that involved in perceiving uniformities in behavior, describing recurring patterns, identifying the determinants and yet reconciling this effort and its underlying premises about the roots of behavior with the liberal, democratic faith in man's individual capacity to determine his own ends, to think rationally, and to reach individual and creative decisions. On this faith rests the political structure of rights, the machinery of the democratic electorate, the party system, and the values of the constitutional democratic state whose political process we are concerned to describe and analyze.[98]

Herbert Simon inherited these concerns—and this dilemma—and he sought to reconcile individual agency with social forces in his studies of the organizational influences on decision-making. As Simon's work reveals, effecting this union of science and democratic values was no easy task.

Interdependence also had implications for the organization of the social sciences. The leaders of the turn-of-the-century effort to professionalize social science, such as Woodrow Wilson in political science and Albion Small in sociology, had believed that social science ought to mirror the society it studied and thus needed to become simultaneously a more specialized and a more interdependent endeavor. They argued that the complexity of this specialized, interdependent world proved the necessity of expert knowledge, for the modern world was too complicated for the untrained amateur reformer to understand. Similarly, this complexity made it impossible for anyone, even an expert social scientist, to encompass it all. Authoritative knowledge, in their view, was necessarily the product of a community of experts—in Thomas Haskell's phrase, "the community of the competent."[99]

The leaders of turn-of-the-century social science believed that this community of experts had a vital role to play in modern society. To men like William Dunning (Merriam's mentor) or William Graham Sumner, modern society always appeared to be on the verge of whirling apart because of the rapid change and extreme specialization that it embraced and because of the weakening of the cultural and religious traditions that once had unified the nation.[100] The community of experts in the sciences of society could combat that tendency to fragmentation and *anomie* by educating the public into a rational appreciation of interdependence.[101] No greater service could any profession render to society.

But how to organize the organizers? How to integrate these specialized researches into a coherent program and thereby cure the ills of modern society? The Chicago School's answer to these questions was threefold. First, Merriam and his colleagues sought to transform their professional association, the American Political Science Association (APSA), and to establish a new pan-disciplinary association, the Social Sciences Research Council (SSRC). In their eyes, these professional associations needed to take an active role in the coordination of research efforts and the promotion of methodological reform. This would not be a heavy-handed imposition from above; as with all Merriam's reform programs, disciplinary reform would take place through the education of the next generation.[102] By the force of their example and the power of their public exhortations, the leaders of the APSA and the SSRC would show the rest of the social sciences the proper path. If the younger generation needed the approbation of such leaders for professional advancement—they were the ones advising foundation program officers and university presidents, after all—then that was not coercion but leadership.

Among the many examples of such efforts were the three National Conferences on the Science of Politics of the mid-1920s (organized by Merriam through the SSRC and APSA), the development of an intimate relationship between the Spelman Memorial Fund and the SSRC (both as recipient of money and as adviser regarding the disposition of other monies), and the SSRC's attempts to promote methodological sophistication through conferences, grants, and guides such as the SSRC-sponsored handbook of *Methods in Social Science.*[103]

These efforts to reform the social sciences from above sparked a debate about the proper role of the SSRC vis-à-vis the disciplines, a debate that was a perfect microcosm of the larger debate regarding the role of the expert in a democratic society. Many social scientists were concerned that the SSRC's Committee on Problems and Policy—the organization's own planning board—was a "secret government" not democratically accountable to the member societies. Others worried that it had too little power to be able to plan effectively. The same debate would play out time and again in the social sciences, arising whenever a new institution (such as the Ford Foundation or the National Science Foundation) arose with the power to exert broad influence and whenever the government sought to use social science as the basis for social reform, as in the War on Poverty.[104]

Second, the members of the Chicago School tried to coordinate their own researches, at least loosely. The primary institutional vehicle for such coordination (aside from informal discussions at the Quadrangle Club or the bar of the Shoreland Hotel) was the Local Community Research Council (LCRC). Established in 1923 with a grant from the Spelman Memorial (matched by local contributions), the LCRC was an attempt to coordinate an interdisciplinary research program into the nature and problems of urban life.[105] Leonard White chaired the LCRC for most of its existence, and he and Merriam and Beardsley Ruml of the Spelman Fund hoped their leadership would inspire their colleagues to address problems and to use methods of which the committee would approve.

The members of the LCRC controlled access to substantial research funds and were senior members of their departments, so their approval carried more than mere moral force. Still, the LCRC was an effort at coordination, not direction, and it was as difficult for it to deny the request of a senior scholar as it was for a junior scholar or graduate student to ignore its advice. The LCRC thus was another microcosm of the Chicago School's liberal managerialism in action, revealing both its promise and its pitfalls. Although it was a remarkably productive group, as time passed the research interests of LCRC members diverged under the pressures of disciplinary specialization and individual idiosyncrasy. The case of the Department of Economics is particularly revealing: the economists were never very active in the LCRC, nor did they need to be. They had other sources of support and were farther along the path of specialization. They could choose not to follow where Merriam led, and that is exactly what they did.[106]

Third, Merriam and the Chicago School attempted to integrate research and reform through the creation and maintenance of a set of coordinating agencies, such as the ICMA, the PACH, and the other groups headquartered at 1313 E. 60th Street.[107] The "experiment in propinquity" at the 1313 building brought together a set of organizations devoted to communicating information from researcher to practitioner (and vice-versa).[108] The result, Merriam and company hoped, would be to transform every administrative program into an experiment whose results could be incorporated into the edifice of administrative science.

These three methods of integration were closely linked, as Simon's experience reveals: as a graduate student he was supported from 1936 to 1939 by his work at the ICMA, which was housed in a building erected with Rockefeller Foundation money on the SSRC's recommendation according to plans drawn up by the LCRC. It is important to remember, however, that all of these integrative efforts depended on individuals as much or more than institutions. Without the Merriams, Brownlows, Whites, and Rumls to provide coordination by serving on several such organizations at once, the connections could—and would—fade. As we shall see in chapter 7, Simon and other leaders of postwar social science later would make similar attempts at interdisciplinary integration, only to repeat this same cycle on a larger scale.

Subjectivity

Despite the increasing power of impersonal, "rationalizing" forces, such as the market and the large-scale organization, social scientists from the turn of the century onward believed that human behavior was influenced strongly by nonrational beliefs and habits, most notably by religion and other local cultural traditions.[109] To most of the social scientists of the 1880s–1890s, who had themselves been raised to value religion and its civilizing virtues, the importance of such subjective factors in determining behavior and belief was not necessarily a bad thing. For them, the problem had been less subjectivity than its decline: from where would values come in a secular world?

In the early twentieth century, however, the problem of subjectivity was reinterpreted. European social thinkers, such as Sigmund Freud and Vilfredo Pareto, ascribed the horrors of the Great War to the power of the irrational.[110] In America, the fear of unreason was less extreme, but it still led to myriad studies of the social disorganization that came when social groups clung to traditional values in the modern urban environment. The work of the Chicago School of Sociology is a perfect example of such concern: Robert Park, for instance, was fascinated by the problem of "social control" in the ethnically diverse modern city, William I. Thomas studied the adaptive and maladaptive qualities of the culture of the *Polish Peasant in America,* and William Ogburn explored the problems of "cultural lag."[111]

Similarly, in political science the subjective aspects of human behavior fascinated the "other" Chicago School. The startling power of American prowar propaganda and the manifest irrationality of public opinion attracted the interest of both Merriam and Lasswell, for instance.[112] Works like Merriam and Gosnell's *Non-Voting* and Lasswell's *Psychopathology and Politics* revealed a corrosion of the iron faith of the older generation of political scientists: no longer could one assume that expert leadership and democratic politics could be reconciled through the education of a rational public. Now, when the public was not apathetic, it was only because its emotions had been manipulated.

Merriam and Lasswell were not the only political analysts who were dismayed by the irrationality of public opinion. Walter Lippmann's biting *Public Opinion* and his plea for expert leadership, *Drift and Mastery,* were widely hailed as taking a clear-eyed look at how democracy "really" worked.[113] From Watson's behaviorism to Freud's psychology to Pareto's sociology, theories that denied the importance (and even the existence) of the rational will excited the interest of political scientists. It was, as one historian of the period has termed it, the "crisis of democratic theory."[114]

The Chicago School's studies of the nonrational sources of behavior were conducted in hopes of establishing rational social control. Social control, however, can be understood in two very different ways. On the one hand, there is the ideal of social "self-control," in which the goal is to teach the populace to understand and thus to master itself. On the other hand, there is the model of self-control as governance, in which the goal is control of the irrational other by a rational elite.[115] From the 1880s to the 1910s, the former interpretation predominated, forming the keystone of the belief in the necessary unity of research and reform. Beginning in the 1910s, however, the second interpretation came to hold increasingly wide sway, and the Chicago School found in the subjectivity of human behavior an intellectual justification for its instrumental approach to the science of politics: the masses could not be enlightened, but they could be governed.

This preoccupation with the subjectivity of the social actor also raised the problem of the subjectivity of the social *observer*. This problem was critical for social scientists, since their intellectual credibility and social authority rested on their claim to objectivity. Consequently, American social scientists strove to develop techniques for controlling their own subjectivity, repeating the refrain that he who would master others must first master himself.

This was no simple task, for there are several ways in which a person can be objective, and not every type of claim suits every time. One can, for instance, understand objectivity in a moral sense, defining objectivity as the honesty, moderateness, and personal discipline appropriate to the gentleman, as Steven Shapin has shown in his *Social History of Truth*.[116] To many social scientists at the turn of the century, however, such traditional claims to objectivity no longer were sufficient, either for themselves or for the public they hoped to lead.[117]

Two new ways to understand objectivity were advanced in the late 1800s and early 1900s. One way was to base one's claim to objectivity on machinery and measurements, for instruments have no agenda and numbers do not lie—or so they claim. Examples of this approach included the "brass instrument" psychology of Edward Titchener and G. Stanley Hall and the turn to quantitative analysis in the economics of Irving Fisher and Wesley Clair Mitchell.[118] In political science, the creation of survey and polling techniques, the reliance on objective measures such as voter turnout, and the attempts to create measures of administrative efficiency are examples of such efforts to attain this mechanical objectivity.

The second new basis for claims to objectivity was method. As Karl Pearson wrote in his widely read *Grammar of Science* (1895), "The unity of all science consists alone in its method, not in its material."[119] That is, science is objective (and powerful) because of the procedures scientists employ in their attempts to discover and verify nature's secrets. By following a uniform procedure based on rigorous logic, scientists both natural and social could assure themselves and their patrons that not only personal bias but also the chances of fickle nature had been eliminated.[120]

What, exactly was that method? This proved a difficult question. Was the key to the scientific method induction? Deduction? Experiment? Statistical inference? The Chicago School, by and large, sought the answer to that question in some combination of machinery and method. If such a combination could be found, then the social sciences could hope to achieve the same status as the natural sciences. As we shall see, the search for this combination would become vital to Herbert Simon's work. Indeed, one reason why he found the computer so fascinating is that it was both a new machine and a new method. Though he moved from field experimentation to mathematical modeling to computer simulation, Simon continued to ask the same questions his mentors had, only in new languages.

Simon and the Chicago School

Herbert Simon was the quintessential product of Chicago social science in the 1930s. In him were blended deep commitments to research and to reform, as seen in his work for the ICMA and his later work with the California Bureau of Public Administration, the Economic Cooperation Administration (the Marshall Plan organization), and the President's Science Advisory Committee. In keeping with his work as expert administrator, Simon, like his mentors, adopted an instrumentalist view of knowledge's purpose, arguing for a sharp separation of facts and values and developing tools (such as optimization algorithms and computer simulations) to aid decision-makers in their management of large-scale organizations.

Simon's work also reflected the Chicago School's concern with change, interdependence, and subjectivity. In keeping with the ahistorical trend of political science in the 1930s, Simon began by studying processes and structures, not their history or evolution. Similarly, he was fascinated by interdependence, system, equilibrium, and organization, and he habitually adopted the functionalist style of analysis that this fascination inspired. He valued both specialized research and interdisciplinary collaboration, feeling the tension between the two and arguing that the key to resolving it was not so much institutional as individual. Finally, Simon studied the limits to rationality imposed by external influences on individual decision-making, and he adopted a combination of mechanical and methodological objectivity as his guide, finding in computer simulation the perfect fusion of procedural rigor and mechanical discipline.

Simon also shared the conviction of his mentors that the solution to these problems necessarily would require a new way of reconciling individual with collective interests, free will with deterministic science, and democratic politics with expert authority. His was a liberal managerial politics, and like his teachers, he believed that science had a vital role to play in making modern democracy work.

But Simon was more than a distillation of his context. For him, unlike his teachers, the limits to rationality were not imposed by irrational passions but by the simple fact that we always have to make decisions without complete information. For him, the enemy was not emotion but ignorance. The first step toward defeating that enemy was to look beyond political science to see how other sciences dealt with the problem of knowledge. These explorations into other sciences would expose Simon to the sciences of choice in addition to the sciences of control and thus would lead him down a unique path. This path began with the questions and concerns of the Chicago School, but it soon crossed into unknown lands.

CHAPTER THREE

Mathematics, Logic, and the Sciences of Choice

Simon's educational experiences outside political science, coupled with his own unique personality, led him to address the central issues of concern to the Chicago School in new ways. Perhaps most importantly, Simon came to understand the problem of subjectivity quite differently than did his mentors. Marx famously remarked about himself that he had stood Hegel on his head: with regard to the limits to reason, one almost could say Simon stood Merriam on his.

What made Simon different from his teachers at the Chicago School was his faith in the power of human reason, limited though that reason might be. To him, human rationality always was bounded, but that did not make reason ineffectual. The key to governance thus was not to chain the passions but to expand knowledge and amplify reason. Political science therefore should study the social supports for and constraints on rational behavior rather than focusing on the means for manipulating irrational beliefs: decision-making would be Simon's subject of study, not propaganda.

Unlike many of his teachers and fellow graduate students, Simon studied mathematical economics and mathematical biology and was influenced by the philosophy of logical empiricism. While these fields and this philosophy ultimately were compatible with Simon's Chicago School training, his encounters with them set his feet on a new path. That path led toward mathematical social science, and it was more commonly traveled by practitioners of the sciences of choice than analysts of social control. Thus, Simon's early training in mathematical social science gave him a new set of tools and a new perspective. These tools did not replace his other intellectual equipment, nor did this perspective render his other training obsolete. Rather, Simon sought to unite them, continually striving to redefine traditional concepts to fit his emerging new framework. The results were both profound and unexpected.

Simon, Schultz, and Mathematical Economics

From an early age, Simon believed that mathematics was the essential "language of discovery" of science. In the 1930s, that belief put Simon in the minority of social scientists. Ever since the early modern period, social scientists had admired mathematics, but they had cast their longing gazes from a distance. Up close, the prospect of a union between social science and mathematics was less attractive. Human thoughts and actions proved to be extremely difficult to measure and to mathematize, rendering many a social scientist skeptical of the value of mathematics, beyond basic statistical analyses. As a result, the development of a mathematical social science had been prophesied in many a preface but was infrequently attempted and still more rarely achieved.

While for some the union of mathematics and social science was a consummation devoutly to be wished, mathematical competence beyond a quite basic level was *not* required for admission to the profession in any of the social sciences until well into the twentieth century. The Chicago Department of Economics, for example, required only one quarter-long course in mathematics for its undergraduates and only one additional course in statistics for its graduate students in the 1930s, despite the department's close association with the mathematically oriented Econometric Society and Cowles Commission for Research in Economics. As late as the 1950s, the SSRC-sponsored "Summer Training Institutes in Mathematics for [*faculty* in] the Social Sciences" did not expect all their attendees to have had calculus before enrolling.[1]

While the majority of social scientists in the early twentieth century were untrained in advanced mathematics and skeptical of its value for the human sciences, many natural scientists and mathematicians found themselves increasingly drawn to the challenge of applying mathematics to social relations. The physicist and statistician E. B. Wilson of Harvard was one of the most prominent examples: he became an integral part of the Harvard Department of Economics in the late 1920s and even became president of the Social Science Research Council (SSRC) in 1929. A number of graduates of Harvard's economics department in the 1930s, such as the Nobel laureate Paul Samuelson, spoke of Wilson as their "true master."[2]

This move by engineers, mathematicians, and natural scientists into social science, which first became notable during the 1930s, was in part driven by supply. The Depression years were hard times for everyone, and physicists and mathematicians were no exception. Work was difficult to find, especially with the influx of European émigrés increasing competition for already scarce jobs. As historian Philip Mirowski has argued, when these mathematically trained physicists and engineers sought new markets for their skills, they looked first to economics, as much of neoclassical economic theory was modeled on the physics of the late nineteenth century.[3] In addition, the economic crisis of the 1930s focused the attention of intellectuals of every stripe on the problems of the economy. A great many engineers (including one of Arthur Simon's friends) attempted to construct physical models of the economic system, hoping to find ways to make the machinery of the nation's economy function more smoothly.[4]

While this new supply of mathematically sophisticated scientists and engineers played an important role in bringing mathematics to economics, shifts on the demand side were vital as well. A new attitude toward mathematics and quantification characterized the future leaders of the social sciences, especially in economics. These younger economists often welcomed the new interest of physicists, engineers, and mathematicians in their field. As far as they were concerned, whatever these *noveaux economistes* lacked in knowledge of economic reality, they more than made up for in their mathematical skill and analytic rigor.

In addition, large-scale bureaucracies' increasing demands for quantifiable data, coupled with the seemingly eternal need for social scientists to prove themselves to be true scientists, made quantification and mathematization appealing across a wide range of fields in the interwar period.[5] One example of this new demand for quantitative social science by large organizations was the U.S. Department of Agriculture's Bureau of Agricultural Economics.[6] Led by Henry Taylor, in the 1920s and 1930s the bureau not only gathered data on economic activity on a new scale but also supported advanced mathematical analysis of that data.

The Bureau of Agricultural Economics was not alone: insurance companies needed statisticians, large firms needed accountants and controllers, and banks needed investment analysts. When the demands for such mundane mathematics went up, so did support for mathematical economic research, especially when "common-sense" economics proved unable to cope with the problems of the Depression: witness the founding in 1932 of the Cowles Commission for Research in Economics by the wealthy investment banker Alfred Cowles, who had become convinced that traditional economics was failing to solve the problems of the day, especially the collapse of the stock market.[7] The National Bureau of Economic Research was a similar product of the interwar demand for sophisticated quantitative analysis of the business cycle.[8]

As a result of such new demands, economists began to welcome mathematical analysis at precisely the time that mathematically skilled physicists and engineers looked to apply their talents to social problems.[9] Supply and demand called each other into being and then grew in tandem, though both grew more slowly than the advocates of mathematization wished.

Of the social sciences, economics was the discipline that changed earliest and most in its attitude toward mathematics. That change had begun with the "marginalist revolution" pioneered by W. Stanley Jevons, Leon Walrás, and Carl Menger in the 1870s and codified by Alfred Marshall, Irving Fisher, and John Bates Clark in the 1890s and 1900s.[10] Like the revolution in political science launched by the Chicago School, the marginalist revolution in economics was a revolution in the basic aims and concepts of economics. It shifted the focus of economic inquiry from progress and poverty to the efficient allocation of a given set of scarce goods. It also entailed a redefinition of economic terminology so that its basic concepts could be measured and quantified.

The best example of this shift toward the measurable was the marginalist redefinition of value: marginalist economists saw value not as a product of labor (a concept with inherently unsettling implications for late-nineteenth-century capitalism) but as a product of exchange. To the marginalist, a good or service's value was the same thing as its price in an open market. With this shift, the traditional concepts of labor value and use value disappeared from economic theory, leaving only exchange value. (The Marxian concept of surplus value, it goes without saying, also vanished.) As the work of John Bates Clark reveals, this redefinition served the dual purpose of enabling quantification through the use of a common unit of measure (price) and shoring up the defense of the economic status quo.[11] According to Clark, every worker *necessarily* received in wages exactly what his labor was worth: it was impossible for anyone to be underpaid.

By the 1920s, the increased interest in mathematization extended well beyond the bounds of marginalist economics. Even the great rivals of the marginalists, the institutionalist school led by John Commons, Thorstein Veblen, Clarence Ayres, and Wesley C. Mitchell, developed a keen interest in quantification, though the institutionalists were skeptical of the deductively derived mathematical models of the marginalists.[12] The struggle between the marginalists and the institutionalists is often portrayed as being between mathematizers and old-fashioned literary/philosophical economists, but that is only part of the story. The difference between them lay not in mathematics and quantification per se but in differing understandings of the importance of nonrational factors in economic decisions, differences that often manifested themselves in opposing viewpoints about the kind of mathematics appropriate for economic analysis.

The 1930s, then, represent a period of transition in economics. The leading lights of the younger generation of economists, such as Paul Samuelson and Kenneth Arrow, were mathematizers, but their senior colleagues were not. As a result, although the trend toward mathematization in economics was clear, mathematical analysis did not become the defining characteristic of professional economics until after World War II.[13] A case in point is John M. Keynes's *General Theory of Employment, Interest, and Money,* easily the most influential economic work of the 1930s, which contains scarcely a handful of equations.[14]

The Chicago Department of Economics reflected this situation perfectly. Probably the most powerful intellectual figure in the department was Frank Knight, but it was his younger colleague Henry Schultz who was the mathematician—and it was Schultz, not Knight, who inspired Herbert Simon.

Schultz had studied under the pioneering mathematical economist Henry Moore at Columbia in the 1910s and received his Ph.D. in 1916, writing his thesis on the demand for sugar. In 1918–19, Schultz studied in London with the statistician Karl Pearson, developing a thorough understanding of modern statistical theory.[15] At Chicago, he combined detailed empirical work with inquiry into theories of demand and theories of measurement, producing in 1938 his magnum opus, *The Theory and Measurement of Demand*.[16] (Sadly, he died in a car accident soon after its publication.)

In this massive tome, Schultz explored the justifications for defining and measuring several of the fundamental terms of economics, developing operational definitions of his terms in a manner directly inspired by Percy Bridgman's *The Logic of Modern Physics*.[17] Schultz's philosophical debt to Bridgman is not surprising, for Bridgman's text was widely read by social as well as natural scientists. Indeed, the Nobel laureate's philosophy, known as operationalism, was extraordinarily influential in the social sciences in the 1930s and 1940s.[18] Even the (self-proclaimed) antipositivist Talcott Parsons accepted the importance of operational definitions in his *Structure of Social Action*, for example, though he gave the concept his own peculiar slant.[19]

Operationalism was a powerful variation on the positivist theme. Its basic thesis was that all theoretical terms must be defined via the operations by which one measured them: length is defined by the operation of comparing an object with a set of standard markings on a ruler, mass by placing an object on a scale, and so forth. If a concept could not be operationalized—if there was no set of procedures by which its constituent terms could be measured (or at least detected)—then that concept had no place in science. If a concept could be operationalized, then it did have a place in science, meaning that the social sciences could become "true sciences" if only they could define their terms properly.

Schultz viewed such operational definitions as crucial, for "as long as a concept remains unoperational, it is vain to hope that it will yield to the quantitative approach." His definitions enabled a new level of mathematization of theories of demand, and his exhaustive data gathering (supplemented by the work of the Bureau of Agricultural Economics) allowed a new level of precision in the quantification of demand curves for specific products. The combined result, he hoped, would enable the development of a "*Synthetic Economics* [that] is both deductive and inductive; dynamic, positive, and concrete."[20]

Herbert Simon studied with Schultz, taking his graduate seminars as an undergraduate, and he found Schultz's book an exciting example of the possibilities of applying mathematics to the study of human behavior. In fact, Simon cites Schultz as one of the three professors who influenced him most at

Chicago. From Schultz, Simon acquired a thorough understanding of marginalist economics, an unusually sophisticated grasp of modern statistics, and an appreciation for the epistemological problems involved in both measurement and model building in science.

Lotka, Rashevsky, and Mathematical Social Physiology

Schultz also introduced Simon to a book by Alfred Lotka titled *Elements of Physical Biology*, a book that Simon found every bit as stimulating as had Schultz.[21] Simon attested to its importance years later when he wrote:

> In the era B.C. (before cybernetics and servomechanisms) it [*Elements of Physical Biology*] was an important source of education and encouragement for the few souls who had a gleam in their eyes about the prospective mathematization of the social sciences. It had a substantial influence on Henry Schultz and Paul Samuelson, and, I am sure, many others besides myself. As a matter of fact, most of the central ideas that [Norbert] Wiener emphasizes—for example, the relation of entropy to organized behavior—can be found in Lotka, and I have felt some annoyance at the lack of recognition of the latter's contributions.[22]

Lotka was a biologist and statistician who had studied physical chemistry under Wilhelm Ostwald at Leipzig in 1901–2. Lotka wanted to redefine biology the way his mentor Ostwald had redefined chemistry. As historian Sharon Kingsland writes, "Lotka sought to create an entirely new field of science called 'physical biology,' which would analyze biological systems in thermodynamic terms." Just as Ostwald's physical chemistry "emphasized thermodynamic principles and mathematical analysis, so Lotka imagined that physical biology must treat the organic world as a giant energy transformer" whose functions could be measured and described mathematically.[23]

Though Lotka's program for physical biology did not bear the fruit he desired, several aspects of his work had an influence on social scientists in the 1930s. Herbert Simon was not alone in his admiration for the book, and many who sought to mathematize social analysis, such as Paul Samuelson in economics and Howard Odum in sociology/ecology, found inspiration in Lotka's work.[24] Such mathematically inclined social scientists shared Lotka's ambition to include human social behavior within the compass of science and were fascinated by his mathematical analyses of energy flows in equilibrium systems.

They also found appealing Lotka's Spencerian understanding of evolution as producing greater complexity, not just greater diversity, as well as his belief that "Man and machines today form one working unit, one industrial system."[25] In the 1940s and 1950s, this set of ideas—partly derived from Lotka and partly derived from the work of similarly minded engineers, mathematicians, and physiologists, such as Norbert Wiener and W. Ross Ashby—would become the backbone of cybernetics and the systems sciences.[26]

Perhaps the most important legacy of Lotka's work for Simon, however, was his understanding of the role of mathematics in science. Lotka firmly believed that mathematics was not just a tool for measuring or verifying concepts derived by other means.[27] Mathematics was a tool for discovery. It enabled more than just the clarification of concepts; it enabled the finding of relationships that would otherwise remain hidden. Whether his understanding of mathematics was derived from Lotka or merely reinforced by it, Simon certainly shared the view that "mathematics is a language of discovery."[28] This belief would play an important role in his mathematical theory-building work of the late 1940s and 1950s and in his writings on the philosophy of science from the 1950s onward.

Significantly, Lotka's work also reveals that mathematical biology and physiology were as much the models for mathematical social science as were physics and engineering in the 1920s–1930s. Of course, mathematical biology was itself modeled on physics, but the intervening layer of application to biological problems did make a difference. In particular, the mechanization of the understanding of the body that accompanied the mathematization of physiology provided a vital link between the two great systems of social-political metaphors—that the society/state is a machine and that the society/state is an organism—legitimating the application of methods and concepts appropriate to one sphere to the other.[29] Traffic along this metaphorical highway ran both ways: social scientists tried to apply scientific methods to the study of society, thinking of themselves as discovering the basic science necessary for socially therapeutic action, while natural scientists and engineers used the analogy to try to reintroduce the concept of purpose into the mechanical world.[30]

Lotka was one medium for this transfer of ideas, but he was not the only one. The prominent physiologist L. J. Henderson, for example, was a major force in the development of sociology at Harvard in the 1930s. He instilled in a generation of famous students and faculty, including Talcott Parsons,

Robert K. Merton, and Bernard Barber, the importance of understanding societies as organized systems whose parts were linked by relations of functional interdependence.[31] Henderson's Harvard colleague, the physiologist Walter Cannon, took a similar interest in the relationships between the "body physiologic and the body politic."[32]

The Chicago-based physiologist Jacques Loeb, with his mechanistic understanding of the body, also inspired many social scientists, particularly behaviorist psychologists such as John B. Watson and B. F. Skinner.[33] Loeb's mechanism, however, was far more rigid than that of L. J. Henderson or Walter Cannon, for he did not temper his mechanism with holism as they did. Hence, social scientists who followed Loeb's lead, as opposed to Henderson's or Cannon's, struggled to eliminate concepts like consciousness and purpose from their work at the same time that some physicists and engineers (e.g., Norbert Wiener and Erwin Schrödinger) were trying to find ways to incorporate them in theirs.[34] Simon would follow Lotka's path rather than Loeb's, spending much of his career seeking new operational, mathematical definitions for some very old-fashioned concepts, such as political power, causation, purpose, and reason.

Both marginalist economics and this new mathematical physiology had strong ties to nineteenth-century energy theory.[35] In many ways, this is not surprising: much early thermodynamic theory, particularly that related to the conservation of energy, was a product of physiological researches, and energetics continued to be profoundly important to physiology up through the early twentieth century.[36] The physiologist L. J. Henderson, for instance, worked at the Fatigue Laboratory at Harvard, which was a direct institutional product of the late-nineteenth-century understanding of work and fatigue as interrelated social, physiological, and mechanical problems.

Marginalist economics also drew heavily upon classical thermodynamics, using the equation of utility with energy as a license to import the mathematics.[37] It is no accident that E. B. Wilson, the teacher of so many mathematical economists at Harvard, studied with J. Willard Gibbs, the American physicist who codified the mathematics of late-nineteenth-century thermodynamic theory. Wilson's student Paul Samuelson even dedicated his famous *Foundations of Economic Analysis* to Gibbs.[38]

A further connection between thermodynamics, physiology, and mathematical social analysis is that both Henderson and Lotka (not to mention

Henry Schultz) were strongly influenced in their social ideas by the Italian engineer-turned-economist Vilfredo Pareto.[39] Pareto's work appealed to social scientists who had lost faith in the rationality of the public but not in the rationality of science. Henderson fit the bill perfectly: he was so taken with Pareto's ideas that he led a faculty seminar on Pareto at Harvard in the early 1930s, from which emerged Henderson's study *Pareto's General Sociology: A Physiologist's Interpretation.*[40]

Pareto's mathematical analysis of society was based on his engineer's understanding of the thermodynamics of equilibrium systems. For example, one of Pareto's central concepts was that of "ophelimity," now usually called Pareto optimality. Ophelimity was a redefinition of marginal utility in thoroughly thermodynamic terms. Pareto defined a system as having maximum ophelimity when the increase of the ophelimity of any one element in the system necessarily reduced that of some other element (or elements) in the system. Thus ophelimity, like energy, always was conserved in a closed system, making the reallocation of ophelimity a zero-sum game. This concept, it should be noted, has had powerful implications in a wide variety of areas, especially welfare economics, where it plays an important role in the work of Paul Samuelson and Abram Berg.[41]

Pareto's belief that behavior was determined by nonlogical factors, coupled with his rational mathematical analysis of that nonlogical behavior, made for a widespread Pareto vogue in the 1930s. Simon, unlike Schultz, never shared his fellow social scientists' enthusiasm for Pareto, mainly because he thought ophelimity was an ideal with little or no basis in reality: to him, all theories that presumed humans seek optimal rather than "good enough" outcomes were fantasies. Simon did, however, share Pareto's basic goal: the rational analysis of the limits to reason.

Another important source of ideas regarding the potential mathematization of social science along the lines of mathematical biology was the work of Nicholas Rashevsky, a biophysicist at the University of Chicago with whom Simon studied. Rashevsky, like Schultz and Simon, was a great admirer of Lotka. In fact, his reading of Lotka in 1925–26 had inspired him to try to create a new science, which he called mathematical biophysics. Eventually, Rashevsky's quest to create this new field led to his departure from the Westinghouse Corporation and his coming to the University of Chicago to develop a program in mathematical biology. In 1940, Rashevsky succeeded in creating a degree-granting Committee on Mathematical Biology at Chicago, institutionalizing Lotka's dream after his own fashion. For Rashevsky, unlike Lotka, the unity of science lay in the math, not the physics: when Rashevsky and his allies arranged for a reprinting of Lotka's classic text, the title was changed from *Elements of Physical Biology* to *Elements of Mathematical Biology.*[42]

Simon was closer to Rashevsky's students A. N. Householder and Alvin Weinberg than he was to Rashevsky, but he nonetheless lists Rashevsky, along with Schultz, as one of the three professors at Chicago who influenced him most.[43] (We shall meet the third in a moment.)

What appealed most to Simon were Rashevsky's skill at applying sophisticated mathematics to empirical problems, his conviction that organization was the key principle of biological and social behavior, and his conclusion that collective, hierarchical organization was more efficient in most cases than individualistic competition. These concepts would prove vital to Simon's work on organizations, for they helped him see bureaucracies as tools for expanding rationality and efficiency, not yokes on the minds of their members.

In addition, Rashevsky and his students A. N. Householder and Walter Pitts were keenly interested in the central nervous system. They were among the first to recognize that the "essential features of interactions of these units [neurons] are such that they can be imitated by different electrical, or electromechanical devices. Hence, by using proper configurations of a sufficiently large number of such devices, we can formally imitate some functions of the brain."[44] Thus, Simon's close familiarity with Rashevsky's work meant that he was aware of the earliest developments in the use of computing devices to model the brain. Simon did not know it at the time, but the parallels between mind and machine that Rashevsky had begun to explore later would lead him into two entirely new fields: cognitive science and artificial intelligence.

Carnap and Logical Empiricism

The third of the most influential of Simon's teachers at Chicago was the philosopher Rudolf Carnap. Carnap provided his avid student Simon with a coherent, rigorous philosophical foundation on which to build his social theories and a framework for understanding the nature and uses of formal logic and mathematics. Carnap's was one seminar Simon attended eagerly.

Carnap was the central figure (at least to Americans) of the Vienna Circle of logical empiricists (a.k.a. logical positivists) that came to international prominence in the 1920s and early 1930s. The Vienna Circle's official name was the Verein Ernst Mach, and the philosopher Moritz Schlick had organized it in order to encourage the development of Machian empiricism.[45] Its members were an illustrious crew: in addition to Schlick and Carnap, it included Otto Neurath, Hans Hahn, Herbert Feigl, Philipp Frank, Karl Menger, and Joergen Joergensen. They counted among their allies Richard von Mises, Hans Reichenbach, A. J. Ayer, Ernst Nagel, Bertrand Russell, Charles Morris (of the University of Chicago), and Percy Bridgman, among others.[46]

Logical empiricism was a philosophical weapon forged for use against the rising forces of reaction in interwar Europe. The ore from which it was made came from the deep mines of philosophical tradition, but it was shaped and sharpened by the interwar crisis of the West. The central targets of the circle's attack were religion, particularly the conservative form characteristic of Austrian Catholicism at the time, and metaphysical philosophy. The two were closely connected: the first commandment of logical empiricism was to "Exclude metaphysics and limit utterances to those about the given. For example: dispose of the idea of God."[47] On this commandment depended all scientific law—and upon it the logical empiricists would hang all the prophets of "blood and history."

Carnap, Neurath, and their allies saw religion and metaphysics as having produced an emotionally appealing but dangerously false faith in ideas like *Volk* and *Geist*, a false faith that was leading to authoritarian rule. Some philosophical problems might be inconsequential, but the "pseudo problems" of "God and the Soul" were dangerous. As Neurath wrote, "And behind it all stands Hitler ... Here comes God and Religion to the front and ancestral truths and the German Volk and what you need to stab a jewish [*sic*] socialist with a knife between the ribs."[48]

In opposition, Carnap and the logical empiricists developed a razor-like philosophy of which Occam would have been proud. This philosophy was based on the radical empiricism of Ernst Mach and the formal-logical understanding of mathematics developed by Gottlob Frege, Bertrand Russell, and Alfred North Whitehead. The logical empiricists held that the only things we could know for certain about the world were our sensory experiences of it.[49] All we can know is that we perceive something. How then to achieve some kind of agreement about things if we cannot depend on the things themselves to force that agreement on us? In the realm of logic and mathematics (which they understood to be wholly congruent) this was no great problem: logic prescribed a set of operations that were universally valid and so could be used to build up elaborate structures from simple elements. But from where would those simple elements come?

Carnap's answer was that one could make valid statements about one's observations of phenomena if one sheared away the layers of metaphysical preconceptions that are commonly added on to the data of sense experience. These preconceptions often implied a causal relationship that could not be proven: to him, science allowed the understanding of correlation, not causation.[50] In the physical realm, for example, this meant cutting away all claims regarding forces: beyond relative motion, all "forces" were metaphysical concepts. In the social realm, this meant eliminating all claims about the motivations of action. There was no "purpose" beyond observed behavior. These basic, neutral observational statements then would provide the basis for the "transparent construction" of more elaborate statements about phenomena.[51]

Significantly, this radical empiricism treated observations of the behavior of humans and observations of the behavior of billiard balls as equivalent: there was no necessary separation between the social and the natural sciences. According to the logical empiricists, it was only because of the faulty reasoning of metaphysical philosophy that the two were seen as such separate endeavors. All valid knowledge was founded on empirical observation and logical analysis. All science, in truth, was unified.[52]

One of the most important consequences of the logical empiricists' approach was that it shifted the focus of philosophy away from the definition of things toward the definition of rules for talking about things.[53] Because Carnap and his fellow Machians believed that the only criterion of truth was agreement among observers, not agreement with ultimate truth (for that was unattainable), they needed to devise a new means for arriving at consensus. They looked to do so by specifying a set of rules for proper scientific discourse. Philosophy, in their view, was properly the logic of the language of science. Thus, logical empiricism involved a turn toward methodological or procedural objectivity (described in chapter 2, above) at the most fundamental level.

Carnap's analysis of the logical structure of language led him to the conclusion that there were three types of statements: synthetic, analytic, and nonsense statements.[54] Synthetic statements were statements about the world and so were empirical propositions regarding phenomena that were observable (at least in principle). Analytic statements were those that were either true or false by necessity. Their truth-value was an inevitable consequence of the definitions of the terms they contained and the universal logical laws that related them to one another. Mathematics, to Carnap, was the perfect example of a system of analytic statements. The third kind of statement was a nonsense statement. Every statement that was neither synthetic nor analytic, including most traditional philosophy and all theology, he consigned to this third category. Logical empiricism was thus a self-consciously revolutionary philosophy. It was the philosophy to end all philosophy. It was only fitting that it emerged from the ashes of the war to end all war.

Simon absorbed Carnap's teachings eagerly, and, as was his habit, he pursued formal logic on his own, devouring Carnap's *The Logical Syntax of Language, Philosophy and Logical Syntax,* and *Testability and Meaning,* as well as A. J. Ayer's *Language, Truth, and Logic* and "some of Ramsey's essays, Morris's essays, Reichenbach—*Wahrscheinlighkeitslehre,* and Wittgenstein—*Tractatus,*" all by the summer of 1937.[55] This intensive study of formal logic and positivist philosophy had a profound, lasting effect on his thinking. Simon's initial title for his thesis, for example, was "The Logical Structure of an Administrative Science," and his first outline for it lists three questions as central to his work: "What is the logical structure of sentences in Newton's *Principia?* What is the logical structure of the sentences of economic price theory? What is the logical structure of the sentences of Aristotle's *Ethics?*"[56] All science was a unity, from Newtonian physics to economics and even ethics.

Simon, however, never became quite so radical an empiricist, or so enamored of formalization, as Carnap. By the 1950s, when the mathematical tide had reached full flood, Simon often found himself in the odd position of criticizing excessive formalization among mathematicians and logicians who were more interested in elegance than empirical accuracy.[57] Simon always believed that his theories did represent some truth "out there," and he consistently sought to find ways to save certain very old-fashioned concepts, such as causality, purpose, and the reality of social entities larger than the individual (e.g., formal organizations).[58] Carnap and his fellow members of the Vienna Circle had deemed such concepts nonsense, but Simon believed that within them lay a kernel of positive meaning that could meet the empiricists' criteria, if only they were properly defined.

Simon and Mathematical Social Science

In 1937, shortly after completing Carnap's course, Simon wrote his famous professor a letter in which he both praised and critiqued Carnap's book *The Logical Syntax of Language*.[59] This letter, first and foremost, shows that the young Simon possessed an enormous self-confidence, one that he never would lose, not even for a moment. In addition, the nature of his criticism is as revealing as the fact that he had the nerve to commit it to paper: Simon thought that Carnap overvalued formal "linguistic coherence" and undervalued correspondence with facts in judging the propositions of science.

Simon could not agree with Carnap on this point, for to him empiricism always meant matching theories to data. Formal analysis, especially mathematical formalization, was valuable because it helped eliminate the logically inconsistent and the vaguely defined from science. To Simon, however, formal coherence alone was insufficient. The key to science was the integration of mathematical formalization and empirical testing: "In empirical science the final test is not mathematical elegance or a priori plausibility, but the match between theory and data. I certainly learned that lesson somewhere, ultimately overcoming my innate Platonism and armoring myself against the aesthetic lures of neoclassical economics, so responsive to mathematical elegance and so indifferent to data."[60] Similarly, in *Models of Discovery* Simon wrote that, "having done much missionary work for mathematics in fields that had been largely innocent of it, I now find myself reacting to a surfeit of formalism more than to a deficit of it, especially in economics, statistics, and logic."[61]

Simon's letter to Carnap also reveals how much of the logical empiricist outlook he had adopted. To Simon, terms had to be operational, hypotheses had to be testable, and theories had to be rendered in a formal language, preferably mathematics, for social science to be truly scientific. Holding these beliefs made Simon an unusual social scientist; that he never doubted they were attainable made him unique.

These goals were attainable because for Simon, unlike the logical empiricists, the enemy was not metaphysics but ignorance, not theology but cognitive incapacity. If ways could be found to use science to expand social knowledge and to organize human understanding, then research would lead to truly fundamental reform.

CHAPTER FOUR

Research and Reform

In August 1939, Herbert Simon and his wife Dorothea set off for California, embarking on a new phase in Simon's career and in their life together. The journey was certainly a dramatic way to begin:

> The Burlington Railroad from Chicago to Denver connected with the Denver and Rio Grande Western. From Denver the train, nosing south to Pueblo, Colorado, pierced the Front Range through the Royal Gorge, then plodded steadily northward again up the valley of the Arkansas, with the Front Range red in reflected sunlight to the east and the shadowed snow-capped Sawatch Range to the west. For five hours, the locomotives throbbed rhythmically, lifting their heavy load 1,000 feet upward each hour ... One by one, Mounts Princeton, Yale, and Harvard passed in review (three almost identical pyramids, but the latter a few hundred feet higher than its rivals).[1]

The Simons arrived in the old resort town of Glenwood Springs, Colorado, on August 30 and stayed to hike the mountain trails the next day. It was a glorious summer's day, and Simon felt invigorated (as so many midwesterners do) by his encounter with the rawer nature of the western mountains. But that night the silence of the mountains was broken by "a loud, rasping voice." Adolf Hitler, his strident voice blaring from a neighbor's radio, was announcing that his armies had invaded Poland. (Simon volunteered to join the army as soon as war was declared in 1941, but he was turned down because he was color blind.)[2]

When Simon arrived in California he began work at the University of California at Berkeley's Bureau of Public Administration. His job was to direct a study of the administration of state relief programs. In late 1939, the worst of the Depression had passed, but there still were more than a million people on relief in California alone.[3] The Simons were not the Joads, but the dust raised by the weary travelers of Route 66 darkened the bright California sky for them as well.

The clouds of war and depression, however, did not dim the light of the Simons' still-new life together. (Herbert Simon had married Dorothea Pye, a fellow graduate student in political science at Chicago, on Christmas Day, 1937.)[4] Nor did they dull the edge of Simon's confident ambition. Rather, they formed a filter that colored the Simons' daily life and work, investing the mundane with the hue of larger meaning. When Simon left his "tiny cottage" on a hill overlooking San Francisco Bay each morning, he walked down Victoria Street convinced that his studies of government organization and administration were contributing to a worthy cause. When he returned home each night to write his dissertation, even the esoteric problems of philosophy and organizational psychology had an air of present weight and moment. Simon was truly in his element, working at an institution that was both academic and governmental, on problems that were both theoretical and practical.[5] In this context, research and reform were intimately linked.

Yet this linkage, however close, was not full union. Although Simon felt equally at home pounding on the desks of state officials or tapping at the keys of his typewriter, there was a distance between the scholar on the hilltop and the bureaucrat in the city below. That both the scholar and the administrator existed in the same body meant that this distance could be bridged, not that it could be eliminated.

To Simon, the way to bridge this gap was through the creation of a new profession, administrative science. The administrative scientist—prepared with a sound theoretical perspective via academic training, equipped with a body of empirical data gathered through experience and experiment, and interested only in efficient public service—was the key to solving the problems of social disorganization that plagued both the city and the broader world. He would do so by helping to sort facts from value judgments, by distinguishing technical from political issues, and by developing neutral tools and accurate measures that would aid the administrator in his work and make him accountable to his public. To do all this, the administrative scientist would need a broad theoretical framework: research was vital for true reform.

Measuring Municipal Activities

In Simon's view, the keys to creating this new profession were measurement, experimentation, and abstraction. To him, what made experts expert was the possession of an integrated body of abstract knowledge—a theoretical framework—and what made them useful was their ability to apply this abstract knowledge to concrete situations. Measurement was crucial to application, as was experimentation, which had the benefit of improving theory as well as practice.[6]

In keeping with this emphasis on measurement and quantification, Simon's first significant attempt to define both himself and his profession was a study on the measurement of municipal services. This study, begun in October 1936, was overseen by Clarence Ridley of the International City Manager's Association (ICMA) and was supported by the usual Rockefeller-centered network of funds from the University of Chicago's Committee on Social Research, the Social Science Research Council's Committee on Public Administration, and the ICMA itself.[7] The chapters of the report appeared serially in the new journal *Public Management* from February 1937 to February 1938.

The study was a natural outgrowth of the municipal reform movement's interest in creating efficient, rational city governance through the elimination of corruption and waste. It was, however, a product of a far more sophisticated, far more analytic approach to governance than was typical of such heirs to the mantle of Progressivism, for it was a brief on behalf of the science of administration. Through such works, Simon hoped to create a profession that would guard the guardians.

Simon was not the first—or the last—to dream of linking science and government. The last quarter of the nineteenth century and first quarter of the twentieth had seen many such attempts, at first topical and ad hoc, as in campaigns for improving sanitation and public health, then wider gauge, as in the social survey movement and in the creation of municipal research agencies.[8] These latter efforts were intended not only to make science available to government but also to make a science of governance. Nevertheless, administrative science was still in its infancy, in Simon's view, for it had barely begun to develop the measures or the concepts necessary for a theory of administration.

One of the first steps in creating any science is to develop accurate measures, and *Measuring Municipal Activities* begins by defending the importance of measuring municipal government services. "A generation ago," the authors write, "a municipal government was considered commendable if it was honest. Today we demand a great deal more of our public service. It must not only be honest but efficient as well." Of course, "it is probably true that in the present state of our knowledge the citizen can often better judge the efficiency of his local government by the political odor, be it sweet or foul, which emanates from the city hall than by any attempt at measurement of services." Still, progress "has been made and is being made," especially in the development of budgetary tools. "As a matter of fact, progress in the construction of tools has very considerably outstripped their application."[9]

This lack of application was not due to the abstractness or impracticality of such tools, Simon and Ridley argue, but to simple ignorance and mistrust of outside expertise. As they write, "It cannot be emphasized too strongly that these [new] standards are not theoretical concepts devised by academicians. Nor are they intended as playthings for statisticians. They are practical tools by means of which practical legislators and administrators can meet the practical need of choosing between alternative courses of action." Because of the utility of these tools, the authors are hopeful that "perhaps some day the need will be recognized of having at least one person with broad statistical training in the city hall of every city of substantial size." Without such expertise in the government—and appreciation of it by the public—the public would naturally fall back on "one of the oldest" but dangerously "fallacious" means of appraising a government: the tax rate. To Simon and Ridley, a low tax rate usually sent exactly the wrong message to the public about the quality of their government.[10]

Mere measurement, however, was not enough for Simon; counting did not satisfy. To him, there had to be a general theory behind the measures for them to have significance, intellectually or professionally. Thus, after these introductory arguments, Simon and Ridley move on to a description of the "theory of measurement" on which their analysis is based. This theory of measurement begins with three ideas: "(1) that units of measurement are of at least four distinct kinds—measurement of costs, efforts, performance, and results—each with a meaning quite different from that of the others; (2) that before the results of an administrative activity can be measured, its objective must be defined in measurable terms; and (3) that measurements may be used to answer two quite different questions, (a) how adequate is the administrative service and (b) how efficient is the administrative service."[11]

These ideas may seem obvious at first, but they have several significant, nonobvious implications. First, to define three units of measurement in addition to costs is to shift the emphasis in evaluating an administration from its *expenditures* to the *services* it provides, implicitly legitimating the expansion of government. If all one can count are costs, everything looks expensive; if one can measure benefits, however, then many things begin to appear worth the price. Second, to call for the definition of the objectives of government in measurable terms assumes that there are measures for most, if not all, services, and it places, implicitly or explicitly, a higher value on services that can be readily measured than those that cannot. Third, the distinction between adequacy (or effectiveness) and efficiency has implications, often unrecognized, for drawing the line between policy decisions and technical decisions. As Ridley and Simon argue, "service adequacy [unlike efficiency of service] ... involves social and political values as well as technical considerations, [and so] is not a decision which can be delegated solely to the chief administrator and particularly not to department heads."[12]

The authors of this "practical" manual for the "practical" administrator follow this explicitly "theoretical" chapter with a historical one, in which the origins of the various approaches to appraising government are explored.[13] Subsequent chapters go through specific government functions, from fire protection to police to public works to education (and more), providing the "practical tools" for "practical legislators and administrators" that the report advertises, such as sample forms for tabulating measurements of specific services. Why put these two introductory chapters in such an avowedly practical report, though?

These chapters were intended to help legitimate administrative science in the eyes of its two main constituencies: professional administrators and the broader reform public. The articulation of a "general theory of administrative measurement," for instance, was both natural and necessary for exponents of a new academic profession. It was natural because Simon and Ridley believed that what made an expert an expert was not merely an accumulation of experience but the possession of a theoretical framework that enabled him to interpret experience properly—how else could a twenty-one-year-old graduate student believe he had something to teach a career administrator? It was necessary for the same reason: only by asserting command over a body of theory could administrative scientists successfully claim jurisdiction (to use Andrew Abbott's term) over administrative issues and therefore be seen as legitimate authorities by their clientele.[14]

These chapters, particularly the historical chapter, also served to help legitimate administrative science in the eyes of the broader public by demonstrating that administrative scientists' concern with facts and efficiency did not make them insensitive to the values and desires of the public they served. Simon and Ridley took pains to describe administrative science as the ally of democracy, not its foe, arguing that their science "sprang from the need of the citizen in his practical action to decide, 'For which candidate shall I vote?'"[15]

Thus, although Simon's early work in administrative science supports Theodore Porter's argument that quantification is a strategy employed by weak groups (such as social scientists) as a means of acquiring social authority, it also reveals that Simon's generation of administrative scientists viewed quantification and measurement differently than did Porter's nineteenth-century statisticians or twentieth-century actuaries.[16] Simon and his fellow practitioners of administrative science in the 1930s did verge on the obsessive in their measurement of everything from the total amount of burnable property in San Francisco to the miles of street patrolled per week in Chicago. To Simon, however, quantification was a necessary component of a broader drive for abstraction because a number is an abstraction itself and because numbers are so useful in specifying the particulars of the case to which a general theory is to be applied.

Quantification was thus a means toward two quite different ends—theoretical generalization and practical application—but in both cases it was only a means, not the end itself. Simon was very critical of those who mistook the means for the end: he always held that the mark of a truly scientific theory was that it could be expressed in a formal language, such as mathematics, but he was well aware that quantification was not the only path to mathematization.

The distinction between mathematics and quantification was quite important to Simon, and he took the confusion of the two to be a mark of the amateur.[17]

Simon's pursuit of theoretical abstraction certainly seems to have suited the times, for in many fields during the middle third of this century, theoretical sophistication joined and sometimes even supplanted numerical precision as a keystone of claims to professional authority. The postwar rise of the economists and the emergence of the profession of management, for example, coincided with both the development of a new general theory (Keynesian economics, Simonian theory of management) and the creation of new measures.[18] A telling indication of the changing attitudes toward theory came in 1951, when the Cowles Commission for Research in Economics changed its motto from "Science Is Measurement" to "Theory and Measurement."

After these context- and legitimacy-building chapters, the report goes through each major department of city government, describing the various means of measuring the services they provide, from miles of street cleaned to per pupil school expenditures to the number of library books in circulation. In their description of such measures, they emphasize time and again the importance of defining clear, measurable objectives for each department—a call that sounds like nothing more than a plea for simple reason until one realizes that having multiple, (at least partially) conflicting objectives is the natural state for an organization created through political action and legislative compromise.

Not surprisingly, city planning turns out to be the hardest function of government to evaluate, but, as the authors take care to point out, it is perhaps the most important function of all, as it sets the goals and standards for all the other agencies. Here it is important to observe that Ridley and Simon, like Dewey before them, see the city planning agency as vital to truly democratic governance, not as a technocratic threat to democracy. To them, the city planning agency was the place where the myriad desires of the public were to be integrated into a rational plan for the whole community.

This hymn to integrated, rational planning rang sweetly in the ears of other planners, administrative scientists, and reformers. As a result, the budding professional community of administrative researchers and reformers welcomed Simon (as it already had Ridley) with open arms. Through this report and through his work with Ridley on the *Municipal Yearbook,* Simon soon became well known within this small but well-connected community, particularly since Ridley rapidly came to trust Simon to present their work by himself.[19] It was not long before Simon was holding forth "like the young Jesus in the temple" on the measurement of government services to audiences whose members had met their first payrolls before Simon first had met the sun.[20]

The Politics of Public Administration

Recognition brought opportunity, and in the spring of 1938, Samuel May of the University of California at Berkeley's Bureau of Public Administration invited Simon to come out west for the summer to write a proposal to the Rockefeller Foundation for a three-year study of various topics in the measurement of municipal activities.[21] Simon, needing funding, jumped at the chance. By the next summer he was the proud recipient of a grant for $30,000 (spread over three years).

The project officially was under May's direction, but May had many other duties. He granted Simon near full responsibility for the study, and Simon behaved in practice as if "near" and "full" were one and the same. He had a research staff of three (most with Ph.D.s), a statistical assistant, a secretary, and additional staff numbering between fifty and over a hundred, depending on the study. In actual practice, his responsibilities were even greater: Simon effectively directed the operations of two large offices and hundreds of employees, despite lacking any official authority to do so.[22] That he was a de facto but not de jure director of these offices helped make him aware of the fact that authority often is not well described by organization charts. Not bad for someone who had barely more than an outline of his doctoral thesis completed.

Simon's new project suited both his interests and those of his patrons at the Rockefeller Foundation, such as Raymond Fosdick, who became the foundation's president in 1936.[23] The foundation's interest in public administration was a relatively new one, having begun in the early 1930s. Its support for this most applied of social sciences was quite risky, for it could lead to political entanglements the foundation wished to avoid. At the same time, however, it was a natural outgrowth of the liberal, managerial politics that foundation officials such as Fosdick and Beardsley Ruml shared with the rest of the national reform elite.[24] To them, the chaos of the Depression and the corruption and inefficiency of government (at all levels) were different manifestations of the same basic lack of intelligent coordination. The economy, the government, the whole society were disorganized; the solution to the problems of the day, therefore, was efficient, rational organization. The way to provide that organization was through the expansion of centralized administrative capacities, both public and private.

While proposals for administrative reorganization in the interest of efficiency always sounded like neutral, apolitical calls for reason to their sponsors, they could provoke fierce reactions. As Harold Seidman, John Mark Hanson, Hugh Heclo, Samuel Hays, Louis Galambos, and others have shown, although we tend to think of the U.S. president as being in control of the executive branch, it has become enormously difficult for the president to exert control over the bureaucracy. Early in the twentieth century, central control over budget-making was thought to be the key to ensuring that the policies set by the president (and his staff) would be implemented by the bureaucracy; hence the increasing importance of the Bureau of the Budget (later the Office of Management and Budget) and the Executive Office of the President. Congress and the various agencies, however, typically have resisted such centralizing measures precisely because they amplified the power of the president over the bureaucracy and so reduced their own.[25]

At the same time Simon was working on his California studies, a number of prominent scholars (many of whom had connections to the "1313" organizations in Chicago) were called in by President Roosevelt to form the President's Committee on Administrative Management. Led by Louis Brownlow of Chicago's Public Administration Clearing House, this committee labored long and hard to produce exactly the proposal for reorganizing the executive branch that FDR had known they would produce. They proposed a number of reforms, all of which would have had the effect of centralizing executive authority in the hands of the president and of making that authority felt more swiftly and surely throughout the ever-growing federal bureaucracy. This report provoked a political firestorm, and most of the proposals died in the conflagration. The only ones to survive were those that did not require Congressional approval.

The Rockefeller Foundation's leaders, from its president Raymond Fosdick to important advisers, such as Charles Merriam and Beardsley Ruml, were charter members of the liberal managerial reform elite, as were the vast majority of the leaders of the Social Science Research Council (whose bills the Rockefeller Foundation paid). Hence it is no surprise that in 1933 the SSRC created a new Committee on Public Administration, headed by "1313" veteran Louis Brownlow, with the blessings of the foundation.[26] Simon's proposal on behalf of the BPA reached the foundation before its leaders had been burned by their support of public administration as a field and so was received enthusiastically. In California, however, Simon soon would find that politics and the science of administration were far more closely linked than either he or his patrons would have liked to admit.

Simon's organizational home out west was the University of California–Berkeley's Bureau of Public Administration. The BPA (which still exists as the Berkeley Institute for Governmental Studies) was one of the leading academic centers for policy research in the 1930s.[27] The BPA's guiding philosophy was similar in many ways to that of the ICMA and the other "1313" organizations in Chicago, though its mission was local, rather than national, in scope. The BPA focused exclusively on state and local governmental issues in California and was linked closely to the California state government. Simon, for example, once worked with the director of the BPA on a report on water issues in the Central Valley for a legislative committee and helped draft the annual report to the state legislature of Berkeley chancellor Gordon Sproul.[28]

The BPA had been created in 1919 in a final postwar surge of Progressivism, and its mission was to perform studies of use to California governments. The BPA received its funding from foundation grants (particularly the Rockefeller Foundation) and from the state (via the university). Its director during Simon's years was Samuel May, a political scientist with a long history of political involvement.[29] May, like most of the BPA staff, was well to the left of center politically, a fact that would cause difficulties for the organization and for various staff members from time to time. All of the staff members seem to have weathered such disturbances safely, however, protected by their being one step removed from actual governance.

For example, Simon remembers his colleague Milton Chernin being hounded by conservative university officials. He recalls with great satisfaction that Chernin survived such attacks and became dean of the School of Social Welfare and a revered member of the university community at Berkeley. Similarly, Simon mentions—with a note of pride in his defiance—that on his departure from the BPA he was given a multivolume set of Marx's *Capital*, which he resolved to display on his office shelf, vowing to move to New Zealand if ever it became politically necessary to remove them. Simon's own politics at the time were quite liberal: in the 1940s, for example, he joined the NAACP and attempted to organize a Committee on Academic Freedom of the APSA in order to combat anticommunist hysteria.[30]

The primary members of the BPA staff with whom Simon worked were William Divine, Myles Cooper, Victor Jones, Ronald Shephard, Milton Chernin, and Frederick Sharp. The connections Simon made at the BPA would serve him well in his later career: Jones, for example, helped Simon get his first academic position at the Illinois Institute of Technology in 1942, and Shephard later would be an important connection for Simon at RAND.[31] But far more important than the professional utility of these connections was their personal quality. This was a young, confident, talented, and ambitious group, and Simon learned a great deal from his colleagues, particularly from Shephard. It was also a lively group, and Simon clearly enjoyed the male camaraderie of his research team.

Simon's BPA project on measuring governmental services was, in actuality, three projects. The first was a "field experiment" in public administration conducted for the State Relief Administration. The second was a study of fire risks and fire losses in the Bay area. The third was a study of the *Fiscal Aspects of Metropolitan Consolidation*. In these studies, Simon learned a great deal about the virtues of experimentation, the perils of politics, and the problems of control. We can learn from them as well, for they reveal much about Simon's goals, methods, and habits of thought.

Simon's training in political science at the University of Chicago had been nonstandard, to say the least. This trend continued throughout his California years, setting the pattern for his later life and work. Instead of being discipline-centered, his training was task- or problem-centered. In the course of his work at the ICMA or BPA, Simon would encounter various problems, all having to do with the challenges of governing a city, none having ready answers within the traditional bounds of public administration. He then would seek ways to solve these problems through wide reading in several disciplines as well as extensive conversation with colleagues trained in other fields.

As a result, Simon came to see disciplines as toolkits from which one instrument or another might be selected to address a given problem. The solutions to the problems of governance could come from a variety of fields, from statistics to sociology, economics to psychology. The administrative scientist, therefore, would have to be able to bring all these disciplines' tools to bear on the problems of administration. A well-run research team could be interdisciplinary, but in Simon's view, being interdisciplinary was best achieved by an individual with diverse training and a broad perspective.

Once in possession of these tools, Simon applied them to the particular problems of particular cities, but he always did so with an eye to drawing generalized conclusions about the value and applicability of these tools to other situations. At times, this drive to generalize could be overwhelming, pushing the original goal of solving the problem at hand into the background. Still, Simon never forgot that there was a practical problem that needed solving at the root of his investigations, though it sometimes was an effort to force himself to come back down from the heights of theory.

In addition to being problem-centered, Simon's training also was largely informal. This is not to say that it lacked rigor; rather, when Simon searched a discipline for tools he gave it a thorough ransacking. Yet when he studied a discipline to see what it had to offer, he did so largely on his own terms and so was not socialized into that discipline as other practitioners might be. He developed his skills in mathematics, economics, and statistics, for instance, through independent reading and through long conversations with his friend and colleague (and subordinate!) at the BPA, Ronald Shephard. From Shephard, a student of the statisticians Jerzy Neyman and Griffith Evans, Simon learned a great deal about the state of the art in statistical inference and economic model-building.[32] One of the new generation of economists trained in the 1930s who were as much mathematicians as economists, Shephard wrote his dissertation on the usefulness of simplified models in the analysis of complex systems, such as economic systems. While Simon did not seize on Shephard's ideas about the value of simplified models right away, he did see great potential in them and stowed them away in his mental toolbox.

Conversations with a friend and colleague teach a subject in a very different way than do lectures at the feet of a professorial master: while he became skilled at economics, Simon never became a professional economist. The same also can be said of Simon with regard to sociology, psychology, statistics, and computer science. Even his original disciplinary affiliation with political science was unusually weak, as his successive redefinitions of himself as organization theorist, cognitive psychologist, and computer scientist indicate.

The profession with which Simon did identify—at least at this time—was that of administrative scientist. Significantly, this occupational ideal was both academic and professional: the administrative scientist undertook research into general principles in order to solve the particular problems of specific clients. Hence, Simon's work as an administrative scientist was strongly client-centered, as well as problem-centered. He needed to produce practical results that would help the ICMA and the BPA solve their clients' problems. He could try to educate his clients (and his patrons), redefining their problems to better suit his research goals, but such education had its limits, especially since the fledgling field of administrative science lacked the cultural authority of law or medicine.

To Simon, who desired to have a present effect in reforming the world about him, such constraints were less limitations than they were opportunities, but because he wished to probe deeper and more freely, such client-centered work would prove frustrating. In the course of his thesis work, Simon's fascination with research would overcome his interest in reform. When the time came, he would choose an academic (peer-centered) rather than a professional (client-centered) career.

Simon's first study for the BPA was titled *Determining Work Loads for Professional Staff in a Public Welfare Agency.* The study came about after the California State Relief Administration's Division of Planning and Research, then headed by Milton Chernin, requested that the BPA assist them in a study of staff work loads. The SRA had been established in 1933 as an emergency agency whose purpose was to "assist in relieving distress caused by unemployment." It supplemented the federal WPA relief program by providing aid to the "employable unemployed" who were in excess of the state's federal quota. The SRA was a statewide organization, and it provided both direct relief and work relief through its own network of district offices. It was independent of the State Department of Social Welfare and thus was only a smallish patch in the crazy quilt of relief programs. Even so, the agency's annual budget in 1940 was roughly $25 million, and its total caseload ranged as high as 114,000 per month in that year.[33]

The study's purpose was to determine how many social workers were needed for the most efficient operation of the SRA program. A more innocuous, apolitical request for technical analysis could hardly be imagined—or so it would seem. The preface to the report, by Frank Hoehler, president of the American Public Welfare Association, gives a hint as to the politics surrounding this request: "While administrators have known that an adequate staff effects real economy in the administration of a relief program, they have had to contend with the notion that the less an agency spends for administration the more effective is its work."[34] This study, he offers, shows that this notion is false. Hence, the report could be taken to show that increased state aid—and an increased state bureaucracy—were necessary for efficient operation in the area of relief. These were fighting words at the time, for during the entire period of the field experiment, the SRA was under severe attack.

California politics at the time was volatile, to say the least.[35] Despite the prominence of radical programs of reform, such as Upton Sinclair's unmistakably socialist platform (the EPIC plan, which had won him the Democratic nomination for governor only a few years earlier), California conservatives were quite powerful. Sinclair had been defeated, after all, and it was not until 1938 that an avowed New Dealer (the Democrat Culbert Olson) was elected governor. Olson encountered fierce opposition from conservative state assemblymen, especially after a new Speaker (Gordon Garland) came to power in early 1940. Garland opposed what he saw as "the increasing trend toward collectivism" in California and the nation. The SRA was the focal point of the power struggle between Governor Olson and Speaker Garland, who believed that the "SRA is being used for the development of the Communist program."[36]

Governor Olson called a special legislative session in February 1940 to reconsider the state's appropriations for relief. As Simon notes in the report, "During this session, the SRA was the center of continuous controversy."[37] Simon did not exaggerate. A special legislative committee, headed by Garland, was appointed to investigate "subversive activities" within the SRA. The report of this committee stated baldly: "This committee offers no apologies for its refusal to join those who blandly offer guaranteed 'cures' for unemployment. Such schemes are worse than the condition they would fail to correct, because they tend to array class against class . . . Proponents of these schemes are deliberately working, more or less openly, toward a revolution . . . Despite repeated statements to the contrary from the Administration, the SRA is being used for the development of the Communist program . . . Communists, fellow-travelers, and the so-called 'intelligentsia' still hold responsible positions in the SRA."[38]

A second special session followed hard on the heels of the first, and in this session a bill was passed transferring SRA activities to the counties. Governor Olson vetoed the bill, but appropriations for the SRA, and for state relief in general, were reduced, procedures were changed, and eligibility requirements for aid were stiffened.[39]

A key issue in all of these battles was the size of the state relief bureaucracy, which conservatives in the state legislature saw as the embodiment of creeping socialism. Prefiguring later attacks on the welfare state, they recognized that attacks on a bloated bureaucracy were acceptable to the broad electorate, while attacks on the services those bureaucracies were supposed to provide were not. Led by Speaker Garland, these conservatives won a 15 percent cap on administrative costs in relief programs in the February special session, forcing personnel cuts in many agencies, including the SRA.[40]

Simon and his colleagues on the study (including Chernin, who had left the SRA to join the BPA after requesting the study) hoped that a careful, dispassionate analysis of where the true economy lay in administration would help to counter the longstanding belief that less is more in government. Even more, they hoped to perform a model field experiment in social science, one that could be repeated in a variety of locations on a host of issues. Such a model experiment, like the renowned Hawthorne experiments, would be crucial to the development of a true science of administration—and to building credibility for that science.[41]

It is no surprise then, that Simon and his co-authors devote the bulk of the report to the description of experimental procedures and analytic methods. Their concern with experimentation even overshadows their interest in the specific issue at hand, for the report's conclusions have more to do with social experimentation than with the delivery of relief services. The political battles surrounding the SRA, for instance, are described mainly in terms of their impact on the experiment's validity, not on the provision of relief.

In order to attempt to control for these disturbances, the experimenters argued for (and won) exemptions from many policy changes for the two district offices under their control. Similarly, the researchers took "special care" to ensure that changes that could not be avoided were "introduced in as orderly a fashion" as possible. "Every attempt was made to preserve normal operating conditions. The workers were assured that personnel would be maintained according to normal quotas in the experimental districts, and non-mandatory changes in procedure were withheld until the conclusion of the experiment."[42]

Such precautions raised the specter of "Hawthorne effects," a term used to describe the situation when membership in the experimental group is in itself a significant factor in influencing behavior (especially morale). The precautions also revealed the remarkable differences between real life and the experimental situation, even in a field experiment: anyone who has worked in a large organization knows full well that the "external" world of higher-level politics and policy can intrude on daily tasks. Such intrusions may be infrequent, but no organization exists for very long without encountering some such crisis. Hence, dealing with ordinary tasks in extraordinary circumstances is a *normal* part of organizational life. To control for such intrusions, even when dealing with seemingly purely technical questions, introduces a significant element of unreality into the experimental situation (as anyone who has sympathized with Dilbert will recognize). This may seem an abstract point, but one of the most persistent weaknesses in organization theory—and systems science more generally—was exactly this inability to deal with such "extraordinary," but normal, disruptions.

Simon and his colleagues were well aware of such potential Hawthorne effects, but they took away from their experience on this study a very different lesson than Elton Mayo and Fritz Roethlisberger, the lead researchers in the Hawthorne study, might have. To Simon, "The many delays and problems caused by the difficult situation of the SRA at the time of the study brought out strikingly the absolute necessity for centralized control over all phases of the experiment."[43] Instead of moving toward incorporating such "uncontrollable" disruptions in their analyses, Simon and his colleagues (particularly Shephard and Sharp) sought to take another step away from the messiness of real life. If control over all phases of the experiment was absolutely necessary, then the laboratory environment was the best—and perhaps the only—place for social experimentation.

Analysis and Control

The city was the focus of Simon's research, however, and it was difficult to see how the problems of social organization in the city could be brought under control, still less into the laboratory. The crucial question, then, was this: were there ways to gain control over a situation *analytically* that one did not control *materially?* Experimentation, with its tight coupling of theory and observation, cause and effect, was still Simon's research ideal, but by the end of 1940 he increasingly began to pursue other paths to analytic control of the problems of the city.

Simon saw three broad strategies for establishing analytic control over an unruly world. First, he could follow the model of mathematical economics and statistics, gathering such measurable data as could be found and working back from such observational data to the underlying patterns of causes and effects beneath the surface world of behavior. We can call this the path of *statistical inference*. Second, he could begin the theoretical work of developing analytic tools and concepts that, by virtue of their internal coherence and broad congruence with common experience, would provide greater clarity and insight into administration (and organization more generally). We can call this the path of *concept formation*. Third, he could undertake the philosophical work necessary for understanding science itself, attempting to understand how artificial constructs like theories, models, or experiments relate to the real world. In particular, such a theory of theories might reveal when and where it was permissible to isolate aspects of the larger whole for closer analysis. We can call this the path of *applied epistemology*.

Simon followed the first of these paths in his final two studies at the Bureau of Public Administration and the latter two in his dissertation. In the first of these studies, an analysis of *Fire Losses and Fire Risks,* Simon and his colleagues collected data on property values, insurance rates, insurance claims, and fire losses for the Bay area. They then used this data to create a unit measure (fire losses per $1,000 valuation) by which to assess the effectiveness of the various municipal fire departments. This report, published in 1942, reveals that Simon had pursued his economic and statistical self-education aggressively in the course of the study: although "the application of modern statistical theory to the problems of Chapter IV [of the report] is almost entirely due to Dr. Shephard," when Shephard left the study before its completion, Simon was able to complete the statistical analysis on his own.[44]

By the end of his study on fire risks, Simon was confident enough of his mastery of economic and statistical analysis to turn his attention to a larger scale analysis of the problems of the city. The subject of this final study was one near and dear to the heart of every city planner and civic reformer: metropolitan consolidation. Big cities then, as now, were in serious financial difficulty. In part, this was due to the movement of wealth from the city to suburban areas where low property taxes could buy a high level of government services. Conversely, the areas most in need of government services were often the ones with the fewest resources available to pay for them. Every big city's metropolitan region was thus a patchwork of varying tax rates and service levels. In addition, although social and economic issues showed no regard for municipal boundary lines, political leaders certainly did. In the eyes of planners, political divisions were preventing people from addressing the problems of the region as the wholes they truly were.[45]

BPA director Samuel May echoed these arguments in the foreword to Simon's study of the *Fiscal Aspects of Metropolitan Consolidation,* writing that: "students of government have been concerned with the waste implied in multiple government units for one metropolitan area, and have long advocated some solution involving integration or consolidation. The practical difficulties, however, of overcoming long-established local interests and loyalties have prevented much actual experimentation." May went on to note that the "study does not assume consolidation is necessary" and "does not attempt . . . to deal with the political problems involved in integration." (An odd silence for a political scientist, and one we will return to in the conclusion to this chapter.) In May's view, though, the study did lead to a "reinterpretation of the impact of consolidation on economic interests and political forces." What this reinterpretation meant as regards "the problem of democratic control of metropolitan government" was left for the citizens themselves to decide.[46]

Despite such qualifications, the thrust of the study was clear: there were striking differentials in the services provided by the various municipalities, differences which did not "represent any reasoned decision on the part of citizens or their councilmen." Rather, "the existing level of service must be admitted to be on the whole an historical accident." Rational economic analysis, on the other hand, revealed that "metropolitan consolidation in the San Francisco Bay Area is highly desirable." In addition, "it would be practicable and desirable to 'level' governmental services throughout the areas," though it would be "undesirable to abolish existing differentials in property taxes" in the process.[47]

While Simon's overall conclusion regarding the desirability of consolidation supported the conventional wisdom of planners, it rested on different grounds than did the typical argument for consolidation. Instead of arguing that differentials in local property tax rates were inequitable, Simon contended that the need for metropolitan consolidation came from the need to eliminate differences in *service levels,* not tax rates. In fact, in his view, it was "preferable that existing tax differentials not be disturbed."[48]

Simon tells his readers that he was able to reach this surprising conclusion because he had two tools in his possession that others had not had: "(1) a theory of the incidence of the urban property tax [elsewhere termed 'the general theory of tax incidence'] . . . and (2) techniques for measuring the *service side* of the metropolitan revenue-expenditure picture."[49] Significantly, both of these tools were, in no small part, the products of his own labors.

In preparation for this study, Simon had undertaken a thorough review of the economic literature on taxation, which was itself a specific branch of the more general theory of supply and demand (particularly the supply and demand for capital). His review was thorough enough that he was able to publish his conclusions on the subject in the *Quarterly Journal of Economics* in 1943. The article, "The Incidence of a Tax on Urban Real Property," noted that there were several schools of thought as to who really paid a tax on property. These differing schools all understood the same broad set of economic laws to be in operation, and (with some small exceptions), they all applied these laws in logical fashion to the facts. Yet somehow they reached divergent conclusions. From this analysis, Simon drew the conclusion that their differences were rooted in different basic assumptions that shaped their understandings of which economic laws were relevant.[50]

The argument in this article typifies Simon's approach to problems throughout his career. He begins with a set of conflicting claims, traces them back to their roots in divergent assumptions about human nature, and then evaluates these assumptions to determine which set best describes the world. At every step in his analysis, mathematics proved to be an essential tool. The final product was a general theory, in this case, of the tax incidence.

In *Fiscal Aspects,* Simon first applied this general theory to data gathered by the ICMA on taxes and services for the Chicago metropolitan area (by way of illustration) and then to data the BPA had gathered for the Bay area. Simon's calculations revealed that equalizing services would require a considerable increase in spending by all areas except the central city and the wealthiest suburbs. These increased costs were so large that, were they to be paid for out of the local property taxes, the increased taxes would be extraordinarily heavy burdens, even amounting to a "complete confiscation" of property values in certain townships.[51] Such an outcome was, of course, unacceptable on many levels. The only solution, then, would be to pool property tax resources areawide (or to use a different revenue source than the despised but unassailable property tax). The question was then, if the region were to pool its property tax resources in order to equalize services, should not the tax differentials between local communities be leveled as well?

Simon admitted that this logic was powerful politically but argued that it reflected a misunderstanding of economic forces. Because of the phenomenon of *capitalization,* inequalities in property tax rates affect only the owners of the property at the time the tax was imposed (or rather, at the time it was widely anticipated). This curious result was due to the fact that the future selling price of the property would be lowered to reflect the added costs imposed by the tax: the amount present owners lose to higher property tax rates, they have, in effect, already gained via a lower purchase price.[52]

Thus, the intellectual tools of the expert had led to an unexpected result with immediate practical implications. General theory had been applied to specific cases and had produced concrete results. Even more, it had produced results that revealed the advantages in both equity and efficiency of rational organization in contrast to historical and political "accident." The "planning approach in public economy" thus offered solutions that other, inexpert, non-interventionist approaches could not.[53]

Indeed, in the course of this study Simon became such a strong advocate of rational planning that in 1941 he began to argue that "There does not seem to be any valid reason why the revenue-expenditure process in governmental agencies need by characterized by less 'rationality' or 'free choice' than the private revenue-expenditure process."[54] Similarly, he held: "There is no *a priori* reason why the community should select the competitive market as the institutional means of organizing its activities, any more than it should select a governmental organization. It is only after the institutional framework has been constructed that the 'rationality' of consumers' behavior in the market has any meaning in economic theory—and the 'rationality' of an administrator in implementing a social value scale through public expenditures would appear to have exactly the same meaning."[55]

Theory and Practice

Curiously, this approach to the problems of the city reversed the linkage between research and reform that Simon himself advocated: research had revealed the proper goal—consolidation—but was silent regarding the political means by which this goal could be made real. This was a significant silence, especially for a political scientist committed to practical solutions. The cause of this silence was the disjunction between the scientific desirability and the political feasibility of consolidation.

This disjunction troubled Simon, as his essay "The Problem of Practicability in Applied Political Science" reveals. Written a few years after the completion of *Fiscal Aspects* (and prompted by one of his many run-ins with the planners in the School of Architecture at the Illinois Institute of Technology), this essay took metropolitan consolidation as the best example of a conflict between the idealized prescriptions of administrative science and the realities of urban politics. Lamenting that "it is an unfortunate fact that there is very often (not always) an inverse relation between the intrinsic merit of a plan and its realizability," Simon argued that "social change of a really fundamental sort will be attainable only when theories of implementation arrive at the same level of sophistication as our understanding of what a desirable solution, if attainable, would be." A crucial step in the development of such theories of implementation, Simon held, was the "recognition of the important psychological barriers to logical behavior."[56]

Perhaps the most vital of such barriers to logical behavior was that between local and larger interests—between the individual and the group. As Simon found time after time in his work at the BPA, what was rational for taxpayers in Oakland might be rational no longer if they became (or thought of themselves as) citizens of the metropolitan region instead. How these bounds to rational choice were set and maintained, how they affected the decisions made within them—these were crucial questions of the broadest importance, and to them Simon turned each night when he climbed the hill at the end of his day's work at the BPA.

CHAPTER FIVE

Homo Administrativus, or Choice under Control

Simon's work at the International City Manager's Association and the Bureau of Public Administration set the pattern for his career as an administrative scientist. From his thesis, *Administrative Behavior*, to his groundbreaking text in *Public Administration*, he continually sought to demonstrate the importance of organizations and expert knowledge in the struggle to master the complexities of modern urban life. He was guided in this effort by the belief that the planning process could improve democratic accountability. He was sustained by the conviction that if the proper framework could be constructed for administrative science, research and reform would be complementary, rather than competing, interests. Thus, while he often was frustrated by the strength of what he saw as the forces of interest, ignorance, and inertia, he never lost faith in the power of organized intelligence, especially his own.

To Simon, understanding decision-making was the key to bridging the gap between research and reform. The study of decision-making struck him as both fascinating and crucial, for it had the potential to link research and reform, democracy and expertise, reason and emotion, choice and control. Through the study of how people make decisions, for example, one could come to understand the roles that facts and reason (as opposed to values and emotion) played in human affairs. Similarly, through the study of decision-making, one could get at the ways individual action was—or was not—determined by social forces.

These abstract issues of facts and values and decisions had concrete meaning for Simon. There were few issues of greater practical import than the administration of public relief during the Depression, for example. But reconciling theory and practice in a science of decision would require making a difficult decision: was the administrative scientist to serve the people or their leaders?

Understanding the mechanisms and contexts of decision-making, as Simon discovered, was as useful for governing others as it was for governing oneself. Potentially, these were two quite distinct ambitions. The clients of the International City Manager's Association and the Bureau of Public Administration were undoubtedly public-spirited, but they were not the public. Similarly, there might well be an identity of interests between honest administrators and the public they served, but to Simon's generation of political scientists that identity of interests no longer could be assumed. How to reconcile efficiency and expertise with accountability and democracy? This was a crucial question for Simon and for others who shared his belief that only the more efficient organization of society could resolve the crises of the day.

The link between research and reform was the starting point for all Simon's work in public administration, and though his thesis wandered far into the fields of philosophy, the city always was its ultimate destination. For instance, in *Administrative Behavior,* Simon's chief criticism of existing theories of administration was that they were worse than useless as guides to practical action. His argument was particularly pointed with regard to the four basic "principles of administration": "1. Administrative efficiency is increased by a specialization of the task among the group. 2. Administrative efficiency is increased by arranging the members of the group in a determinate hierarchy of authority. 3. Administrative efficiency is increased by limiting the span of control at any point in the hierarchy to a small number—say six. 4. Administrative efficiency is increased by grouping the workers, for purposes of control, according to (a) purpose, (b) process, (c) clientele, or (d) place."[1]

These principles were enshrined in the collection of *Papers on the Science of Administration,* compiled by Luther Gulick and Lyndal Urwick for the use of the President's Commission on Administrative Management.[2] The principles appear relatively simple and clear, but they are far from it, Simon argues. Noting that every attempt to implement them has led to controversy, he asks "How is it possible for such disagreement among experts to arise concerning the application of the most fundamental principles of administration?"[3]

The problem, of course, was with the principles. They were vague, contradictory, and logically unsound. Specialization, for instance, "is not a condition of efficient administration; it is an inevitable characteristic of group effort." The real problem of administration is "not to 'specialize,' but to specialize in that particular manner ... which will lead to administrative efficiency." Similarly, the principle of the unity of command actually contradicts the principle of specialization, for the whole purpose of specialization is to bring multiple sources of expertise to bear on a decision.[4]

The other principles fare no better. Each is revealed to be logically confused and empirically invalid, a mere "proverb" rather than a principle of administration.[5] In Simon's view, such confusions were endemic in the profession, and all administrative science was suffering from a consequent "lack of reality." Perhaps most damning, administrative science typically dealt with "symptoms" rather than with the underlying causes of bureaucratic disease. Though seemingly abstract and theoretical, administrative science in fact "lack[ed] a theory" altogether, for "no comprehensive framework has been constructed within which the discussion can take place." Simon's analysis of the principles of administration was thus a clear "indictment" of the current state of the art.[6]

Indeed, his criticisms were so thorough that he was forced to ask, "Can anything be salvaged" from the wreck of administrative theory? His answer to this rhetorical question is unexpected after such an assault: "As a matter of fact," he replies, "almost everything can be salvaged."[7] The key to this rescue operation was the conceptual reorganization of the field: "Before a science can develop principles, it must possess concepts. Before a law of gravitation could be formulated, it was necessary to have the notions of 'acceleration' and 'weight.' The first task of administrative theory is to develop a set of concepts which will permit the description, in terms relevant to the theory, of administrative situations."[8]

Concepts alone were not enough, however. For Simon, a concept only had meaning in the context of a broader theoretical framework. If a new theoretical framework were to be constructed, then useful concepts like specialization, hierarchy, authority, and efficiency could be given precise definition and thereby be preserved. Existing data and hard-won practical experience likewise could be saved by being reinterpreted in terms of this new framework. The stars of old would remain, but they would take on a different meaning as parts of a new constellation.

Thus, in *Administrative Behavior,* Simon pursued the second and third of the paths to analytic mastery of the problems of the city mentioned in chapter 4: the development of the analytic tools and concepts necessary for understanding administration in general (concept formation), and the building of a philosophical framework to undergird such a science of administration (applied epistemology). In Simon's view, only with this philosophical and conceptual framework in place could a profitable program of empirical research be launched, and only then could the administrative scientist have a legitimate claim to jurisdiction over the problems of organization. A reformed administrative science thus was the key to linking social research and social reform.

Sources for Simon's Synthesis

The primary sources for this theoretical framework were the works of Talcott Parsons, Edward C. Tolman, Chester Barnard, and John Dewey.[9] Simon's debts to the first three are not hard to discern: he cites Barnard, Parsons, and Tolman in his preface as the three greatest direct influences on *Administrative Behavior.* The influence of Dewey requires only slightly more effort to discover, for he cites Dewey repeatedly throughout the text, especially in his discussion of the psychology of decision, habit, and attention, the conceptual core of the thesis. In addition, Dewey's belief that science advances through the reconciliation of seeming opposites rather than through the victory of one view over another also played a significant role in Simon's thought.

Though Simon's thesis was a text primarily devoted to the analysis of other texts, it was no idle intellectual exercise. Simon's own experiences in organizations formed the background against which the ideas of these authors had to be tested and represented one of the main sources of his examples in the text. (The other major examples of organizational decision-making he uses come from manuals of military procedure and from studies of behavior in similarly controlled environments, such as rats running mazes and people playing chess.) In *Administrative Behavior,* Simon forged of these elements a synthetic theory of human behavior in organizations that he hoped would provide the foundation for a positive science of the individual in society.

The four main textual sources for Simon's ideas had much in common. From Tolman and his "purposive behaviorism" to Parsons and his "analytic realism," these social analysts all sought to create a philosophy that was rigorously scientific yet less rigid than the positivism of the Vienna Circle and thus, they believed, better adapted to the explanation of human behavior. A crucial element in all of these approaches was the acceptance not only of the occasional necessity of using terms that referred to unobservable entities but also of the reality of the entities that those terms represented, so long as those entities produced regular relationships among things that could be observed. For example, purposes are unobservable, as are most psychical phenomena, but to these four, that did not mean that they did not exist: all one had to do was to define purpose as a mechanism that transformed one (observable) state into another in a uniform manner. To justify this view they used the example of atomic physics: one never *saw* an electron or a photon. One did see, however, the regular effects produced by them.

There were other, more material, connections between these men as well. Parsons and Barnard corresponded often, for example, and had many discussions when Barnard made his frequent visits to Harvard. In addition, Barnard cited Dewey and Tolman, as well as Parsons, many times in his best-known book, *The Functions of the Executive*. Similarly, though Tolman left Harvard long before Parsons arrived, his work was a source of inspiration for Parsons and other members of Harvard's Department of Social Relations in the late 1940s and 1950s, as is evidenced by Parsons's request that Tolman write a chapter for *Toward a General Theory of Action* setting out the psychological underpinnings necessary for any general theory of social behavior.[10]

Simon found these writers' theories and philosophies very appealing, though he treated them as starting points for theory construction rather than as frameworks within which he would work. The philosophy Simon built upon these foundations combined the instrumentalist approach to knowledge characteristic of the Chicago School with a blend of positivist operationalism and Parsonian analytic realism. To this mix, Simon added a desire for synthesis and a preoccupation with choice, free will, and purpose. The keys to the theoretical synthesis he created were the concepts of organization, hierarchy, and function, and the hinge on which it turned was decision-making. The sum of these views was an early form of Simon's bureaucratic worldview: it was a way of seeing the world that placed choice firmly under control.

Simon's analysis of decision-making in *Administrative Behavior* begins with a discussion of the philosophy of science, for he presents his theory of decision as derived from his philosophy of science, and his theory of administration as derived from his theory of decision.[11] The primary source of his views on the philosophy of science is not hard to discover: citing Charles Morris, Rudolf Carnap, A. J. Ayer, and Percy Bridgman, Simon states that "The conclusions reached by a particular school of modern philosophy—logical positivism—will be accepted as a starting point, and their implications for the theory of decisions examined."[12]

Probably the most important of these conclusions is that only statements that are empirically testable, at least in principle, belong in the realm of science. Nontestable statements (when they have any meaning at all) are value judgments, and between facts and value judgments there lies an impassable barrier. Following Bridgman, Simon also holds that the operational definition of terms is crucial to making statements clear and testable. Quite simply, "Concepts, to be scientifically useful, must be operational."[13]

Operationalism, like logical positivism, had been developed as a means of unifying science. As noted in chapter 3, the logical positivists believed firmly in the unity of science, and many of the leading members of the Vienna Circle wished for nothing so much as the inclusion of the social sciences under the banner of unified science. There was, however, a great obstacle to the application of positivist philosophy to the social sciences: the human mind. The tenets of logical positivism appeared to exclude any statements about unobservable things, such as mind, consciousness, or purpose, leaving only observable behavior as the stuff of social science. When one's topic of interest is the causes of human action, however, it is extremely difficult—perhaps impossible—to exclude the concept of purpose.[14]

In the 1930s, the psychologist Edward C. Tolman had addressed that very problem in a way that was to have a profound influence on Simon.[15] Tolman began his landmark book, *Purposive Behavior in Animals and Men,* by defending the behaviorist approach to psychology, arguing that "all that can ever be observed in fellow human beings and in the lower animals is behavior. Another organism's private mind, if he have any, can never be got at." Similarly, in his chapter on "Conscious Awareness and Ideation," Tolman began with a section on the "Shameful Necessity of Raising [the] Question" of consciousness. Despite his "strong anti-theological and anti-introspectionist bias," however, he found it necessary to introduce concepts such as purpose, cognition, and consciousness in order to explain behavior. Purpose, in particular, was a vital concept because behavior "always seems to have the character of getting-to or getting-from a specific object, or goal-situation."[16]

Tolman permitted terms like *purpose* and *cognition* because he believed that they could be used in a "purely objective" manner if they were defined operationally. The key was to define purpose in terms of "docility" or the adaptation of means to ends: "wherever a response shows docility relative to some end—wherever a response is ready (a) to break out into trial and error and (b) to select gradually, or suddenly, the more efficient of such trials and errors with respect to getting to that end—such a response expresses and defines something which, for convenience, we name as a purpose." Purpose is thus defined by the observable adaptation of a creature's behaviors to the achievement of certain ends, and it is intimately linked to learning and to the selection of behaviors from a certain set of alternatives.[17]

The paradigmatic example of such selective behavior, for Tolman, was the behavior of the white rat in a maze. Such experiments were so central to Tolman's work that he dedicated the book to "M.N.A."—the *mus norvegica albina*—the brave creatures who were "nobly devoting large portions of their daily existences to the running of mazes." The rat's role in Tolman's thinking is also exemplified in his definition of conscious awareness as consisting in a "running-back-and-forth" behavior, and ideation as the mental performance of a "running-back-and-forth."[18]

In *Administrative Behavior*, Simon followed Tolman's lead, arguing that purpose was central to understanding human behavior and that it could be defined operationally in terms of the selection of behavior alternatives. Such an operational definition of purpose, he believed, would enable the creation of a true science of human behavior. This notion of purposeful behavior as being characterized by the selection of alternatives was fundamental, for it provided an avenue for observing the "choice which prefaces all action." Choice, understood as decision-making, would be the keystone for the reconstruction of administrative science.[19]

Simon's interest in decision-making fit in well with his personal conviction that "the only real certainty" in life lay in the necessity of accepting "the burden of personal ethical choice."[20] He was not alone in connecting decision-making with such moral and philosophical issues. For example, Chester Barnard, the philosophically minded president of New Jersey Bell, began his masterwork, *The Functions of the Executive,* with a discussion of free will, ask-ing, "What is an individual? What do we mean by a person? To what extent do people have a power of choice or free will?" Barnard, like Simon, tried to have it both ways, writing that "an individual human being is a discrete, separate, physical thing ... however, it seems clear that no thing, including a human body, has individual independent existence." Similarly, "at the same time that a power of choice is always present ... the person is largely or chiefly a result-ant of the present and previous physical, biological, and social forces" he has encountered.[21] Thus, "Free and unfree, controlling and controlled, choosing and being chosen, inducing and unable to resist inducement, the source of authority and unable to deny it, independent and dependent, nourishing their personalities, and yet depersonalized; forming purposes and being forced to change them, searching for limitations in order to make decisions, seeking the particular but concerned with the whole, finding leaders and denying their leadership, hoping to dominate the earth and being dominated by the unseen—this is the story of man in society told in these pages."[22] The same could well be said of the "story of man" told in the pages of *Administrative Behavior.*

After his philosophical introit, Simon begins his analysis of the psychology of decision-making by arguing that although the term *decision* usually implies conscious deliberation and *choice* does not, it is not possible to draw a clear line between decisions and choices. Hence, he uses the term *decision-making* to refer to "any process whereby one behavior-alternative is selected to be 'acted out,' from all the alternatives which are accessible to choice."[23] Thus, his study of decision-making encompasses both rational, conscious decisions and nonrational, unconscious choices.

This seemingly small shift in terminology turns out to have weighty impli-cations, for Simon uses the positivist injunction to "assume nothing" about the unobservable to legitimate the inclusion of the subjective within the ambit of behavioral science. Decision-making, he argues, involves "both conscious and unconscious, rational and nonrational elements." Here the parallel between *Administrative Behavior* and Talcott Parsons's *The Structure of Social Action* is quite clear: both Simon and Parsons developed philosophies that val-idated their attempts to bring the subjective aspects of human behavior into the realm of scientific analysis.[24]

Indeed, the similarity between the basic argument of *Administrative Behavior* and *The Structure of Social Action* is striking: in *The Structure of*

Social Action, Parsons attempts to solve the problem of the nature and scope of economics through a newfound understanding of the problem of social order. Parsons distilled from his Harvard environment a philosophy (analytic realism) that legitimated social science as a true science, that defended theory and abstraction as essential to science, and that validated his own particular social theory as an attempt to bring the subjective into the realm of scientific analysis.

In keeping with this philosophy (and presented as logically derived from it), Parsons articulated a "voluntaristic theory of social action" in which he attempted to describe the ways society shapes the actions of individuals without denying agency and free will to the individual. To this end, Parsons described society as an integrated system of action governed by a set of common values embodied in that society's institutions. This philosophy of order and system put economic behavior—the paradigmatic example of rational behavior—in its place: it defined the nature and scope of economic rationality by categorizing the economy as one subsystem, one set of institutions and values, within the larger social system.

Drawing on Parsons as well as Tolman, in his thesis Simon argues that the first thing to understand about decision-making is that it is not random. Decisions are "integrated" and "interdependent" because they are purposeful. This statement applies to the myriad small decisions one makes every day, as well as to the most momentous of choices, for "the minute decisions which govern specific actions are inevitably instances of the application of broader decisions relative to purpose and to method." The concept of purposiveness thus "involves a notion of a hierarchy of decisions—each step downward in the hierarchy consisting in an implementation of the goals set forth in the step immediately above."[25]

Hierarchy was an important concept for Simon, not only in *Administrative Behavior* but also in his later work on complex systems. Fittingly, the concept was important to him on many levels. On the intellectual level, it played an increasingly important role in his understanding of experimentation, particularly in his later theories of "near decomposability" and in his analyses of causality. For example, Simon later would argue that the causal relationships between the elements of a system of equations can be discovered by organizing them into a hierarchical structure that mirrors the causal structure of the world that the system of equations describes. The concept also was significant in Simon's postwar work in computer science: it is no accident that the basic structure of files in a computer memory is called its "hierarchy" and that Simon conceives of complex programs in terms of nested feedback loops.[26]

The concept of hierarchy also was useful to one interested in carving out a space for a new science. The concept was used by Tolman in *Purposive Behavior,* for example, as a way of delimiting the bounds of psychology versus physiology in the system of disciplines: "For us [purposive behaviorists], behavior has emergent patterns and meanings [and thus] is more than and different from the sum of its physiological parts."[27] Parsons also used the concept of a hierarchy of levels of action in *The Structure of Social Action* to separate sociology from psychology, in addition to using it as a key aspect of his analysis of social systems.[28] Simon, as was his wont, went Tolman and Parsons one better, depicting administrative science as the discipline that studies the structure of the hierarchy itself. (Whether that put administrative science at the top of the hierarchy of disciplines or outside it was not exactly clear.)

Hierarchy, of course, is a concept with clear political implications. Simon, like many in his day, associated organizational hierarchies with the military, big business, the centralized state (such as Prussia), and the Catholic Church, none of which was regarded as democratic. Hence, one of the crucial questions for Simon as political thinker was how hierarchical structures could fit into a democratic society. As we shall see in later chapters, Simon would struggle with this question for years, tackling it head on in his work on the definition of power and authority and obliquely in countless corners of his oeuvre.

Facts and Values, Means and Ends

Simon uses the concept of hierarchy to go a step beyond Tolman in his analysis of purpose in decision-making. Simon argues that because decisions are integrated in a hierarchy of goals, they involve the selection of ends as well as the choosing of means to achieve them. Here, theory returns to the aid of philosophy, revealing the relative nature of "factual" or "value" judgments. To Simon, "in so far as decisions lead toward the selection of final goals" they are "value judgments." In so far as they "involve the implementation of such goals" they are "factual judgments."[29]

Since every level in the hierarchy of an organization selects goals for those below, a judgment is thus very likely to be *both* a factual judgment (an implementation of higher-level goals) and a value judgment (a selection of goals for lower levels to implement). Only at the very top, where few goals are given, and the very bottom, where there are no more underlings to influence, is one likely to see judgments that are unequivocally factual or value judgments. Thus, whether a judgment is a factual or a value judgment depends upon the framework in which a decision is considered.

The interconnection between fact and value goes even deeper, for a decision itself always comprises both fact and value judgments. To Simon, "a decision is analogous to a conclusion drawn from a number of premises—some of them factual premises, and some ethical premises." This is true of every decision: all decisions are based on facts and values, though whether the decision as a whole is understood by an outside observer as a value or a factual judgment depends on whether the observer sees it as selecting or as implementing larger goals.[30]

While the decision itself shifts as a unit from being a factual to value judgment when the observer's perspective shifts, the premises from which a decision is drawn can be separated permanently and unambiguously into factual and value premises. To Simon, a factual premise is, quite simply, a proposition that is empirically testable. As a result, a factual premise is a factual premise from any perspective, unlike a factual judgment, which depends on the observer's perspective and on the actor's intentions. The reason why the premise is accepted as a basis for decision is not relevant in determining whether a premise is a factual or a value premise: a factual premise may be taken on authority (and so not be tested) just as a value premise may be misunderstood to be a factual statement.

Simon freely admits that in the real world, "problems do not come to the administrator carefully wrapped in bundles with the value elements and the factual elements neatly sorted." Though his use of the language of logical positivism might lead one to suspect otherwise, this inevitable intermingling of fact and value in every decision is one of the chief conclusions of his analysis of decision-making, and the elucidation of its implications is one of *Administrative Behavior*'s main purposes.[31]

Simon treats factual and value premises as almost perfectly parallel. They have different empirical bases, of course, but they enter into decisions in similar ways. Though he does not go into this idea in detail in his thesis, he suggests that the human mind applies logic to both kinds of premises in much the same way, simply treating value premises as factual premises plus an imperative. To use Simon's example: "It is desirable to surprise the enemy in battle" equals "surprise leads to victory" (factually testable proposition) plus "victory is good" and should therefore be sought (value premise in the form of an imperative). Later, at the Illinois Institute of Technology, he would work with the mathematician and philosopher Karl Menger who also was interested in developing such a "modal logic."

In his discussions of facts and values, Simon shows a clear debt to Dewey, for here the Deweyan desire for the reconciliation of opposites shines through. Facts and values, like theory and practice, are distinct, but they are not antithetical. Rather, they must be conjoined in a new synthesis.[32]

Simon also shared Dewey's fundamental optimism about the potential for achieving such syntheses. Whatever the conceptual divisions that prevented intellectual synthesis, action in the world always required such a union. Since people manifestly are capable of action, such reconciliations of seeming opposites are not merely possible but common. Impossibility existed in the philosopher's world, not the actor's.

One area where this close connection between facts and values is particularly important is in the definition of the proper realm of democratic politics versus that of expert administration, a subject on which Simon cites Dewey repeatedly. Woodrow Wilson and the other leaders of the first generation of "scientific" scholars of public administration had drawn a sharp distinction between facts and values as a part of legitimating an expansion of the expert administrator's domain. Simon, however, uses his novel concept of the roles of facts and values in decision-making to redefine that domain and its boundaries. He is just as concerned that the administrator be subservient to the policy choices of the people and their elected representatives as he is that political calculation be excluded from technical decisions.[33]

As a result, while Simon makes the standard claim that "the distinction of value from fact is of basic importance in securing a proper relation between policy-making and administration," his notion of the proper relation is different from that of the thoroughgoing technocrat. To him, "a democratic state is committed to popular control over these value elements," but "too often the basic decisions of policy are reached by technicians in the agency entrusted with budget review, without any opportunity for the review of that policy by the legislature." The danger that the bureaucrat will exceed his authority is as much to be guarded against as the ignorance of the masses.[34]

To Simon, society requires expert leadership *and* democratic participation; the two are only opposites if one misunderstands their relationship. They are complements, distinct but vital to each other. One key to their union would be the administrative scientist. The administrative scientist would be an expert citizen, a professional democrat, and so could police the bounds of policy and of administration.

To Simon's later critics, this approach did not produce synthesis but confusion. In their view, he misunderstood real democracy and so confused consent with control, describing "domination" while using "democratic language."[35] To do so, as one political scientist argued, was "analogous to a description of a jungle using a theory of a farm."[36] Such criticisms have some merit, for Simon's democratic values usually were filtered through the lens of management. Still, it is a mistake to see his professions of democratic faith as mere window-dressing (or naiveté or self-deception) rather than as integral components of his philosophy.

Such criticisms were natural responses to Simon's compartmental approach to integration, however. To Simon, while factual and value premises were inevitably intertwined in practice, they nevertheless were separable analytically. "Since decisions involve an imperative, or ethical, element as well as a factual element, they cannot be 'correct' or 'incorrect' in an unqualified sense. It is possible, however, to separate decisions into two elements, one of them purely factual, and to apply the criterion of correctness to this factual element."[37]

Here Simon uses Parsons's concept of "analytic realism" to defend the use of analytic objects that do not exist in the real world (the separate, purely factual and purely value-based premises of decisions) in order to understand that world. Parsons, like Max Weber, called such analytic objects "ideal types." Simon does not use the terms "analytic realism" or "ideal type" in *Administrative Behavior,* but his ideas about theoretical constructs—at this point in his career—show a strong, clear debt to Parsons and to L. J. Henderson.

Against the view that "the progress of scientific knowledge consists essentially in the cumulative piling up of "discoveries" of "fact," and that theory consists "only in generalization from known facts," Parsons held that scientific theory "is not only a dependent but an independent variable in the development of science." In *The Structure of Social Action,* Parsons built on this claim, stating that "Not only is theory an independent variable in the development of science, but the body of theory in a given field at a given time constitutes to a greater or lesser degree an integrated 'system.'" Thus, Parsons saw a parallel between the structure of the world and the proper structure of thought: both the world and theories about it must be understood as systems.[38]

Such systems of theories were themselves parts of even larger systems of thought, as "there are no logically watertight compartments in human experience. Rational knowledge is a single organic whole." These integrated systems of theory, in Parsons's view, do more than organize what we know already; they also tell us "what we want to know, that is, the questions to which an answer is needed."[39]

Such theories are all-pervasive, inescapable. "All empirically verifiable knowledge—even the common sense knowledge of everyday life—implicitly, if not explicitly, involves systematic theory in this sense." Citing L. J. Henderson's "Approximate Definition of Fact," Parsons writes that "a fact is a statement about experience in terms of a conceptual scheme" and is thus not a free-floating, unencumbered thing. Science is the product of the organization of facts into such conceptual schemes, and the progress of science is due to the development of more sophisticated, elegant, and parsimonious theoretical systems, not to the discovery of masses of new facts.[40]

This notion of a conceptual scheme is very like Thomas Kuhn's concept of the paradigm—an idea that would play an important part in the postwar behavioral revolution in social science, which Simon helped lead. According to Henderson, before the creation of a systematic conceptual scheme, a body of knowledge is just a mass of mere data and is not a science, just as a pre-paradigmatic body of knowledge is not a science for Kuhn.[41] As we shall see in chapter 11, the behavioralists quite consciously strove to be scientific revolutionaries along Kuhnian lines, believing that their behavioral-functional analyses of social (and economic, political, cultural, and psychological) systems and their new model of man as a rational, but limited, problem-solver finally had lifted them out of the pre-paradigmatic, and thus pre-scientific, state. The connection is not accidental, for Kuhn developed his ideas in a context heavily influenced by Henderson, Parsons, and the cognitive psychologist George A. Miller, who was in turn strongly influenced by Tolman and by Simon.

Simon's assault on the false "principles of administration" reveals that he shared this conviction that theoretical innovation was a keystone of scientific advance. As Simon wrote in response to a friend's comments on his thesis: "In general, your comments express approval of the empirical conclusions stated, but scepticism of the method used to reach them ... I should like to assure you most emphatically that the theory generally led to the conclusions, and not the reverse, so far as the development of my own thinking was concerned."[42] In similar fashion, in 1944 Simon wrote to the chair of the American Political Science Association program committee to ask that a panel on theory and methodology be added: "Perhaps wartime is not the best time to suggest more 'theory,' but before the war there was a depression, and after the war I have no doubt that we will be equally blessed with holocausts and upheavals of one sort or another—and if we political scientists simply sit down and wait for a period of peace and quiet before we return to our theoretical pursuits, I am afraid that we will never return."[43]

As with all things in Simon's work, this defense of abstraction had a practical purpose, for it served to legitimate the administrative theorist's work. As Simon wrote, "if there is no more to administrative theory than accumulated empirical observation, then there is nothing at all that we academicians can do for the practitioners than to record practices and to pass them on." Still, he was "optimistic enough to believe that theory—based on correct empirical premises—has something to contribute to practice, as it does in the other sciences."[44]

Henderson, Parsons, and Simon were not alone in their belief that theory was the engine of progress in science. The 1930s saw an increasing fascination with theory in field after field in both the natural and the social sciences in the United States. This emphasis on theory was new to American scientists, who had a longstanding reputation as careful observers but weak theorists. European scientists, on the other hand, had the reputation of being builders of grand theoretical systems, some buttressed by sound data, some too ethereal to find support in concrete facts.

American social scientists had viewed such theories with a mixture of envy and disdain. If they wanted to be thought of as true scientists, American social scientists, whatever the discipline, went out into the field and measured and counted, or brought things back into a laboratory, again to be measured and counted. This was not a mindless empiricism, of course, and even the most Baconian of fact-gatherers did have some theory to guide him. Nevertheless the obsession with measuring and counting is palpable everywhere in prewar American social science, right up into the 1930s. In many places, such as the monumental *Recent Social Trends,* there is almost a belief that the numbers will speak for themselves if one could just go out and find them.[45]

From the turn of the century well into the 1930s, one saw everywhere in

American social science a parallel distrust of any theory, any concept, any term that left no physical trace to be counted or measured. This was the era of "brass-instrument psychology," devoted to quantifying sensations; of behaviorist psychology, which observed the very material reactions of thousands of white rats to a range of precisely calibrated stimuli; of countless social surveys and polls, which were, again, physical traces of concrete actions; of compilations of numerical data on everything that had a ready measure.[46] It was entirely representative of the generation before Simon that the inscription over the doors of the new social science building of the University of Chicago, erected in 1929, should read, "Where you cannot measure, your knowledge is of a meager and insufficient kind."[47]

Yet after the war a number of the students of this prewar generation, many trained in the 1930s in the rooms beyond that very inscription, came to believe that science was all about constructing models and testing them through simulations. A belief that manipulations of symbols could be legitimate science took hold, especially in psychology and economics, and skill at theoretical abstraction came to be valued in much the same way that skill at devising new ways and novel devices with which to measure things had been before the war. The sociologist Bernard Barber even went so far as to write that: "The history of science, and especially of modern science because of its rapid rate of progress, could be written in terms of the successively greater development of conceptual schemes and of the correspondingly greater reduction in the degree of empiricism in science. Good conceptual schemes, says President Conant in sum, are the essential *cumulative* component of all science."[48] The inscription in the stone of the University of Chicago social sciences building could not be changed easily, but the research practices of those inside certainly could—and did.

The advocates of this new, abstract experimentalism believed themselves to be even more rigorous empiricists than their teachers had been. To them, symbols were real, material things and even a fact so bare as a simple count of voters in an election never came unadorned by theory. They saw themselves as better empiricists because they had to be explicit about their assumptions in order to make their models work and their simulations run. They believed that they were better empiricists because they were experimentalists. They did not just find their facts where nature left them; they produced facts that told for or against their hypotheses.[49]

This new interest in theory had many sources. Among the most important of these were the example of the quantum revolution, which was widely viewed as a revolution led by abstract theories rather than novel facts; the forced migration of European scientists to the United States in the 1930s; and the growing mathematical competence of American scientists.[50] The most influential articulation of this new idea of conceptual change as the driving force behind scientific advance was probably Alfred North Whitehead's *Science and the Modern World,* which helped shape the scientific outlook of an entire generation of American scientists, especially in the human sciences.[51]

The "empiricist temper" of American scientists was not replaced by the new emphasis on theoretical abstraction so much as wedded to it, however. In this, as in so many things, the scientists of Simon's generation sought a third way, seeking a philosophy and a practice that could bring together theoretical abstraction and empirical observation, just as they sought a sociology that could encompass both social control and individual choice, an economics that could embrace both bureaucratic organization and the free market, and a psychology with room for both reason and passion. The changing motto of the Cowles Commission for Research in Economics, the leading center for economic research in America in the 1930s–40s, offers a case in point: in the early 1930s its motto was "Science is Measurement." By 1951, it had become "Theory *and* Measurement."[52]

The Task of Knowledge

In this discussion so far, I, like Simon, have referred to "means" and "ends" and used them as the logical partners of "facts" and "values." In the third chapter of *Administrative Behavior,* however, Simon expresses dissatisfaction with the common forms of means-ends analysis. Though the terms are useful (and he employs them often), Simon finds the means-ends schema insufficient for a full description of decision. Here he returns to his original interest in decisions as selections, describing decision-making in terms of "alternatives and possibilities." Simon also shows his debt to Tolman by stating that "A simplified model of human decision-making is provided by the behavior of a white rat when he is confronted, in the psychological laboratory, with a maze, one path of which leads to food."[53]

Choices are selections of certain actions or sets of actions from all the vast set of possible behaviors, selections intended by the chooser to bring about one (or a set) of possible future states. Drawing on Wittgenstein, and using chess as a recurrent example, Simon describes all decisions as being like moves in a game, where one out of a vast set of possible moves is chosen in order to bring about a new configuration of the game board that is the world. In his view, "the problem of choice is one of describing possibilities, evaluating them, and connecting them with behavior alternatives."[54]

To Simon, "it is the task of knowledge to select from the whole class of logical possibilities a more limited subclass of empirical possibilities, setting up certain functional relations among the variables which describe the possibilities." The "ultimate goal" of knowledge is thus to "determine relations of known facts of past and present with facts of the future, so that a single possibility is consequential on the present."[55] A more instrumental view of knowledge's purpose I cannot imagine.

At first, this task appears to be overwhelming, for it would seem that any rational choice between alternatives must then involve "a complete description of the possibilities consequential on each alternative and a comparison of these possibilities."[56] Here, the problem of interdependence raises its hydra head, making such calculations impossible in all but the simplest and most isolated of situations. Since people manifestly do make choices, however, there must exist some means by which we limit the alternatives we consider and the depth in which we consider them.

In response to this problem, Simon advances the idea that "The human being striving for rationality and restricted within the limits of his knowledge has developed some working rules partially to overcome this difficulty. He uses rules which implicitly assume that he can isolate from the rest of the world a closed system containing only a limited number of variables and a limited range of consequences." Therefore, "rational choice will be feasible to the extent that the limited set of factors upon which decision is based correspond, in nature, to a closed system of variables—that is, to the extent that significant indirect effects are absent."[57]

This assumption is usually—but not always—valid, for "the empirical laws which describe the regularities of nature tend to arrange themselves in relatively isolated subsets." In other words, interdependence exists, but it is usually negligible beyond the bounds of well-defined systems. This fact, he argues, "provides both scientist and practitioner with a powerful aid to rationality, for the scientist can isolate these closed systems in his experimental laboratory, and study their behavior, while the practitioner may use the laws discovered by the scientist to vary certain environmental parameters without significantly disturbing the remainder of the situation." Here lies the kernel of Simon's experimental philosophy and of his later work in "near-decomposability."[58]

The bulk of *Administrative Behavior*, however, is not devoted to the analysis of experiment and of model-building. It focuses instead on the means by which we limit the alternatives we consider. The key factor in this process, according to Simon, is organizational membership. Here is where Simon's theory of decision-making intersects with the problems of administration and organization and where Simon lays the groundwork for turning upside down the traditional understanding of organizations.

To Simon, choice is not possible without limits, and freedom is not feasible without bounds. Organizations enable their members to make decisions by virtue of the fact that they limit the sets of alternatives they consider. Indeed, this enabling of decision is the ultimate purpose of all organization, and organizations are themselves decision-making machines.[59] As a result, the organization is structured in the same broad way that individual choice is structured—in a pyramidal hierarchy, with the administrative hierarchy of superior and subordinates corresponding to the hierarchy of goals and subgoals, from the CEO who decides broad policy down to the operative worker who implements it.

Drawing directly on Barnard's *Functions of the Executive,* Simon argues that organizational membership produces two primary phenomena that limit the alternatives we consider and thereby enable us to make choices: *authority* and *organizational identification.* Authority, in its simplest sense, is the power to make decisions that guide the actions of another. As a member of an organization, the individual "sets himself a general rule which permits the communicated decision of another to guide his own choices without deliberations on his own part." This general rule only applies to decisions within a certain area, the employee's "zone of indifference" (a term he borrowed from Barnard), but it does enable the centralization of "the function of deciding, so that a general plan of operations will govern the activities of all members of the organization."[60]

Significantly, this understanding of authority makes it dependent on the actions of the subordinate, not the status of the superior, helping Simon reconcile the hierarchical nature of authority relationships with his democratic ideals. When a subordinate does not permit his decisions to be guided by those of his superior, "there is no authority, whatever may be the 'paper' theory of organization."[61] Why, then, does the individual sacrifice his power to choose, even within a limited area, if he is not forced to do so?

First, Simon argues, there exists in the broader society a set of widely accepted premises as to the nature of the roles of superior and subordinate in any given organization. These institutionalized roles help individuals form reliable expectations about the behaviors of others, and they teach them to suspend independent judgment within certain bounds. Different organizations may draw those boundaries in different places, but they do not have to teach their members that such bounds exist or what they mean.[62]

Second, the individual chooses to accept the authority of the organization (and of his superiors within it) because he or she identifies with the organization, agreeing with its basic goals and the broad structure of authority it has created as a means toward achieving those goals. The individual who so identifies with the organization "evaluates the several alternatives of choice in terms of their consequences for the specified group" instead of for himself alone. The individual does so for several reasons, of which probably the most important is that identification enables rational decision-making. As Simon writes, "Identification, then, has a firm basis in the limitations of human psychology in coping with the problem of rational choice . . . From this point of view, identification is an important mechanism for constructing the environment of decision [because] it permits human rationality to transcend the limitations imposed upon it by the narrow span of attention."[63]

Organizational identification thus provides a crucial link between the needs of the organization and the needs of the individual. It is a powerful, pervasive phenomenon, and without it all the benefits of rationally coordinated action would be unattainable.

Its effects are not always benign, however, for people tend to identify with the subgoals that guide their immediate decisions and with the subunits of the organizations with which they have the most frequent contact. (Specifically, individuals usually identify with the "unit organization"—the highest level at which organizational goals are translatable into direct action upon the external environment.) This tendency to identify with sub-goals is due to the fact that root premises and basic values often are obscured by the chains of decisions or levels of the organizational hierarchy that stand between them and the decisions involved in daily tasks. When the connection between present decisions and basic goals becomes hazy, then the interests of the different units within the organization can come into conflict.[64]

Simon sees this identification with subgoals and subunits in every area of life, from the behavior of taxpayers who identify with Oakland rather than the Bay area, to that of the fire chief who identifies with his department rather than the needs of the city as a whole, to that of the expert who identifies with his profession instead of the public interest. Such identifications are natural, universal, and dangerous: without awareness of the broader picture, "the integrated busyness of the adult is simply a more patterned busyness than the random movements and shifting attentions of the child. The organized wholes of which it is composed are larger and more complex, but as wholes, of no more significance." Fortunately, the conscious human is able to "ask himself not only what means are appropriate to his ends, but more broadly, 'what kind of world do I want to live in, and which behavior alternatives lead to this kind of world.'"[65]

One cannot ask such questions before making every decision, of course, but one is able to construct an "environment of choice" to guide future decisions. That is, one is able to make plans based on expectations of one's own future choices and the choices of others. Simon makes sure to note that "we have reached here an extremely significant point in the analysis ... It is because the human being can deliberately fashion the psychological environment of his decisions that these can be welded together into an integrated pattern of a high order of rationality." Hence, he argues, we are led "to a concept of 'planned' behavior as the proper means for maintaining rationality at a high level."[66]

Just as making plans for oneself is crucial to maintaining a high level of rationality in individual behavior, making plans for society is crucial for maintaining a high level of rationality in the behavior of that society's members. As Simon writes, "The deliberate control of the environment of decision permits not only the integration of choice, but its socialization as well." Since social institutions "largely determine the mental sets of the participants, they set the conditions for the exercise of docility, and hence of rationality in human society." The highest level of rationality, it follows, is the planning of the institutional environment in which the individual moves.[67]

The most important element of the broader institutional environment, to Simon, is the formal organization. In this belief, Simon again followed Parsons and Barnard, who believed that "formal organization [is] a most important characteristic of social life, and [is] the principal structural aspect of society itself." The formal organization was important because it was a bridge between the "mores, folkways ... attitudes, motives [etc] and the actions of masses."[68] Thus, to Simon, while "administrative organizations cannot perhaps claim the same importance as repositories of the fundamental human values as that possessed by older traditional institutions like the family," nevertheless, "with man's growing economic interdependence, and with his growing dependence upon the community for essential governmental services, formal organization is rapidly assuming a role of broader significance than it has ever before possessed." As a result, "the behavior patterns which we call organizations [have become] fundamental ... to the achievement of human rationality." Indeed, to Simon "the rational individual is, and must be, an organized and institutionalized individual."[69] The bureaucratic organization thus was no Weberian "iron cage," but essential scaffolding for thought.

Even more, not only must the rational individual be "organized" and "institutionalized," rationality (and thus, to Simon, human existence) itself derives significance from its organizational setting. "If human rationality has any ethical meaning, if it is more than a pleasant game, then it gets its meaning, that is, its higher goals and integrations—from the institutional setting in which it operates and by which it is molded." Not surprisingly, the highest level of rationality, the highest level of meaning in human thought and action, "consists in taking an existing set of institutions as one alternative and comparing it with other sets," which is the task of the ideal citizen, the administrative scientist.[70]

Simon's Synthesis: A Science of Administration

Simon's own perceptions of change in the world around him informed his understanding of administrative science and its place in the modern world. He, like so many who lived in the first third of the twentieth century, was deeply conscious of the end of the old and the birth of the new. The modern world appeared to be separated by a vast gulf from the world of even a generation before. A new system was in place, and a new class of people—the administrative class whose decisions he wished to influence—was needed to lead this vastly more complex, interdependent, urban society.

The chief claim of the administrative scientist to be the one to lead the process of social adaptation was that he had mastered the objective analysis of subjective behavior. His behavioristic methods, his operationally defined terms, his logic, and his skill at measuring the heretofore unmeasurable certified the administrative theorist as a true scientist.[71]

Not only had the administrative scientist mastered the procedures that guaranteed objectivity, he now had the theoretical framework necessary to understand subjectivity itself. In *Administrative Behavior,* Simon argued that subjective behavior is simply rational behavior in a "non-rational matrix," and hence subject to the rational analysis of science. He had shown that subjectivity is at once inescapable, essential, and dangerous, and he had argued that the goal of science was not to eliminate subjectivity but to educate and control it.[72]

Emotion, passion, and individuality are strangely absent from Simon's analysis, despite his intense interest in the subjective aspects of behavior. He does not so much dismiss emotion as omit it, leaving the reader to assume that the love between husband and wife, say, is but an identification with the organization of the family or a set of behaviors learned as part of the roles of parent or lover.[73]

Similarly, the individual in *Administrative Behavior* appears to be but a "clean slate upon which the organization writes its desires," as Simon later admitted.[74] Indeed, the individual in Simon's analysis seems little more than the sum of his organizational affiliations. Each person is unique, perhaps, but unique in the way that the intersection of many sets may be a single point, not in the way an individual soul is a unique whole.

This question of how to understand the individual in relation to the group brings us to the problem of interdependence and thus to perhaps the most important question for any social science: are the actions of individuals the actions of atom-like units, or are they the actions of components of a larger system? Many believed the future of not only social science but also society hinged upon the answer to this question.

Simon's answer to this problem was to understand the organization as creating an environment of decision in which free choice took place and had meaning. Some environments were more restricted than others, but no choice was possible without limits. Freedom could not exist without boundaries, and choice would have no meaning if it were not integrated into a larger structure of purposes.

A crucial aspect of this synthesis of choice and control was the idea of levels of interdependence. Just as there are levels to the hierarchy of an organization and to the hierarchy of decision, there are levels of interdependence in the systems of the world. No system is wholly closed over the long run, but all the important interactions for the short term can be captured if one draws the boundaries with sufficient care. As a result, one can go about one's daily tasks according to the routine prescribed by the organization—or one's immediate boss—without worrying about how one's decisions will affect those far away. Eventually, the need to question the premises of such habitual decisions would arise, but such larger questions could be left to those farther up the hierarchy—or outside it, such as the administrative scientist.

In sum, in *Administrative Behavior,* Simon effected a remarkable fusion of scientific philosophy, psychological theory, and political argument. Through the union of positivist philosophy and the psychology of decision, Simon created a new theoretical structure for administrative science and a new rationale for an active—but accountable—government. The keystone of this new structure was the concept of decision-making as a selection among alternatives, a selection based on premises both factual and value-laden. The linchpin of this political argument was the necessity of planning the individual's "environment of decision" in order to enable him—and society—to act rationally. These were powerful ideas, and they became important parts of a broader intellectual-political movement that sponsored the creation of both the "administrative state" and the "systems sciences" that were vital for this state's operation. But before Simon (or his ideas) could do their work in the world, he would need to find entree into a broader world than that of municipal reform.

CHAPTER SIX

Decisions and Revisions

Herbert Simon returned to Chicago twice in 1942, the first time in the spring to defend his thesis and the second, in the fall, to take up his new position as assistant professor of political science at the almost equally new Illinois Institute of Technology (IIT). The thesis defense went smoothly, for the most part. According to Simon, the only real difficulty "on the exam derived from my stubborn positivism. The examiners, especially devout Catholic Jerry Kerwin and British Labourite Herman Finer, found it difficult to believe that one could not *prove*, from self-evident premises, that Hitler was a bad man. And if one couldn't prove it, what right had one to believe it? I doubt whether they bought my positivist explanation that choice begins with faith in value premises, not with proof of their correctness."[1]

Though Simon passed the exam with no trouble, he did leave with a nagging sense that he was speaking a different language than his former mentors. At IIT, however, Simon found a community that did speak his language. For example, he was able to institute a calculus requirement for political science majors, though how he "got this particular piece of whimsy through the curriculum committee" mystified him.[2] Simon enjoyed the support that IIT gave his scientific ambitions, and he would spend the rest of his career pursuing behavioral science in technical schools.

At first Simon found that "teaching constitutional law to engineers is, as you might surmise, not the most exhilarating pursuit that can be imagined," leading him to wonder, "How does one instill curiosity into people? Without it, this 'educational' process is a continual struggle between a student who is trying to get by and a teacher who is trying to catch him at it."[3] But after a short while, Simon grew accustomed to his surroundings, feeling that "engineers, on the whole, are not a bad bunch to teach," provided that the instructor explained things in a way they could understand: "If students are engineers, they will better understand the logic of a Supreme Court case if you can represent it on the blackboard as a wiring diagram for an electric circuit, the switches representing the yes-or-no choices of the court. The wiring diagram for the case of *Marbury vs. Madison* . . . being rather baroque, pleased the students no end."[4]

Despite IIT's technical orientation, when Simon arrived to join the Department of Political Science, it was expected that he would teach some traditional courses such as Constitutional Law and American Political Institutions and Ideas. In addition, due to the strictures of wartime, staff were spread rather thinly: "Whether I will teach political science next semester, mathematics, or chemistry, I don't yet know," the harried junior professor wrote to a friend.[5] Simon resolved to "proceed cautiously, conservatively, and traditionally for the next year or two—but not indefinitely," and threw himself into the work with his typical intensity. In the end, he found that reviewing the classics of American government, while "not particularly scientific," was "great fun" nonetheless.[6]

Simon rapidly became dissatisfied with the standard texts on American government. After conducting a review of the leading dozen texts, he found them all to be "inconsistent" and "vague" in their definitions of key terms such as *democracy* and *liberalism*.[7] Simon found such theoretical vacuity disturbing because, to him, it made the texts useless for understanding current political issues. As he wrote, "The aim of studying political theory is not to construct a taxonomy of political theorists" but to help one make political decisions today.[8] Political theory, like all science, should be a tool for improving decision-making.

Simon's experiences teaching broad, traditional topics in political science and his dissatisfaction with the existing texts helped inspire him to extend the framework he had developed in his thesis. As a result, the text on *Public Administration* that he and his colleagues Donald Smithburg and Victor Thompson wrote near the end of Simon's stay at IIT took a very broad perspective on administration. *Public Administration* defined its subject so broadly as to make it virtually congruent with all human activity involving more than one person, stating at the outset that "When two men cooperate to roll a stone that neither could have moved alone, the rudiments of administration have appeared."[9]

There were many movements between Simon's first version of *Administrative Behavior* and the writing of *Public Administration,* however, many resources in the IIT environment for him to draw on, many irritants for him to react against. Of such stimuli, the three most important were the "problem of working out a raison d'être with the faculty of the architecture department," the mathematical ferment of the Cowles Commission for Research in Economics, and the intellectual excitement of the seminar in the philosophy of science he organized with the mathematician Karl Menger.[10] The combined effect of these experiences was to convince Simon that his framework for understanding decision-making needed to be elaborated in new ways, using new techniques. His thesis had incorporated elements of the sciences of choice into a framework set by the sciences of control, and that would remain true of his later work as well, but he increasingly found he had to expand and alter that framework for it not to be a Procrustean bed. Choice could not be controlled easily.

IIT and the Perils of Planning

In 1940 the Armour Institute of Technology and Lewis College joined to form the Illinois Institute of Technology. Like the University of Chicago, IIT was intimately tied to the Chicago urban reform community and to national philanthropic foundations. IIT's president, Henry Heald, for instance, became president of the Ford Foundation in the mid-1950s. (Simon's long-distance mentor, Chester Barnard, became president of the Rockefeller Foundation in 1948, meaning that Simon had top-level connections to the two most important private patrons of social science in the postwar years.) Unlike the "1313" organizations, however, IIT's connections primarily ran through architectural rather than political circles: James Holabird of Holabird and Root sat on IIT's board, for example, and he played an active role in university affairs, helping to bring Ludwig Mies van der Rohe to head IIT's renowned School of Architecture.[11]

This was a significant addition for Simon, since Mies van der Rohe and his disciples, such as Ludwig Hilbersheimer, were city planners as well as architects. As noted in the discussion of *Fiscal Aspects* and of *Administrative Behavior* in chapters 4 and 5, Simon was a strong advocate of "the planning approach to political economy." Indeed, his researches into administrative behavior had been undertaken in part to show that "There is no *a priori* reason why the community should select the competitive market as the institutional means of organizing its activities, any more than it should select a governmental organization."[12]

Nevertheless, Simon found the kind of planning embraced by Mies van der Rohe, Hilbersheimer, and the other members of the School of Architecture to be both autocratic and unrealistic. As he wrote to a friend, "they teach these people a very Utopian variety of city planning" in the School of Architecture. As a result, Simon found that the students in his courses on the "Administrative Aspects of Planning" and the "Economics of City Planning" often became "confused" and "discouraged" when he informed them that the kind of planning the architects advocated was not legal in a democracy.[13]

Quite simply, Miesian planning was not Deweyan planning, and Simon found himself reacting strongly against this aspect of the IIT environment. As he later wrote, "In this setting, I felt less like a teacher than a missionary—one preaching not to tolerant pagans but to true believers of another faith."[14] Reading the course outline for the Economics of City Planning, one can easily imagine Simon, the sage veteran of California's political wars, chiding his students for their ignorance of political and economic realities. An ironic stance for a confirmed theorist, perhaps, but one that well suited Simon's taste for poking holes in received views.

It is no coincidence that Simon saw economics as the antidote to autocratic planning, for economics had a long tradition of valuing—indeed, of nearly worshiping—the free market. Simon, however, was no neoclassical economist. In his course on the Economics of City Planning, for example, he sought to synthesize free choice and collective responsibility, arguing: "There is no sharp or complete dichotomy between the planned and the unplanned city. City planning . . . never completely displaces the mechanisms that give structure to the unplanned city, but merely supplements them . . . Planning, far from substituting any central control for these individual decisions, simply modifies them and limits them by establishing certain regulations with which the actions of the individual must conform."[15] As always, for Simon, choice existed and was meaningful, even though it existed within bounds.

Simon's studies of city planning (and his experiences in California) convinced him that "the mechanism that brings about the pattern of the unplanned city is fundamentally an economic one."[16] Although no decision was purely economic in nature, the economic aspects of decisions had principal importance in producing urban patterns. This conviction, combined with his growing admiration for the mathematical sophistication of economics, led Simon to join eagerly the biweekly discussions of the Cowles Commission, then in its heyday.

Economics at the Cowles Commission

Simon's experiences as a member of the Cowles Commission had a powerful effect on him. The commission's atmosphere was immensely stimulating, and he found in its members kindred spirits, fellow believers in the possibilities of bringing mathematical rigor to the social sciences. Also, on a practical level, the connections he made there to a rising community of mathematical economists, statisticians, and applied mathematicians served him throughout his career, leading to his first contracts with RAND and the Office of Naval Research (ONR). In addition, its seminar series became his model for high-level intellectual discussion, and its organization became his model for a research institution.

From its creation in 1932 until its move to Yale in 1956, the Cowles Commission for Research in Economics was perhaps the single most important institution in the mathematical metamorphosis of modern economics.[17] It was created in 1932 by the wealthy investment analyst Alfred Cowles, who, in the wake of the crash of 1929, came to the conclusion that stock analysts did not know what they were doing. In his view, better data and superior statistical knowledge were necessary before economic analysis could be of any use to the practical businessman.

Through various connections (primarily the mathematician Harold T. Davis), Cowles was introduced to the fledgling Econometric Society in 1932. The Econometric Society had been formed in 1930 by a small group of mathematically minded economists. Led by Ragnar Frisch, Charles Roos, and Irving Fisher, the society operated on a shoestring budget, able to provide little more than moral support to its worldwide membership of around eighty. Not surprisingly, Fisher hardly could believe his good fortune when Cowles rode in from the wilds of Colorado offering an annual budget of $12,000 for a journal and a research institute. After brief deliberations, Fisher and the society took Cowles up on his offer, and the newly formed Cowles Commission housed and financed the society, its journal *Econometrica,* and a small number of research personnel. The operating staff of the society, the journal, and the commission remained intimately intertwined until the 1950s, and *Econometrica* became the primary vehicle for the postwar "econometric" revolution in economics.[18]

In 1939, the commission (plus the society and *Econometrica*) moved to Chicago from its original home in Colorado Springs, as Cowles came to the Windy City to take over his father's business interests. The commission sought affiliation with a university and soon settled on the University of Chicago, which was glad to trade some office space in the Social Sciences Building for access to top-level mathematical economists, especially since the death of Henry Schultz in 1938 had left a gaping hole in the university's program. The commission and the University of Chicago remained closely associated until 1956, with commission research staff commonly holding university appointments and university faculty sitting on the commission's board.[19]

The staff assembled at the commission during its Chicago years was nothing less than spectacular: Theodore Yntema, Tjalling Koopmans, Jacob Marschak, Franco Modigliani, Gerard Debreu, Hendrik Houthakker, Kenneth Arrow, Leonid Hurwicz, Oskar Lange, Lawrence Klein, and, of course, Herbert Simon were all active members during this period. In addition, Milton Friedman, George Stigler, Trygve Haavelmo, and Ragnar Frisch were frequent participants in its discussion series, along with many other distinguished members and guests. The lists above include no fewer than ten future Nobel Prize winners in economics, a tally unmatched by any research school in any science, except for the Cavendish Laboratory during Ernest Rutherford's heyday and the Manhattan Project.[20]

The Cowles Commission, in some ways, looked both backward and forward as regards the development of economics. Over half of its research staff held advanced degrees in mathematics, statistics, or physical science, representing the influx of the mathematically trained from other disciplines into economics in the 1920s and 1930s. At the same time, however, over half of the members also had advanced degrees in economics or the other social sciences, representing the increased level of mathematical sophistication among those trained primarily in economics.[21] (Both numbers are over half because several staff members had advanced degrees in both mathematics and economics.)

Fifteen of the seventy-eight members of the research staff during the late 1940s had been members of planning organizations before the war, and dozens more worked for the armed forces in various planning-related occupations during the war. This connection to governmental planning agencies continued into the cold war era; in 1951 ten members already were working under contracts for RAND, which had been created in 1948 in order to help the Air Force plan weapons development, logistics, and strategy. Many more would follow suit later in the decade, Herbert Simon among them. Simon thus was far from the only member of his generation who moved from the problems of planning economic systems to the problems of planning defense systems (and back again), which helps to explain why RAND's defense analysts were so eager to apply their new skills to the War on Poverty in the 1960s: in many ways, they were returning to their intellectual roots.[22]

The Cowles Commission's purpose was straightforward. Its mission was the "conduct and encouragement of research in economics, finance, commerce, industry, and technology, including problems of the organization of these activities, and of society in general. Its approach is to encourage and extend the use of logical, mathematical, and statistical methods of analysis."[23] The mindset of the commission's members is captured well by a song sung at one of the group's annual parties. (To the tune of "the American Patrol"):

We must be rigorous, we must be rigorous
We must fulfill our role.
If we hesitate or equivocate
We won't achieve our goal.
We must investigate our systems complicate
To make our models whole.
Econometrics brings about
Statistical control!
Our esoteric seminars
Bring statisticians by the score,
But try to find economists
Who don't think algebra's a chore.
Oh we must urge you most emphatically

Decisions and Revisions   127

To become inclined mathematically,
So that all that we've developed
May some day be applied![24]

One can detect a certain wistfulness here in the recognition that "much of the work of the Cowles Commission is of an abstract nature, and that many of its fruits are not likely to be reaped in the immediate future." Nevertheless, its members believed that their work was "connected in a very real way with the fundamental problems of a free and democratic society. It is by learning to predict in detail the consequences of *general* economic and social policies that we will be best able as a society to achieve desirable objectives without resort to direct controls over individual economic behavior."[25]

The specific problems the commission took as its focus all were related to the problems of matching theory to data, problems that Simon had taken an interest in because of his work with Ronald Shephard at the Berkeley Bureau of Public Administration (see chapter 4) and Henry Schultz (see chapter 3). In particular, under Marschak's leadership the commission's staff undertook studies where (1) the theory involved was a system of simultaneous equations, (2) the equations involved random terms, (3) the data were in the form of time-series, and (4) the data referred to aggregates, not to individuals.[26]

These four criteria described a wide range of mathematical models, but they were particularly relevant for problems of planning and prediction in complex, interdependent systems. Thus, the chief pursuits of the commission members during this period—activity analysis, linear programming, game theory, and studies of the "identification problem"—all were linked by their importance for planning, inference, and prediction. How to predict the decisions of others, how to decide what the true state of the world is now and what it will be tomorrow, how to plan one's own strategies in response—these were the central questions for the commission members, and their work would form the foundation for the postwar sciences of choice, particularly for operations research (OR).[27] Simon, for his part, included all of these under the rubric of decision theory.

One key obstacle to effective planning and prediction was the problem of technological change, as the sudden advent of nuclear weaponry demonstrated all too clearly. The dramatic effect of the atomic bomb on the war and on postwar geopolitics led many to believe that it marked a sudden break; humanity had crossed the Rubicon of history. The physicist Ralph Lapp, for example, wrote that "physicists have split the century—and all time—in half; history is bisected into pre-atomic and atomic eras."[28] Would nuclear power bring similar, startling transformations in the economic realm? In the economist's language, did technological changes merely shift production functions one way or the other, or did they reshape the curves entirely, perhaps even creating new ones? Perhaps even more importantly, could the effects of such changes be predicted and thus planned for?

These questions formed the basis for Simon's first research projects for the Cowles Commission.[29] In a series of papers on the economic aspects of productivity increases for technological change, Simon combined his abiding interest in the patterns of the city with this newer concentration on technological change, attempting to discover the implications of productivity increases for rural-to-urban migration. In a fashion that became entirely typical of Simon, this article began a train of inquiry that eventually led him to investigate a host of apparently dissimilar phenomena that nonetheless were produced by similar mechanisms. This article, for example, led him to analyze the size distribution of cities, which is described remarkably well by a formula devised by the statistician G. Udny Yule. Simon later found that the Yule distribution also correlated well with the size distribution of business firms and with the frequencies of word usage in English. (For the curious, the reason for the similarity is that the Yule distribution is produced in situations where there is a "Matthew" effect: those that have tend to get, and they do so in proportion to how much they have already.)[30]

Another product of this interest in the problem of technological change was a report on the *Economic Aspects of Atomic Power,* written by Jacob Marschak, Sam Schurr, and Simon.[31] In this report, the authors took on the popular belief that nuclear power would provide "energy too cheap to meter."[32] Contrary to such atomic hyperbole, their conclusion was that nuclear power was *not* going to have a dramatic impact on the economy; only where there were no inexpensive fuel sources available would it have a significant effect. Their prediction has turned out to be quite accurate.

This report, like much of Simon's work before the publication of *Administrative Behavior* in 1947, appears to have sunk without a trace into the ocean of policy analysis. Still, it was a valuable experience for Simon, for his participation linked him to a new community of increasingly powerful players in science and in politics. On the strength of his expertise in the economic implications of nuclear energy, for example, Simon later became a board member of the Nuclear Science and Engineering Corporation, where he became friends with Eugene Zuckert, who became secretary of the Air Force soon after.[33]

In addition, the Cowles Commission was Simon's original point of contact with RAND and the Office of Naval Research (ONR)—and thus with the postwar military-industrial-academic research complex. This connection was particularly important for Simon because he, unlike so many of his colleagues, was not involved in defense research during World War II. Without Cowles, Simon would have had a much harder time breaking in to the inner circles of behavioral science. In 1949, for instance, Simon began work on a study of the "Theory of Resource Allocation" on a subcontract from the Cowles Commission of an original contract from RAND, and he worked on study of "Decision-Making under Uncertainty" on a subcontract from the commission of an original contract from the ONR.[34]

These contracts came after Simon left Chicago, however. During his years at IIT, his membership in the Cowles Commission was more important to him as a source of intellectual stimulation than of patronage. His exposure to mathematical model-building, formal analysis, statistical inference, and all the tools of the trade of postwar mathematical economics was crucial for his intellectual development, for in these he discovered a new language in which to address the problem of achieving analytic mastery of nonexperimental situations: the language of the sciences of choice.

Simon's association with mathematician Karl Menger also spurred his investigations of models, inferences, and decisions. Menger, the son of the pioneering marginalist economist Carl Menger, had participated in the Vienna Circle and was a distinguished expert in mathematical topology. In addition, he was interested in some of the same problems related to statistical inference and mathematical model-building that animated Simon and the members of the Cowles Commission. One of Menger's students before the war had been Abraham Wald, the brilliant creator of statistical decision theory.[35]

In 1947–48, Simon, Menger, and a few of their colleagues at IIT organized a yearlong faculty seminar on the philosophy of science.[36] The philosophical orientation of the seminar was decidedly positivist and operationalist: the series began with a "general talk on the logical structure of an empirical science, introducing the idea of operationalism," followed by talks on modern logic, axiomatization, and problems in theoretical model-building.[37]

Simon's own contribution to the seminar was a novel analysis of a weighty problem in the philosophy of science: mass. Mass has been a central concept in Western science since the seventeenth century, but Simon did not believe it ever had been defined in a philosophically rigorous manner. In Simon's analysis of mass, we can see that a thoroughgoing operationalism can be used to legitimate, as well as to cut away, conceptual abstractions. In a talk on Newton's laws which he wrote for the seminar, for example, Simon went Ernst Mach one better in the operational definition of terms like *mass* and *force*. In this lecture, Simon began by showing that Mach's own definition did not meet operationalist criteria. The ground swept clear, Simon then set about creating one of his own that did.[38]

Significantly, in this lecture, Simon defined mass in terms of the relationships between objects in a system. A single object alone does not have mass. Mass only exists in relation to other objects, for the mass of an object can be defined operationally only if it causes certain effects in the other elements of the system. Hence, Simon concluded, understanding the causal relationship was necessary for the elemental process of defining terms, just as it was for guiding practical action in the world.

The temporal juxtaposition of Simon's interests in the problem of causality, the operational definition of terms, and the challenges of drawing inferences from statistical data eventually would lead him to connect them in ways that would have profound implications for his understanding of science, experiment, and the possibility of intervention in complex systems.[39] In particular, Simon (and others) later would employ this operational approach in order to legitimate the reintroduction of terms like *mind* and *purpose* to experimental psychology, as well as to justify the use of computers to simulate human cognition.[40] The outcomes of this line of research lay some years in the future, and Simon could not have known the ends to which he would apply his ideas about operational definition or causality, but he already had developed a keen sense of what sort of projects might lead to something useful. Like a skilled chef, Simon kept his larder well stocked with things he thought might go well together in a new recipe someday.

From *Administrative Behavior* to *Public Administration*

In 1945, Simon decided that the time had come to revise *Administrative Behavior* for publication. He knew it would need some thorough rewriting in order for it to be accessible, but he was so confident of the work's merits that he launched a rather ambitious direct-mail campaign, running off two hundred copies of his thesis on the departmental mimeograph machine and sending them to anyone he believed ought to be interested in his work.[41] His original mailing list has not survived, but it is clear from the responses he received that he did not define his target audience by discipline. While a majority of responses were from political scientists, Simon also received commentary from psychologists, economists, sociologists, mathematicians, and businessmen. The method to this madness? Outside the obvious target group of specialists in administration, the respondents all were interested in decision-making, the problems of organization (writ broadly), and the application of mathematics to human behavior. Simon later called this group of people studying the psychology of decisions and the properties of organized systems an "invisible college."

As might be expected from such a selected audience, the responses Simon received were broadly positive. There was wide agreement with his argument that decision-making was the heart of administration, strong sympathy for his attempt to distinguish (but also to link) facts and values, and general applause for his unmasking of the false principles of administration.[42]

Nevertheless, the preliminary edition of *Administrative Behavior* received sustained criticism from even this friendly audience. Many found it too abstract, too formalistic, and too functionalistic, arguing that it did not take into account personal motivations and emotions. Even ardent functionalists such as the political scientist James McCamy felt that in *Administrative Behavior* the individual disappeared into the organization and that emotion had vanished in a puff of reason.[43]

By far the most significant single critique of Simon's thesis was Chester Barnard's. Barnard believed that Simon's was the "first good book on administration," so he took particular care in reviewing the manuscript, sending Simon first six and then nineteen more pages of detailed comments.[44] Simon was very grateful for this thorough critique, writing to Barnard that this "is exactly the kind of analysis I have needed, and haven't gotten even (or perhaps I should say 'particularly') from my thesis committee."[45]

In his letters, Barnard expressed strong agreement with Simon's general approach, reassuring him that while "what you have written may be, as you call it, 'armchair philosophizing,'" it was "not philosophizing in a vacuum."[46] Such theoretical work was vital, in his view, for (citing Alfred North Whitehead) "it is impossible to get anywhere with bare facts without a theory since without a theory there is no basis for determining what facts are relevant and evidential."[47]

Still, Barnard did find some serious flaws in the preliminary version of *Administrative Behavior.* In his view, Simon was trying to "produce a 'physics' and at the same time to solve the riddle of the universe."[48] This cryptic critique boiled down to four main criticisms: Simon's thesis (1) was inconsistent in its use of the terms *rational* and *efficient,* (2) did not take into account the enormous amount of uncertainty involved in most decisions, (3) did not pay sufficient attention to the processes of communication within the organization, and (4) did not take a properly neutral political stance. Simon took all four criticisms to heart.

On the first point, Barnard observed that Simon mixed together several perspectives on rationality, sometimes defining it from the point of view of the individual decision-maker, sometimes in terms of some outside observer's assessment of the efficacy of actions in achieving certain ends, and sometimes in terms of an Olympian judgment of the rationality of the whole environment that shapes an individual's decision. In particular, Simon made "the test of rationality the adaptability of results of decisions rather than behavioral characteristics. In this sense a beehive is a consequence of extreme rationality," an understanding of rationality that flew against common usage of the term.[49] Similarly, it appeared inconsistent for Simon "to speak of a 'rational decision' consisting of nonrational elements and made unconsciously."[50] These problems were severe enough for Barnard to warn, "you certainly have to get this [definition of rationality] straightened out or some critic will have your liver out on the chopping block and begin throwing it to the dogs."[51]

Regarding uncertainty, Barnard pointed out time and again that it was often "impossible ever to determine which alternative was or would have been the better," even after the fact.[52] Decisions commonly are made on the basis of estimates and speculations because "knowledge of the future is at best extremely doubtful."[53] As a result, people treat "hard facts" differently than they do estimates and speculations when making decisions. To call such speculations "factual premises" thus overemphasizes the scientific, logical aspects of decision-making. This overemphasis on the exact and the rational, Barnard believed, led Simon to make some false assertions, such as the claim that a rational person usually can reach only one decision, given the values and the facts of the situation: "This 'unique answer' trick you pull several times," he noted. In his view, however, two equally rational people often arrive at different estimates of the same situation. The world was not so deterministic as Simon believed: "I recall my surprise when I first learned that quadratic equations had two sets of roots," he commented dryly.[54]

As was fitting for the president of New Jersey Bell, Barnard was convinced of the importance of communication in organizations, and he criticized Simon for taking communication for granted. To Barnard, communication was not a simple passing of information up and decisions down the hierarchy. Rather, it was a complicated process that could take place through formal channels or through informal ones, inside or outside the hierarchy. Also, in his view communication was fundamental to the structure of authority of an organization. To attempt an analysis of the "anatomy of an organization" without studying its nervous system, therefore, was to leave out its brain.

Finally, Barnard took Simon to task for his frequent reference to problems of "social value"—that is, problems related to evaluating the rationality of the system as a whole. "In view of the strong ideological currents now running," he wrote, "I think you weaken the influence of a book of this character by anything that suggests a lack of scientific objectivity and complete neutrality."[55] Barnard took particular issue with Simon's conclusions about the possibility of introducing elements of public economic planning into a free economy, asking, "Are you trying to write in the scientific mode, or in that of social philosophic speculation?" In the same vein, he chided Simon that "In a book which contains your Chapter III you have no right to abandon defining your terms [such as social cost or social value]."[56]

Despite the bite of these criticisms—questioning Simon's scientific resolve must have cut to the quick—Simon paid close heed to Barnard's comments. Replying to Barnard, Simon wrote that, "In line with your suggestions the chapter on 'Efficiency' will be thoroughly revised, material on communications will be added, and likewise material on the influences upon decision of the informal organization." In addition, he agreed that the thesis could be "revised to avoid taking a stand on current economic issues," especially "those pages which sound like a defense of 'New Deal' economics."[57]

Simon paid attention to the other critiques as well, and, in the end, the manuscript was thoroughly rewritten. As he told one friend, "By the unanimous demand of my reading public I am eliminating the 'possibilities' from Chapter 3 and the rats from Chapter 4, and bringing the critical material on the principles of administration forward to a new Chapter 2."[58] In addition, Simon, true to his word, included more material on communications and the informal organization and removed much of the "political" language. The mathematical appendices were cut as well.

Such a thorough revision "required a great deal of soul-searching," for it involved removing several parts of the thesis that Simon had felt were fundamental.[59] He did preserve his essential theoretical framework, but its philosophical underpinnings were submerged. The resulting book, published in 1947, thus was simultaneously less philosophical and less "political." It had softer edges and so was much more readable, much more acceptable to a broad public. It was not, however, a perfect work, and the tensions inherent in his understanding of rationality, of the individual and his relation to the group, and of theory and its relation to practice lived on beneath the surface.

Despite these unresolved tensions, *Administrative Behavior* received several very positive reviews upon publication. In particular, it was admired by political scientists, such as James Fesler and John Millet, who were sympathetic to the emerging "behavioral" approach to political science.[60] Chester Barnard's laudatory foreword also helped it achieve recognition. For its first five years, however, it did not sell as well as Simon had hoped. Beginning in 1952, though, its sales began to rise notably.

Administrative Behavior's rise to prominence in the early 1950s is a curious story, for its acceptance was due as much to its critics as to those who sang its praises. Alongside its positive reviews, the book also received some dismissive and critical responses.[61] It did not set the world afire. Beginning in the early 1950s, however, it came to be associated with works such as David Truman's *The Governmental Process* and Harold Lasswell and Abraham Kaplan's *Power and Society* as a symbol of the "behavioral revolution" in political science.[62] Opponents of the behavioral trend, such as Herbert Storing and Henry Morgenthau, often took aim at Simon, perhaps under the mistaken impression that the relative unknown would be an easier target.

Matters came to a head in 1952 when Dwight Waldo penned a review article for the *American Political Science Review* that was strongly critical of Simon and another new analyst of administration, Peter Drucker.[63] Both Simon and Drucker replied in the next issue. Simon came out swinging, arguing that "the faults of Waldo's analysis are characteristic of the writings of those who call themselves 'political theorists' and who are ever ready to raise the battle cry against positivism and empiricism." Writing "love me, love my logic," Simon clearly took offense at Waldo's comments, stating that "A scientist is not ...

flattered by being told that his conclusions are good, but do not follow from his premises."[64]

Simon went on to lecture Waldo, chiding him for misunderstanding *Administrative Behavior's* argument in particular and positivism in general. He concluded with a sharp rebuke:

> Philosophy is a serious study. The history of human error demonstrates repeatedly that philosophers will inevitably reach the conclusions they wish to reach unless they subject themselves to a merciless discipline of rigor ... Quite apart from whether Mr. Waldo's premises are right or wrong, I do not see how we can progress in political philosophy if we continue to think and write in the loose, literary, metaphorical style that he and most other political theorists adopt. The standard of unrigor that is tolerated in political theory would not receive a passing grade in the elementary course in logic, Aristotelian or symbolic.
>
> If political philosophers wish to preserve democracy from what they regard as the termitic borings of positivism, I suggest that as the first step they acquire a sufficient technical skill in modern logical analysis to attack the positivists on their own ground. Most of the positivists and empiricists of my acquaintance will then be likely to receive them more as allies in the search for truth than as enemies.[65]

Simon's vigorous reply, combined with the growing acclaim for his other work, helped renew interest in *Administrative Behavior*. As historian John Gunnell writes, in his reply to Waldo, "Simon enunciated an account of scientific explanation to which behavioral political scientists would routinely subscribe within the next few years."[66] Thanks to Simon's newfound notoriety and to the growing acceptance of "behavioral" political science, the sales of *Administrative Behavior* began to rise steadily. By the end of the 1950s, the second edition had become a staple of business school and public administration curricula. It has remained so ever since and is in its fourth edition as of 2004.

Administrative Behavior was not the only text of Simon's to become a standard. As noted earlier, Simon's dissatisfaction with existing texts on American government led him to try his hand at writing a synthetic work on public administration. Simon did not attempt this task alone but worked with his IIT colleagues Victor Thompson and Donald Smithburg. Intellectually, however, he clearly was the lead member of the trio, as can be seen by the authors' focus on organizational influences on decision-making and their adoption of Simon's notion of a decision as being separable into fact and value premises.[67]

The early drafts of the chapters of *Public Administration* received much the same response that *Administrative Behavior* had: praise for its theoretical sophistication and its focus on decision-making, criticism for its tendency to depict the individual as wholly shaped by the formal organization.[68] In its final form, though, *Public Administration* addressed directly many of the problems *Administrative Behavior* only touched on, devoting chapters to "The Individual and the Organization," "Group Formation," "The Communication Process," and "Formal and Informal Controls" over behavior.

As the chapter titles reveal, *Public Administration* was an ambitious project, far broader in scope than its title would indicate. It was, in fact, a text in fundamental sociology and social psychology, as applied to the problems of administration. It so emphasized "underlying sociological and psychological phenomena [as opposed to] specific rules and 'know-how'" that the authors were anxious about its reception.[69] Smithburg, for example, wrote Simon in early 1950 wondering if the "dum-dum repetition of Weber, Durkheim, Parsons, Barnard, and Simon," though unavoidable because they were "the only sources," might not give offense to "all those obscure characters" who had written on public administration in the past.[70] In a similar vein, Smithburg also worried that parts of the text would not be "understood by the students, let alone the faculty" in most schools of public administration.[71] A panel convened to discuss *Public Administration* at the annual meeting of the American Political Science Association (APSA) in 1951 seemed to Smithburg to confirm all his fears: "The stupidity of the panel shocked me. The points that bothered me were the usual two: 1) a complete failure to understand the fact-value problem (Ex: Waldo claimed that we had re-introduced the policy-administration dichotomy under new names); 2) the general feeling that our approach was somehow immoral . . . These goddamn political scientists never got an education . . . only Shipman and Wengert seemed to understand what we were driving at . . . I was depressed, however, as to how far political science is behind the other social sciences."[72]

Smithburg was not alone in feeling unappreciated; Simon kept lists, inscribed on the inside covers of his books, of people who *should* have cited his work but had not done so. He also was known for sending unsolicited reprints of his articles to those he thought needed to read them and for circulating an annual list of his publications to colleagues (so that they might request any reprints of interest to them). Since Simon averaged between ten and fifteen publications per year throughout the 1950s and 1960s, such friendly gestures could be quite intimidating.[73]

In the end, however, their novel perspective did find a receptive audience, particularly among the younger generation of political scientists. As the reviewer for the *Journal of Politics* noted: "For their part, Simon, Smithburg, and Thompson do the job for public administration that Jeremy Bentham did for Blackstone's Commentaries. Nothing is sacred, and all principles suspect. They explore for fresh insights the heady studies of industrial sociology, nonrational behavior (after Freud), social anthropology, and philosophy of science. They have struck pay dirt, too."[74]

Philosophy, Politics, and Administrative Science

Simon had struck pay dirt. By 1950 he had produced not one but two masterworks and was a journeyman no longer. He had developed a synthetic, positivist philosophy as the foundation for his work, and he had built a theoretical structure upon that foundation, enabling him to address the key problems of the age: the relation of the individual to the group, the nature of the limits to reason, and the role of the expert in a democracy. One can disagree with Simon's answers, but he certainly was asking the right questions.

As the prominence of terms such as *organization, hierarchy,* and *planning* in Simon's work reveals, this synthetic, positivist outlook had implications for the organization of society, as well as for the organization of social science. Simon's mentor-by-mail, Chester Barnard, made the link explicit:

Men are now dismayed by the evidences of world disorganization . . . However, the present questioning and discouragement do not come, it seems to me, merely from economic disturbances and international conflict. Much more do they arise from a deep conflict of beliefs concerning cooperation itself. There are two beliefs that are far apart, both struggling not only against each other but also against unrecognized limitations. One of them centers upon the freedom of the individual and makes him the center of the social universe . . . The second extreme faith is adulatory and optimistic. It places its emphasis upon the order, the predictability, the consistency, the effectiveness, of untold myriads of concrete acts that are cooperatively determined . . . Those who speak from this point of view are likely to advocate uncritically a vast regimentation, an endless subordination ... And so we find ourselves again with the very problem with which we began; for the issue between these faiths, I think, is unconsciously centered upon the old question of free will and determinism ... I found [this issue] not as an abstract question unrelated to the daily lives of men, but as one evident in the collapse of actual cooperation and in the moral disintegration of living men and women. Scarcely a man, I think, who has felt the annihilation of his personality in some organized system, has not also felt that that same system belonged to him because of his own free will he chose to make it so.[75]

To Barnard, and to Simon, the connection between the abstract problems of free will and determinism and the concrete problems of social disorganization could not be more plain. They were not alone: from city hall to the White House, from Main Street to Wall Street, democratic planning, rational coordination, and expert guidance were seen by many from the 1930s through the 1960s as the answer to the nation's ills—and as the logical consequences of a properly scientific understanding of society. Simon agreed. His politics were as thoroughly bureaucratized as his scientific worldview.

Yet Simon, Barnard, and their audiences in business and government strongly wished to retain a role for free, rational individual choice, both in their theories and in their policies. This demand for experts who believed in management and free choice, in the system and the individual, helps explain why the discipline that has had the most success in claiming a role in public policy-making after the war should be economics; of the social sciences, it was the discipline that retained the strongest faith in the virtues and powers of rational individual choice, even as it provided the tools necessary for managing enormously complex systems, from multinational firms to national economies.

Furthermore, the rise of the economists as key advisers to policy-makers coincided precisely with the emergence of two strands of economics that meshed with the bureaucratic worldview: game theory and OR, which were useful for planning everything from weapons procurement to production scheduling, and Keynes's general theory, as elaborated and transformed into the "neoclassical synthesis" by Paul Samuelson. The decline of Keynesian theory since the 1980s and the adoption of a far more atomistic, choice-centered style of economic analysis (one that assumes an omniscient rationality rather than the bounded one Simon described) has produced a tension between what economists do (develop and use tools for managing economic systems, from production lines to firms to nations) and what they believe (individual economic actors are free to choose, order emerges spontaneously in a free market).[76]

As might be expected of a philosophy dependent on the union of opposites, there were profound tensions in the bureaucratic worldview and related programs for society. The systems that Simon, Parsons, Barnard, and colleagues studied all had equilibrating mechanisms—else they could not survive long enough to be studied—but where was the need for expert guidance in a self-maintaining system? Was it really possible to blend the objectivity of descriptive science with the prescriptions of applied science? Did bounding rationality integrate it with the nonrational or did it wall it off? Were individuals merely sets of functions or roles or were they something more? If one defined purpose, mind, emotion, and the like in operational terms, did they still refer to something meaningful, or did it render them empty? Are there human qualities that are meaningful precisely because they cannot be measured?

Despite such problems, Simon and those who shared his outlook believed that they had created a practical philosophy that would help guide the individual and society through the tempests of depression and war. Like Barnard (and Plato before him), Simon believed that though this science was far from perfect, "in a storm there must surely be a great advantage in having the aid of the pilot's art."[77]

In 1949, Simon left IIT for Carnegie Tech, for there he saw an extraordinary opportunity to build an institution that could teach that "pilot's art."

CHAPTER SEVEN

Structuring His Environment

In 1946, Herbert Simon won the chair of IIT's Department of Political Science, taking the helm at the (relatively) tender age of thirty. As the vigorous young chair of a department clearly on the way up, Simon became a favorite of IIT president Henry Heald and an active member of the IIT community.[1] The IIT environment suited him well, and Simon thought he might well stay there for a good while, at least until the Harvards and Chicagos came calling. (He never doubted that they would.) In early 1949, however, Simon received an offer he couldn't refuse. It did not come from a big name university but from a small-ish technical college, the Carnegie Institute of Technology (CIT), located near the smoky heart of Pittsburgh. While CIT was no Ivy League school, it, like Simon, had great ambitions and the means to realize them.

After a few months of consideration—and increasingly generous offers—Simon decided to leave Chicago for Pittsburgh, trading Hyde for Schenley Park.[2] His new title was head of the Department of Industrial Administration, one of the two core departments in CIT's brand new Graduate School of Industrial Administration (GSIA). His salary ($10,000) was large enough for Herbert, Dorothea, and their three children to purchase a nice home on the

suburban slopes of Squirrel Hill, about a mile from CIT's yellow-brick campus, which Simon came to prefer to Mies van der Rohe's IIT.[3] Simon would
spend the rest of his career at CIT (later Carnegie Mellon University), preferring to stay even when Chicago, MIT, and Harvard did come calling. Chicago's
offer was especially tempting: in 1954, Morton Grodzins, Chicago's dean of
social sciences, offered Simon a "super-professorship" in which Simon could
"define [his] own teaching role and load" and "roam in the way Frank Knight
has in the past."[4] Despite the appeal of returning to his alma mater in triumph, Simon stayed at CIT.

It was good for CIT that he did, for Simon and the GSIA were the keys to
putting Carnegie Tech on the map as a major research university. The school
was well aware of their significance, rewarding Simon and GSIA deans G. L.
Bach and Richard Cyert with leading roles in broader university affairs. Cyert,
for example, became president of CMU in 1972 after a decade as dean of the
GSIA. Similarly, Simon was made trustee for life in 1972 and was given an
endowed professorship created specifically for him.[5] One of his later colleagues describes CMU as being "made in Herb's image."[6]

Although the GSIA proved to be a bridge, not a destination, its environment played a major role in shaping Simon and his work, just as he played a
critical role in defining its mission. In particular, the GSIA environment fostered his move from administrative science to cognitive science, and it provided a model for his efforts to reform the social sciences.

Although much about Simon and his work changed during his years at the
GSIA, certain things did remain (relatively) constant. Perhaps the most
important of these constants was Simon's vision for the social sciences.
During his years at the GSIA, Simon pursued a consistent strategy intended to
reform the social sciences through the creation and interconnection of a network of centers for interdisciplinary social research. The achievement of these
goals, in Simon's mind, depended on two things. First, research in the more
advanced areas of the social sciences needed to be expanded, particularly
research in the interdisciplinary region characterized by the mathematical,
behavioral analysis of decision-making in social systems. Second, regular connections to a diverse set of powerful patrons needed to be developed, both to
enable work in this area and to legitimate it.

In Simon's view, the available resources—in both dollars and men—needed to be concentrated in a few top-notch programs. These programs, in turn,
needed to be linked to each other in a network of regular communication and

loose coordination of research. Simon advocated the concentration of resources because he believed that there were only a few men in the social sciences who were true scientists. (A few did not mean none, however: Simon resolutely defended the merits of the work done by the best social scientists, even as he turned his acid criticism on those who preferred "weaker medicine" than his mathematical, behavioral approach.)[7]

Simon sought to link such centers in a loosely coupled network because he believed in the virtues of coordinated, but not directed, research, just as Charles Merriam had at the University of Chicago. In keeping with his own theories of planning, Simon wanted to structure the environment of his fellow researchers' decision-making so that interdisciplinary work would be the natural—but not the only—choice. As a result, Simon tried to create incentives for researchers to choose to work together, especially across disciplinary boundaries. Such coordination by incentives worked best when the community members shared a sense of mission—and when no one group within it had a steady, singular pipeline of resources distinct from those available to the rest of the community. When goals diverged and when funding streams began to flow in deep, well-defined channels, Simon and his colleagues started to go their own ways.

The first step in making Simon's vision for the social sciences a reality was to build the GSIA into a leading center for interdisciplinary social research. As there was neither an established network of patrons accustomed to supporting research in this area nor an existing market for the products of such centers, Simon and his allies at the GSIA faced real challenges. They had to display all the talents of the Schumpeterian entrepreneur, bringing together interests and resources into a new organizational pattern while at the same time creating a market for this new organization's products.[8] This was no small task, and it entailed many risks. Simon, however, was confident that his ambitions suited the times. In his view, the environment at CIT—and of the social sciences generally—was fit to be shaped in his image.

Organizing diverse interests required reconciling divergent goals. For a time, Simon's goals and those of the other major stakeholders (such as the Carnegie Tech's administration and the GSIA's economists, students, and patrons) were similar enough that they could be made to reinforce each other. There were tensions between the various interests from the start, however, tensions that broke along the same fault lines Simon had been trying to bridge since his days at Chicago: was the GSIA to be dominated by the sciences of choice or control? Would it be a center for research or for education and reform?

As the GSIA grew and flourished, the divisions grew as well. Eventually, Simon came to believe that a new constellation of interests and resources was necessary for him to achieve his goals. The result was a new interdisciplinary program in Systems and Communication Sciences, which soon evolved into Carnegie Mellon University's renowned Department of Computer Science.

Although this story is of a specific individual and his institutional endeavors, a story not only colored but also shaped by the local environment, it has much to teach us about broader issues as well. Not only does it help us understand Herbert Simon and his role in the creation of a pioneering research enterprise, it also sheds light on a pervasive phenomenon in postwar American science: the creation and dissolution of interdisciplinary research communities. Accordingly, this tale of Herbert Simon and the GSIA strives to link the local and the global, just as Simon did.

Two Visions

When Herbert Simon came to Carnegie Tech in 1949, he found a technical institute in the process of transforming itself into a research university. Although it had been created in order to provide a practical education for mechanics interested in advancing themselves, over its first forty years, CIT—like its technical school brethren—had progressively expanded its educational program to include more and higher-level science. The first major steps along this path came in 1912, when CIT became a full-fledged degree-granting institution, and in 1922, when Thomas Baker assumed CIT's presidency and began to encourage the establishment of research laboratories on campus.[9] The biggest changes in CIT, however, came after Robert Doherty took the reins in 1936.

Doherty had been trained as an electrical engineer at Cornell and had been dean of engineering at Yale before coming to CIT, and he hoped to instill in the Carnegie Tech faculty some of the same spirit that animated such research universities. In addition, his experiences working at General Electric and teaching at Yale had convinced him that professionals in general, and engineers in particular, were too narrowly trained. In his view, professional education needed to "place less emphasis upon routine know-how and mis-cellaneous technical information, and much more upon [the] intellectual skills [and] fundamental knowledge ... essential to coping with practical sit-uations."[10] For Doherty, like Simon, fundamental research was necessary, not antithetical, to developing a broad perspective.

In keeping with this educational philosophy—which Doherty quite accu-rately described as "a major break from education tradition"—in 1940 he put forward a set of guidelines that quickly became known as the "Carnegie Plan" for engineering education.[11] According to this plan, Carnegie Tech would strive to cultivate: "1) the ability to think independently ... 2) the ability to ... deal with the *whole* problem or situation and not only part of it ... 3) the ability to learn from each new experience ... [and] 4) competence and inter-est in dealing with the responsibilities of citizenship."[12] In short, Carnegie Tech would train engineers to be professionals and citizens, not mere techni-cians.

One of the key elements of this unusually broad vision of engineering edu-cation was instruction in the social aspects of engineering. Doherty believed that training in economics, for example, was vital, since engineers always must design with a cost constraint. Similarly, he knew that successful engineers typ-ically advanced to management positions where technical skills alone were insufficient. As a result, Doherty committed himself to developing programs in economics and management as a part of the Carnegie plan.[13]

After the war's end, Doherty and his provost, Elliot Dunlap Smith, decided that the time was right for taking this next step in the Carnegie plan. So, in early 1946 they hired economist G. L. (Lee) Bach and charged him with revi-talizing economics at CIT. Bach had received his Ph.D. from the University of Chicago, and like Simon, he had absorbed the characteristic Chicago convic-tions that interdisciplinarity was vital to innovation, that social science should be linked to public service, and that social science must be theoretically sophisticated to be useful as a guide to practical action.

In keeping with these views, Bach staffed his department with economists who combined theoretical skill with experience in applying theory to practi-cal situations, using his Washington connections to recruit men with practical government experience: ten of his first fifteen hires had worked for the gov-ernment during the war.[14] His first hire was W. W. (Bill) Cooper, for example, whom Bach knew from Cooper's days as chair of an interagency task force on the government's wartime statistical systems.[15] Cooper was another Chicago

Ph.D. and was closely connected with the Cowles Commission for Research in Economics, counting Bert Hoselitz, Franco Modigliani, and Herbert Simon among his friends. All three of these prominent scholars followed Cooper to CIT in the late 1940s and early 1950s.

In 1947, with the reform of economics at CIT well underway, Doherty turned to the problem of management training, asking Bach to take over the program in industrial engineering and charging Provost Smith with developing a program in personnel management. Bach assigned Cooper the task of revising the industrial engineering curriculum, and he soon devised a new two-year program on "Quantitative Controls in Business" to accompany Provost Smith's new courses.[16] With these new arrows in his quiver, Doherty approached William Larimer Mellon in late 1948 about the possibility of his endowing a new school of industrial administration.

Doherty envisioned this school as having a five-year program leading to a bachelor's degree in engineering and a master's in administration. Such a program would constitute another marked break with educational tradition, for at this time the MBA was a very new thing. Fewer than ten thousand master's degrees in business administration or management were granted nationwide in the *decade* of the 1940s, whereas close to 100,000 MBAs were awarded *every year* by the turn of the millennium.[17] The GSIA is a major reason for this swell, for the institution Bach, Simon, and Cooper created at Carnegie Tech rapidly became *the* model business school during the explosive growth of business education in the 1950s and 1960s.[18]

Mellon's endowment also would entail a break from another kind of tradition, as the Mellons usually invested their philanthropic dollars in the University of Pittsburgh, not Carnegie Tech.[19] The novelty of the plan appealed to Mellon, however, and he was easily persuaded that engineers made the best managers and that management education was both vital and sadly deficient in America. In an intense but discreet courtship, Doherty capitalized on these shared sentiments and on Mellon's growing dissatisfaction with the University of Pittsburgh. In early 1949 the relationship was consummated, with Mellon agreeing to donate $6 million for the founding of a new school of industrial administration. This was a remarkably large gift for the time; among business schools, only Harvard had ever received such a gift.[20]

Doherty appointed Bach dean of the new school, and Bach's first hire was Herbert Simon, then known only as a promising young political scientist with interests in public administration and economics. The addition of Simon to the staff was a signal event for the GSIA, for Simon would change the school's goals even as he helped it achieve them.

Doherty and Mellon envisioned the GSIA as a center for professional training in management. They "valued its educational program above all else, and they expected the school's students to be its greatest product."[21] Herbert Simon, however, had a different vision, one that was very much a product of his education at the University of Chicago and his later experiences at the ICMA and BPA. As earlier chapters have shown, these experiences had taught Simon that reform programs needed to be based on a body of fundamental knowledge and theory. Too many people, in his view, still were guided by "the Proverbs of Administration," not by the findings of administrative science.[22] Research into the fundamentals of organizational sociology and social psychology, particularly the ways organizations influenced the decisions of their members, was necessary for true reform.

In addition, Simon's aptitude for mathematics and his fierce positivism had led him to an ever more rigorously mathematical, behavioral approach to the study of human thought and action, an approach he promoted with the relentless energy of the true believer. His association with the Cowles Commission had only strengthened his faith, and he was more than ready to profess his creed on a new stage to a larger audience.

For Simon, the combined result of these ideas and influences was a linked set of three basic beliefs that structured his ambitions, both intellectual and institutional. First, he believed that the sciences of choice and control were beginning to converge on a shared set of questions and methods, all related to the study of the "atomic phenomena of human behavior in a social environment."[23] The nexus of this convergence lay in the problem of choice: how does the social environment affect how individuals make decisions? In 1949, Simon still thought that the common conceptual language in this interdisciplinary nexus would be the language of the sciences of control, but he had a growing appreciation for the power of the analytic tools of the sciences of choice and hoped to use these new tools to answer his old questions.

Second, Simon believed that work in this interdisciplinary region necessitated a synthesis not only of choice and control but also of theory and practice, research and reform. One important consequence of this belief was the idea that research should be problem-centered; that is, it should focus on solving problems with tangible, testable, real-world implications in addition to having significance for theory. This "problem-centered" research was not simply applied research by another name; rather, the gradual elaboration of ever more sophisticated, ever more general theory was every bit as important as a focus on real-world problems.[24] Problem-centered research would be a synthesis of abstraction and action, just as the experiment is a synthesis of theory and empirical practice. To Simon, theory only could advance by being put to the test in experimental situations, and problems only could be solved through the advance of theory.

Third, Simon felt that the proper way to conduct such interdisciplinary, problem-centered research was to have teams of researchers working on projects funded by external contracts and grants at a small number of model research centers. The ideal center would conduct both empirical and theoretical work and would bring together specialists from several disciplines to work on projects funded primarily by foundations and military research agencies.[25] Work would be initiated through concern with practical problems, and it would move from there to the theoretical and fundamental before returning to a practicable solution. Happily, all three components of this vision reinforced each other: a focus on problems would appeal to patrons with concrete problems to solve, and because real-world problems do not respect disciplinary lines, it would encourage the "reintegration of the social sciences" that Simon so ardently desired.

Simon was not the only one who had such a vision; across the country a great number of interdisciplinary centers for social research were born in the first twenty years after the war. Some of the most famous were the Harvard Department and Laboratory of Social Relations, the University of Michigan's Institute for Social Research and Mental Health Research Institute, MIT's Center for International Studies, Columbia's Bureau of Applied Social Research, the University of Chicago's National Opinion Research Center, the Air Force's RAND, and the Ford Foundation's Center for Advanced Studies in the Behavioral Sciences. These centers were merely the tip of the iceberg: a 1968 study by the National Academy of Sciences on the state of the behavioral and social sciences counted over four hundred institutes for social research, almost all of which had been created since World War II and nearly three hundred of which involved multiple disciplines.[26] These centers depended on grants from private philanthropies or contracts with federal (primarily military) agencies, and, as a rule, they attempted to fuse fundamental and applied research.

The GSIA, as defined by Doherty and Mellon, would not have been such a center, nor was Bach inclined to deviate quite so far from its founder's goals. Simon, however, saw in the GSIA the potential for creating a center for inter-disciplinary social research.[27] He guessed, rightly, that the school had the potential to draw on a variety of significant resources and that Bach could be persuaded to side with him on the issues that mattered most.

Part of the bait that had lured Simon to Carnegie Tech in the first place were the promises that he would be appointed head of the Department of Industrial Administration (one of the two basic divisions in the GSIA), that he would have a free hand in hiring and firing within his department, that he would be a member of the school's three-man executive committee (and so would have a voice in all policy and personnel decisions), and that he would serve as the school's representative on the Carnegie Institute's external com-mittee overseeing the school's operations.[28] These appointments gave Simon unusual status within the GSIA, as is indicated by the school's catalog where he is listed, out of alphabetical order, at the top of the page next to the dean.[29] Later he was named associate dean, and he eventually served a year as dean, which he did not care for. He much preferred being the power behind the throne to occupying it himself.

Simon employed his official powers vigorously, hiring a young, talented, and like-minded faculty of social psychologists and organization theorists to staff the Industrial Management program. His hires included such now well-known names in the field as George Kozmetsky, Harold Guetzkow, James March, Allen Newell, and Richard Cyert, who later became dean of the GSIA and president of Carnegie Mellon. Simon also used his negative authority willingly, securing the removal of several faculty members (parti-cularly economists) who did not see eye-to-eye with him. This formal author-ity was magnified when both Doherty and Mellon died suddenly in the summer of 1950. The new president, Jake Warner, was not as interested in management education as Doherty, and he left Simon and the school to their own devices.

In addition to this formal authority, Simon wielded great informal power. His informal authority stemmed from many sources, from his personal influ-ence with Bach and other CIT administrators, to the table-pounding vigor with which he promoted his vision for the school, to his obvious intellectual prowess. A number of the staff, particularly in Industrial Management, were rather in awe of him, writing memos to him peppered with lines like, "Gee, Herb, you're a genius," which went down well.[30]

In addition, in the school's second year "an informal three-man elected faculty committee" was established "as an informal research steering and coordinating group," with Simon at the helm. Although there was "no thought that this research group will dictate what research projects should be undertaken by individual faculty members," it did "participate actively in decisions on the allocation of S.I.A funds to research purposes" and "on [decisions about] the School's acceptance of outside contracts for research work."[31] Thus, Simon was able to position himself at the gateway between external resources and internal policies and programs. As every master strategist has found, controlling the interfaces—the points where the local connects to the global—brings wealth and power. Simon had learned his organizational sociology well.

Mobilizing Support: Patrons of the Revolution

This larger world of support to which Simon sought to make connections was changing rapidly. The most obvious change was one of scale, for the scale of resources available for social science research after World War II was many times greater than before the war, supporting vast increases in the numbers of social scientists, the amount and sophistication of equipment available, and the scale and scope of research projects. While the numbers involved may seem small to historians used to reading about the billions appropriated for research on nuclear energy or electronics, they produced stunning changes in the social sciences nonetheless. Take psychology, for example, the field Simon soon would enter: in 1947, the American Psychological Association (APA) had 4,661 members, making it the second largest professional association in the social sciences. Ten years later, it had tripled in size, to 15,545. Another ten years saw another ten thousand new members join. A psychologist who received a Ph.D. in 1945 thus might join a department of five or ten as an assistant professor only to have thirty, forty, fifty or more colleagues by the time he or she became a full professor. Specialization in research followed in step with the increase in numbers: a psychologist with an interest in language, say, soon became a psycholinguist and then a specialist in language acquisition in early childhood.[32]

The sources of this flood of funds were as new as their scale. While the continued prominence of the Rockefeller Foundation (RF) provided some measure of continuity, it no longer was the dominant patron in the network of support for social science. Indeed, one major difference between the prewar and postwar patronage systems for social science was that there was no single dominant patron in the 1950s and early 1960s. No single institution could set the tone the way the RF had in the 1920s and 30s. The closest thing to such a leading patron was the Ford Foundation, which dwarfed the RF, but even the Ford Foundation had to share center stage with the military research agencies, especially the Air Force/RAND and the Office of Naval Research (ONR). In addition, by the early 1960s the National Science Foundation (NSF) and National Institutes of Health (especially the National Institute of Mental Health, or NIMH) had become major patrons of the behavioral sciences as well.[33]

The program officers at these diverse institutions did share some common goals, however, particularly those at the Ford, Carnegie, and Rockefeller foundations and at the ONR, Air Force/RAND, and army research agencies.[34] (The federal patrons of relatively pure science, such as the Defense Advanced Research Projects Agency [DARPA], the NSF, and NIMH, were a different story, which will be discussed in chapter 11). During the 1940s, 1950s, and early 1960s, the program officers at the foundations and the military research agencies all supported social science that was behavioral-functional in approach, mathematical in technique, problem-centered in focus, and interdisciplinary in organization.

This shared agenda can be seen in the career of Merrill Flood, a mathematician who, like Simon, was fascinated with decision processes. Flood was a key figure in a number of military research agencies, and his ability to move easily from one agency to another reveals how widely these four basic goals were shared. Flood ran the army's Fire Controls Research Office of the Army during World War II, sponsoring work by Norbert Weiner, Marston Morse, Albert Tucker, and other pioneers of operations research (OR) and cybernetics. After the war, he became the chief civilian scientist of the War Department, in which capacity he sponsored research on OR, game theory, linear programming for logistics analysis, and other mathematical approaches to human behavior. Flood later moved to the ONR, where he ran its Behavioral Models Project, an important source of funding for mathematical behavioral science in the mid-1950s.

Simon first met Flood in the summer of 1952 while attending a conference (hosted by RAND) on "The Design of Experiments in Decision Processes." This conference revealed that Simon was not imagining things when he saw an interdisciplinary convergence on the topic of decision-making in social systems. Many of the participants, such as Gerard Debreu, John Nash, John von Neumann, Frederick Mosteller, and Howard Raiffa, either were or went on to be major figures in economics, sociology, or psychology. The conference also revealed the sources of support for those working in this area: of the thirty-seven participants, twenty-one were sponsored by RAND, seven by the ONR, thirteen by the Ford Foundation, and three by the Cowles Commission. In addition, seven speakers reported on research that had been funded by more than one of these four agencies. Both the ONR and RAND, for example, funded the University of Michigan's Behavioral Models group, just as they supported Simon's group at Carnegie Tech.[35]

Simon's institutional goals thus fit his patrons' agendas perfectly. The fit was so perfect that it can be difficult to tell whether Simon adapted himself to suit his environment, whether he shaped that environment to suit himself, or whether he was selected by it. Was he an opportunist? A visionary? Or did he simply happen to be a peg of exactly the right size and shape? His outward actions could be interpreted in each of these ways. In the end, though, Simon appears to have been a fortunate evangelist rather than a clever careerist: he found his institutional environment fit to be shaped in the ways he wanted to shape it.

Simon's rise to influence in the new patronage system of postwar social science began shortly before his arrival at Carnegie Tech. In 1949, Quincy Wright invited Simon to join the American Political Science Association's (APSA) Committee on Research, headed by Harold Lasswell, one of Simon's former teachers at Chicago.[36] This was only the first of many committee assignments for Simon, who took to committee life like a fish to water.

It was no accident that Simon's introduction to the world of associations, agencies, and foundations came through a committee position. Many scholars have noted that the ability to manage large projects became an important part of the ambitious scientist's repertoire of skills in "Science After '40," but postwar science was a world of committees as well as projects, which historians have not fully appreciated.[37]

Committee work can be team work; that is, it can be project- or mission-oriented. Committees frequently serve coordinating rather than executive functions, however, and they often are used to assure the stakeholders in an issue that their interests have been represented. Such coordinating committees are vital linking mechanisms in the modern world of specialized, task-oriented bureaucracies, and one who serves on multiple committees often becomes not merely a communications link but also a broker—one who creates indirectly by bringing the right people and interests together.

Skillful brokers, such as Simon, are immensely valuable, not merely to their constituencies but to the whole network of organizations they link together. As a result, they often find themselves in great demand—a demand that they help to create, of course. Simon, for example, stated to me several times that "I only served when asked," regarding the many, many policy-making committees on which he served. Strictly speaking, he was telling the truth; he never had to lobby for the positions he wanted. He always was the obvious choice: no one else on committee X also was a member of committee Y, not to mention committee Z, which he well knew.

Simon took his committee assignments quite seriously and used them to his advantage. He was a master of committee politics, knowing when and how to take a stand, throw a fit, or turn on the charm. Even more, Simon used his memberships on multiple coordinating committees to great advantage. Simon's many committee memberships enabled him to serve as a broker—and even as ambassador plenipotentiary—coordinating the policies of multiple agencies while advancing his own interests and agenda.[38]

The importance of Simon's abilities as committeeman and broker can be seen vividly in his relationship with the Ford Foundation. By 1951, Simon's work on the APSA Committee on Research and his persuasive programmatic statements in *Administrative Behavior* had attracted the attention of Bernard Berelson, the head of the Ford Foundation's new program in the behavioral sciences. Berelson found Simon's philosophy and energy appealing, and he turned to Simon frequently for advice about the direction the Ford Foundation's program should take. This relationship was formalized (to an extent) by Simon's appointment to several of the official advisory committees the foundation established in its early years.

Membership on one of these advisory committees was a sure sign of either existing or coming prominence. Membership on more than one such committee defined the inner circle of the social sciences. Simon was a member of four official committees and a number of ad hoc groups, which was remarkable given Simon's relative youth. Only Paul Lazarsfeld, Robert K. Merton, Ralph Tyler, and Thomas Carroll—the organizational elite of the social sciences during the 1950s and 1960s—served on more of these committees than did Simon.[39]

Simon's association with the Ford Foundation was significant in many ways, for it played an extremely important role in the social sciences during the 1950s. The Ford Foundation's resources and ambitions may have been greater than those of the other civilian patrons of the social sciences, but Berelson and the other program officers at Ford shared the widespread beliefs that research should be oriented toward solving problems, that it should adopt mathematical, behavioral approaches, and that it should be interdisciplinary. Similarly, its internal pattern of coordination through interlocking committee memberships mirrored the structure of the patronage system for behavioral science as a whole.

The Ford Foundation had been established in 1936, but its history really began in 1948, when the Ford family assigned a vast amount of their stock to the foundation in an attempt to lessen the bite of inheritance taxes.[40] This sudden influx of money meant that the foundation became the world's largest private philanthropy almost overnight, commanding greater resources than the Rockefeller and Carnegie philanthropies combined. The old organizational form and mission of the foundation were not sufficient to handle responsibilities of this magnitude, so it organized a Study Committee on Policy and Program to devise a structure and a purpose to direct its funds.

H. Rowan Gaither, who had just organized the new RAND Corporation in Santa Monica for the Air Force, chaired this committee.[41] Gaither's experience setting up RAND and his wartime experience as assistant director of the Radiation Laboratory at MIT had given him a thorough knowledge of Washington science policy circles. He knew that the Ford Foundation, despite its vast resources, could not compete with the federal government as a patron of research and development in the natural sciences. It would have to make its presence felt in other areas.

America's new global position, the emerging Soviet threat, and the stunning power of recent scientific advances (especially the atomic bomb) taught Gaither that the United States needed a better understanding of itself and the world beyond its shores. The foundation, he believed, could make its mark best by sponsoring programs intended to advance peace, democracy, and economic development. The Gaither committee's report, therefore, advocated the organization of the foundation into five program areas: "the establishment of peace, the strengthening of democracy, the strengthening of the economy, the improvement of education, and the better understanding of man."[42]

In 1951, political scientist Bernard Berelson was appointed program officer for the fifth of these areas, and he soon set to work devising a strategy for the development of the "behavioral sciences." He chose the term *behavioral science* for several reasons, only one of which was the oft-cited desire to avoid the confusion of social science and socialism. He chose the term primarily because he desired to promote a particular kind of social science—mathematical, behavioral social science—and because he wished to avoid channeling its funds through the traditional disciplinary structures.[43]

After consulting with Simon and his others advisers, Berelson circulated a draft "Plan for the Development of the Behavioral Sciences," in which he set forth his agenda. Agreeing with the Gaither committee that "the critical problems which obstruct advancement in human welfare and progress toward democratic goals are today social rather than physical in character," Berelson stated that the "goal of the program is to provide scientific aids for use in the conduct of human affairs." Since the program's ultimate goal was the solution of social problems, not the mere increase of abstract knowledge, "the program [did] not fall within any one conventional field of knowledge, and traditional academic disciplines as such [were] not included or excluded. On the contrary, the program [was] interdisciplinary and inter-field."[44]

Unfortunately, Berelson argued, there was in the social sciences "a tendency toward diffuseness, unorganized specialization and lack of cumulation and integration." The social sciences were "too isolated from their neighbor disciplines," and traditional university departmental structures only made the problem worse. Similarly, theory and empirical observation were poorly integrated, which was detrimental both to social science and society, for "only if theory is thus translated into action will democracy outperform its totalitarian competitors."[45]

In keeping with these views, Berelson outlined a series of problem areas to serve as focuses for research, such as "Political Behavior," "Values and Beliefs," "Social and Cultural Change," "Formal Organization," "Communication," and "Behavioral Aspects of the Economic System." All these problem areas demanded interdisciplinary attack, Berelson argued, and they required the active leadership of foundation officers. In his view, the Ford Foundation would initiate action in these areas through specific programs and projects while it would respond to proposals in other areas. It thus would select and commission work in certain fields, conducting "inventories" of "tested propositions" in specific problem areas, such as organization theory, for example.[46]

The centerpiece of Berelson's proposed plan was the creation of an interdisciplinary "advanced education and training" institute to be "staffed by a small group of absolutely top men." This institute would have a permanent staff of ten to twelve, with a smaller number of visiting faculty, and it would train the leaders of the next generation of behavioral scientists. In Berelson's words, "the objective is to create *the* center for training and research in the behavioral sciences."[47]

Simon was one of the select group who received this draft plan for review. He was pleased that Berelson and the foundation were interested in encouraging interdisciplinary social research of real scientific merit, and he applauded Berelson's calls to integrate theoretical and empirical work.[48] He strongly opposed many of Berelson's specific proposals, however. While Simon agreed that the "training of a moderate number of first rate people is . . . more urgent than [producing] a large number of merely competent people," he was skeptical of Berelson's proposals to initiate activity in a few chosen fields.[49] In addition, he thought the idea of trying to create *the* institute for behavioral science was sheer "Megalomania!"[50]

To Simon, picking research areas and topics was unlikely to produce good results in the long run, and the foundation was far better off giving $1 million each to five or ten schools than it was putting all its eggs in one basket, especially since he would not be the one holding that basket.[51] In short, to Simon, Berelson's plans involved "Too Much Masterminding!" He preferred a simpler plan, based on three principles: "1) Big Money. 2) Few Strings. 3) Long Time and Good Men."[52] Simon wanted the Ford Foundation spend its money on talented individuals, enabling them to build programs around themselves.

In a similar vein, although Simon did want to encourage coordination and integration across the social sciences, he did not think that a single center or a series of commissioned projects were the ways to buy such integration. To him, commissioning work was fine for agencies with concrete, applied interests (like defense agencies), but he believed that the Ford Foundation's role should be to support more basic, purer research. In his view, "Program Five [behavioral sciences] resources should be jealously guarded for fundamental work whose application is a very long run proposition."[53] The idea of commissioning fundamental work in such areas was a fantasy.

Simon also believed that the key to integration in the social sciences was to "integrate people!"[54] That is, real interdisciplinary work required "interdisciplinary people," not just the assembly of a team of specialists.[55] (His difficulties in working with the GSIA's economists at precisely this time doubtless strengthened this conviction.) The foundation could best serve the cause of interdisciplinarity, therefore, by enabling individuals to broaden themselves and by rewarding them when they did. It could do so by selecting "good men" with broad ambitions and "turning them loose" to learn what they needed, wherever it was to be found, however long it took. Summer training institutes in various skills and techniques (especially mathematics), travel grants to visit and learn from leaders in other fields, and flexible, long-term grants were the keystones of Simon's plan for promoting interdisciplinary work.[56]

Simon was not Berelson's only adviser, of course, and Berelson had his own agenda. As a result, the Ford program in the behavioral sciences did not hew precisely to Simon's views. A Center for Advanced Study in the Behavioral Sciences was established, for example, even though Simon opposed it. However, due in large part to Simon's influence, the center little resembled Berelson's original idea of an advanced training institute.[57] As O. Meredith Wilson later wrote (to Simon): "The memory of your role in the advisory group that first recommended the creation of the Center is bright indeed. I reinforced my impression that in the local oral tradition, you are one of the heroes. You are credited with preventing the Center from becoming a place where inexperienced postdoctorals would sit at the feet of superannuated seniors."[58]

Travel grants and summer institutes were established as well, with important benefits for many individuals. "Inventories" of selected problem areas were commissioned, again contrary to Simon's wishes, but due to his influence they tended to be defined as "major projects of theory construction" rather than "inventories."[59] Simon even led one inventory team, under the principle that if it was going to be done, it might as well be done right. This inventory developed into *Organizations,* one of the most important works in organizational theory of the postwar era.

In the end, the Ford program in the behavioral sciences did not lead but rather followed—and funded—leaders like Simon.[60] While this was not Berelson's plan, it is no small thing to have enabled leaders to lead—or to have played such an important role in selecting those leaders. Simon, for one, used the influence he gained through his association with the foundation to advance his agenda (and himself) to heights he could not have reached without such support.

Team Work

Simon's intimate connections to influential patrons aided his efforts to build the GSIA into his ideal research center. At the same time, his status outside the GSIA advanced as it became a model center for behavioral science research, exemplifying the qualities he and his patrons sought to promote: problem-centered (and therefore interdisciplinary), mathematical, behavioral-functional analysis.

The most direct source of Simon's pervasive influence at the GSIA, however, was not his ability to win funds. Rather, it was his unique position as an integral member of all of the major team research projects undertaken by the school's staff in its first five years. His multiple memberships, combined with his forceful personality and captivating intellect, enabled him to coordinate work across projects and to inspire a sense of shared commitment to his vision of behavioral science.

The first five major projects reveal the diversity of the sources of support for Simon's kind of science: (1) a subcontract via the Cowles Commission for work on the RAND corporation's project on the theory of resource allocation; (2) a closely related study for the ONR on decision-making under uncertainty; (3) a contract with the Air Force and the Bureau of the Budget for work on Project SCOOP, a study of intrafirm production planning and control; (4) a study done for the Controllership Foundation on the comparative merits of decentralization and centralization in budgeting; and (5) an inventory of organization theory funded by the Ford Foundation, which led to the book *Organizations,* co-authored with Simon by James March and Harold Guetzkow.[61]

The Air Force and the Cowles-RAND contracts were signed by early 1950, and they not only covered the research expenses of the projects but also included sufficient overhead payments to pay for "nearly all the faculty time . . . allocated to research planning and development during the year ahead."[62] This was quite fortunate, since $6 million in endowment didn't go as far as it used to, and the school's new building already was running over budget.[63] Despite such unexpected expenses, these and the other contracts, which soon followed, enabled the school to dispense with charging its master's students tuition for its first few years. They also supported a very high staff-to-student ratio, enabling the faculty to spend the majority of their time on research. To give an idea of the sums involved, which were quite extraordinary for the social sciences at the time, one of Simon's colleagues, as part of an attempt to relieve him of his administrative authority, suggested in 1951 that Simon be set up with a $100,000 annual research budget—ten times his salary—in exchange for his letting slip the reins of power.[64] Instructively, Simon refused.

Such plentiful resources enabled the school to attract talented researchers and to equip them well, with the result that the staff quickly gained considerable recognition, particularly for their work in economics and in organization theory. Many business schools, from MIT to Chicago to Stanford, followed the GSIA's model, often hiring GSIA staff to lead the way. Bach played a particularly important missionary role in this regard, as he was intimately involved in two influential reports on business education: Robert Gordon and James Howell's *Higher Education for Business,* produced for the Ford Foundation, and Frank Pierson's *The Education of American Businessmen,* produced for the Carnegie Corporation.[65] These "Flexner Reports for Business Education" went along with a $35 million program of grants to business schools by the Ford Foundation, and both reports described the GSIA as a model business school. Simon did his part to promote the GSIA model through his friendship with Thomas Carroll (dean of the University of North Carolina Business School and program officer at Ford Foundation responsible for the business education program) and his participation on the SSRC's Committee on Research on the Business Enterprise.[66]

These contracts did more than just pay the bills, however; they structured the life and work of the school at a basic level. In the GSIA, people were organized into research teams, in sharp contrast to the prewar period when individuals conducted most social science research. One indicator of this trend is that the percentage of papers with multiple authors published in the flagship journals of the various social sciences leapt dramatically after World War II.[67] The GSIA's research teams were interdisciplinary, and they met frequently (usually weekly) to discuss progress. Project staff wrote an endless series of memos and updates, and project leaders wrote annual or semiannual reports for their patrons and for the dean, who compiled them for the school's annual report to the president.

One consequence of the school's project organization was that team leaders and particularly energetic or forceful team members had the opportunity to exert a strong influence on the work of the whole team. As head of two of the five major early projects and lead hell-raiser on the others, Simon's ideas about decision-making, bounded rationality, and system organization came to characterize the work of much of the school's staff. As a result, a clearly definable Carnegie Tech style (or even "doctrine," as some called it) was born. The effects of a prolonged exposure to Simon could be dramatic indeed, as a letter from R. Freed Bales of Harvard's Laboratory of Social Relations reveals. After a brief visit to Carnegie Tech, Bales wrote: "Don't tell anybody the horrible truth about the effect you had on me. My ideas are even crazier than when I came ... About to abandon the stochastic approach, wavering away from the multiple weighted factor approach, and teetering toward the grundlich-gosh-awful-fundamental-complex-information-processing-and-problem-solving-monster approach. My philosophy seems to have been affected."[68]

Another result of this project organization was the product orientation of the staff. Although RAND and the ONR were not as narrowly results-oriented as other military research agencies, which typically had a strong task-and-technology focus, their patronage did encourage the school's staff to think in terms of creating readily identifiable products—that is, of producing not just publications but theorems, models, discoveries, and easily packaged techniques, the more novel the better.[69] For example, the GSIA staff not only studied production control systems but also developed a marketable set of techniques for "quadratic programming," going "linear programming" one better.[70]

One final way in which these contracts and grants structured work at the school, and one that is often overlooked, is that they brought with them connections to the world beyond Carnegie Tech, which was just as Simon intended. Program officers at RAND, the Air Force, the Ford Foundation, ONR, and the like were involved with many projects at many sites, and they, quite naturally, wanted to ensure that their charges communicated with each other. As contract workers for RAND, Simon and some of his colleagues were invited out to Santa Monica, not just for weekend conferences but also for entire summers. Similarly, the Cowles staff came to visit CIT on many occasions, delivering papers, holding conferences, and meeting with their counterparts for hours on end.

In a similar vein, the Ford and Rockefeller Foundations—partly at Simon's behest—sponsored travel grants for researchers to study at various centers of research in the interdisciplinary area of convergence Simon had identified. A steady stream of visitors came to Pittsburgh on such grants to discuss common interests, to assist in setting up experiments that they had run elsewhere (Simon believed that a tradition of replication needed to be created in the social sciences), and to participate in seminars.[71]

In addition, beginning in 1953 the Social Sciences Research Council, with money from the Ford Foundation, organized a series of six-to-eight week intensive summer seminars for promising young faculty.[72] All of these seminars promoted a mathematical, behavioral-functional approach to social science, and several of them explicitly focused on Simon's interdisciplinary area of convergence (today called the "systems sciences" by historians). This was no accident, as Simon played a major role in organizing the summer seminar program and put together two of the most famous of the seminars, including one held at RAND in 1958 that has become near legendary in the world of artificial intelligence (AI) and cognitive psychology.

The 1958 summer seminar on simulation techniques was a classic example of Simon's ability to bring together diverse resources in the service of his synthetic vision. Simon and his GSIA colleague Allen Newell organized this seminar under the SSRC's aegis with funding provided by the Ford Foundation and the RAND Corporation. As an indication of how close relations were between Carnegie Tech and RAND, RAND continued to pay Newell's salary throughout the 1950s, even though he moved to join Simon at the GSIA in the fall of 1954. Ford and RAND were similarly closely connected: H. Rowan Gaither played a vital role in organizing both, to the extent of arranging for Ford to give RAND $1 million in start-up money in 1948.

Planning for the seminar began in the summer of 1957, when Simon and Newell (both running yet another summer seminar for the SSRC, this one on organization theory) found that there was growing interest in the computer simulation aspects of their work on decision-making. They suggested to their friends at the SSRC and the Ford Foundation that a summer seminar specifically oriented around teaching younger scholars the techniques involved in computer simulation would be valuable.[73] Simon, who would be elected to the SSRC's board of directors the next year, already was a prominent member of the SSRC and was well connected to the Ford Foundation, as noted above. Through his good offices, Ford (via the SSRC) agreed to pay the attendees' travel expenses and the SSRC provided its official sponsorship and some administrative assistance. RAND provided the facilities and made a number of its staff members available to the seminar to offer technical support. The various sponsoring organizations divided the costs of housing the attendees among them, with RAND paying the lion's share—not a trivial sum for an eight-week seminar with more than thirty participants.

Such seminars, projects, and connections tied Simon, his colleagues at Carnegie Tech, and researchers at other centers into a self-aware community. This community lacked a formal structure, but it made up for this lack with frequent, intense communication, collaboration, and competition, the use of a common mathematical language and behavioral-functional approach, and the possession of a shared sense of mission regarding the importance of understanding decision-making in social systems.

This shared sense of mission did not unite everyone at the GSIA, however. Simon quickly became the dominant intellectual figure in the school and its most skilled administrative politician. His role was so significant that it soon began to cause strains among the faculty. For example, Simon was a forceful advocate of a "behavioral" approach to economics and management and an increasingly sharp critic of the assumptions of "omniscient" rationality and maximizing behavior that were central to neoclassical economics. Indeed, he was "prepared to preach the heresies of bounded rationality to economists, from the gospel of *Administrative Behavior,* Chapter 5, in season and out."[74] The economics staff did not like being told how to do economics, of course, and tempers often flared.

These differences first came to a head in summer 1951, after the stormy departure of a young economist named David Rosenblatt. As Simon recounted in a memo to Bach in 1951, Bill Cooper had come to him saying "the economics staff was now in a state of complete demoralization, principally because I have terrorized them into thinking that they must learn social psychology under pain of administrative reprisals." Cooper, Simon relates, went on to lament Simon's "Machiavellian skill as an administrative politician," noting that Simon was "one of the few people he had met who could match him in administrative politics"—which was saying something coming from a veteran of the Bureau of the Budget.[75]

The chief point of contention in such early struggles was over personnel. In particular, Rosenblatt's departure had left the "strong feeling that one's advancement rate here depends largely on one's adherence to the official doctrine," and many faculty were afraid to argue with Simon, even about intellectual matters, for fear that he "might interpret disagreement as heresy or stupidity."[76] In keeping with this view, the economics staff often referred to themselves as "vassals," and many felt that "issues raised at staff meetings had been pre-decided" by the triumvirate of Bach, Cooper, and Simon, with most of the votes being two-to-one against the interests of the economists.[77]

Simon viewed these complaints as the normal products of the "stresses that commonly go with rapid organizational change," and he did not believe that the "basic direction" of the school needed changing. In Simon's mind, the economists were "not so much opposed to the special focus of the school's activities as desperately intent on retaining their professional roles as economists."[78] Thus, he concluded, only allowing the economists greater control over their own program could allay their resentment of his power.

The economics faculty at the time did not have an official head, aside from Dean Bach, who did not really represent them, so Simon suggested that a committee composed "solely of economists" be put together to guide the new graduate program in the field. If that were not sufficient, a department head for economics could be created, though that would be less desirable because of the danger of him being "captured" by the economists, thus becoming a "center of division and a hindrance to further progress in an interdisciplinary direction."[79]

These changes, coupled with the departure of the most discontented staff members and Simon's conscious (but not entirely successful) effort to be less confrontational, did help smooth over the differences between his group and the economists. The tensions remained beneath the surface, however, and the greater the autonomy the economists gained, the less like Simon's vision the school became.

Similarly, faculty members who were more concerned with practical undergraduate and master's level training than with theory or research also felt pressured to conform to Simon's emphasis on research. These stresses became increasingly prominent after 1954, when the school doubled the size of its master's class, largely to pay for staff increases.[80] Melvin Anshen, a professor of management, grew steadily more dissatisfied with the priority given to research and abstract knowledge, eventually venting his frustrations in a long memo. "Some of the older graduates," he wrote, believed that "the faculty [were] learning more and more about less and less."[81] In addition, many graduates were "not finding opportunities to put their quantitative skills to work" and wished that they had studied "practical human relations skills" instead.[82]

These battles between economics and industrial management and between research and practical training were in large part the products of the inherent tension between Simon's vision of the GSIA as interdisciplinary research center and the school's official charter as a business school. Such tensions could be creative as well as destructive. For its first decade the shared sense of being "leaders of the revolution" helped the staff use their differences as spurs to innovate, both in research and in teaching. In its second decade, however, this internal dynamic changed and the divisions between the factions grew.

In the GSIA's first decade, Simon played a key role in maintaining a sense of common purpose. Although his sharp opinions could drive people apart, his ability to engage the economists on a high level and his membership in so many of the school's team research projects served to unite the staff. On another level, Simon's rapidly rising prestige outside the school and his authoritative position within it made it clear that working with him was a far surer path to success than was opposing him. (The chair of the Psychology Department, as we shall see in chapter 11, learned this lesson the hard way.)

In the late 1950s, however, Simon's role in the GSIA began to change. His research agenda shifted dramatically after 1955, as he became ever more focused on using computers to study cognition. He withdrew from the GSIA's other research projects, and no one at the school was willing (or able) to take over his coordinating role.

At the same time, the main topics of research of the GSIA economists—operations research and game theory—moved from the margins to the center of economics during the 1950s, increasing their sense of connection to their parent discipline. In addition, Franco Modigliani, Jack Muth, Charles Holt, William Cooper, and others developed steady sources of support distinct from Simon's projects, and the differences between the behavioral scientists and the economists grew sharper as the latter's autonomy increased. In the end, Simon would find he needed a new set of partners—and patrons—to perform the kind of interdisciplinary research he wanted to do.

Variation on a Theme

In this story of Simon and the GSIA we see a familiar pattern, one common to a wide range of institutions and disciplines in postwar America. Time and again, interdisciplinary research centers were founded and flourished, often in close contact with similar centers, together forming a multidisciplinary, multi-institutional research community supported principally by military research agencies and by the Ford, Rockefeller, and Carnegie Foundations.[83]

Such centers and communities often were spectacularly productive, but they could be unstable as well. That instability was caused by a dependence on the shifting agendas of program officers at mission-oriented funding agencies and the efforts of unusual individuals who could broker the interests of many groups. When agendas changed, or when such individuals moved on to other things, communities like the one oriented around Simon's interdisciplinary "convergence" usually dissolved into their component disciplines, perhaps changing them in the process, or coalesced into new disciplines—such as computer science—that fit traditional university structures.

Centers like the GSIA repeated this pattern, either reproducing traditional disciplinary divisions within themselves, devolving into university departments, or separating from the university world entirely and reorganizing themselves as business enterprises. Which path they followed correlated well with the nature of their financial support: institutes with multiple, problem-oriented patrons tended to maintain an interdisciplinary focus, while ones with one or two dominant patrons tended to form discipline-based departments aligned with those patrons' interests, especially if the patrons were discipline-oriented patrons of (relatively) "pure" science, such as the National Science Foundation, DARPA, or National Institutes of Health.[84] Finally, institutes that supported themselves by producing marketable products or services tended to become businesses, as in the case of MITRE (a spin-off of MIT's Lincoln Laboratories) or the Systems Development Corporation (a spin-off of RAND's Systems Research Laboratory).[85]

Historians of science and technology are used to studying such communities when they become disciplines and such centers when they become permanent departments or laboratories or firms. We like to have a concrete endpoint to provide an implicit teleology, and we like formal organizations because they produce extensive paper trails. Informal research communities—the interdisciplinary area of convergence Simon identified—are more troublesome. We know they exist because we can observe patterned relationships among researchers who share a sense of common endeavor. But because such communities are often unstable and evanescent, there is a tendency to treat them as failed or "immature" disciplines, especially if they do not organize themselves formally. There is some justification for this view, for leaders who seek an enduring presence for their community (or greater personal power within it) strive to institutionalize their vision, as Simon did at Carnegie Tech.

In many cases, however, viewing such informal communities as "imma-ture" disciplines is not the best way to understand them. These communities and the networks of extradepartmental, interdisciplinary centers in which they found homes have been among the most important sources of innova-tion in postwar science. They thrive precisely because they provide an essen-tial counterbalance to the otherwise insistent pressures toward specialization that pervade the academic world.

Herbert Simon's career at Carnegie Tech illustrates both community- and institution-building on this interdisciplinary model. From the beginning, he sought to build the GSIA into an interdisciplinary research institution organ-ized around the mathematical, behavioral-functional study of decision-making in social systems. At the same time, he sought to link the GSIA with other such centers, forming a community united by a common perspective, a common language, and a common network of patrons. In these tasks, the challenge always was to instill that perspective and teach that language with-out imposing formal central direction. The most common means of doing so was to create a network of interlocking memberships on research teams and coordinating committees.

Simon's ambitions suited the times, for the absence of any dominant single source of patronage for the social sciences meant that those prospered who were skilled at bringing together a variety of interests: scientific entrepreneurs on the one hand and brokers on the other. Those who were more comfortable with a stable, orderly system, or those who wished to be left alone, found themselves on the outside looking in, puzzled and alarmed at how complicat-ed their world had become. Simon saw this complexity as opportunity. Indeed, he was fond of telling his graduate students that they "needed to learn to live with chaos for a while."[86] Good advice, perhaps, but Simon was com-fortable with chaos because he was always sure that an order lurked within. Similarly, when he embraced leadership rather than formal control, it was because he was confident that when he found the Truth, others would see the light. If they did not—or even if they did—there was always another pattern to find, another programme to found, another set of patrons that would fund.

CHAPTER EIGHT

Islands of Theory

At Carnegie Tech, Herbert Simon worked to build new institutions, new communities, and new relationships in the social sciences. He strove to create a new environment for social science research, one that would select for inter-disciplinary, mathematical, behavioral social science without imposing stan-dards by central fiat. This changing environment had a profound effect on Simon's intellectual work, with the communities that Simon helped to create shaping him in turn.

Just as the key to Simon's efforts at institution-building was the mobiliza-tion of varied sources of support, the story of his idea-building hinges on his attempts to bring together concepts, practices, and techniques from varied fields. The two main fields he sought to integrate were organization theory and decision theory. Although he saw these fields as connected in important ways, they were quite distinct pursuits circa 1950. Organization theory, for example, drew primarily upon institutional sociology and social psychology. Decision theory, by contrast, encompassed game theory, utility theory, and statistical decision theory, all of which were the province of econometricians. Organization theory, in short, was a product of the sciences of influence and *control,* while decision theory was the product of the sciences of *choice,* especially rational choice.

When Simon came to Carnegie Tech in 1949, he already had an idea of what the union of these two fields would look like. The product of their integration would be a mathematical, behavioral-functional, and empirical social science, one that could hold both choice and control within its compass.

Simon also had a clear vision of the road he thought would lead him to this goal. The journey, however, soon took on a life of its own, as journeys often do. As he traveled, his destination changed, subtly. Problem-solving became the object of his quest instead of decision-making. So understood, his goal appeared to be best approached via a new path: the construction of formal models of how individuals solve problems.

The high road to this destination was the new science of machines. This science, most commonly called "cybernetics," gave its practitioners a lofty and exhilarating view. From this abstract aerie, the sciences of choice and control appeared to be but channels in a larger stream, lanes in a broader highway. That broader way was the functional analysis of adaptive systems, an approach that transformed the study of decision-making in organizations into the study of system survival. The new model of the human would be neither the perfectly rational *homo economicus* of the sciences of choice nor the wholly plastic *homo administrativus* of the sciences of control; rather, it would be *homo adaptivus,* the active problem solver of finite, but real, powers.

The key obstacle Simon would encounter along this path was the chasm that lay between formal models and empirical experience. In the computer— and, above all, in the concept of the *program*—Simon would find a way to bridge that chasm. Once he crossed that bridge, he entered a strange and wonderful new land.

The Sciences of Choice and Control

Both the sciences of choice and those of control had experienced remarkable growth during the 1940s. Talcott Parsons's structural-functional social theory, for example, had been blended with Freudian ideas about the individual's internalization of group values, with anthropological investigations of "culture and personality," and with social-psychological studies of attitudes, aspirations, and the adjustment of the individual to the group.[1] The confluence of these ideas was spurred by the military's interest in questions of morale, leadership, and propaganda, producing a boom in research on small groups.[2]

This confluence led many, particularly those affiliated with the Harvard Department of Social Relations and the University of Michigan's Mental Health Research Institute, to anticipate the development of a unified social science based on the functional analysis of social systems. Talcott Parsons, for example, believed that this development was imminent. In 1944 he wrote, "a very big scientific development has been rapidly gathering force," stating that he would stake his "whole professional reputation" on it being "one of the really great movements of modern scientific thought."[3] Similarly, James G. Miller, the spokesman for the Mental Health Research Institute, the home of the journal *Behavioral Science,* believed it was time to move "Toward a General Theory for the Behavioral Sciences," arguing that the basis for such a theory would be "general behavioral systems theory."[4]

While the greatness of this movement may be debatable, its strength—and the confidence of its exponents—during the 1940s and 1950s cannot be doubted. Parsons's *The Social System* (1951), Robert K. Merton's *Social Theory and Social Structure* (1949), and George Homans's *The Human Group* (1950) set forth closely allied theories all based on the functional analysis of the individual as a component of a larger social system. Clyde Kluckhohn and Henry Murray's *Personality and Culture* (1948) and Eliot Chapple and Carleton Coon's *Principles of Anthropology* (1947), two very important texts in the remaking of anthropology after the war, hewed to a similar line, as did political scientist David Easton in his landmark *The Political System* (1953). Even the most important texts in mainstream economics of the late 1940s, Paul Samuelson's *Foundations of Economic Analysis* (1947) and *Economics: An Introductory Analysis* (1948), were predicated on the assumption that the economic actor was a component of a highly interdependent system and thus was not entirely free to choose.[5]

The sciences of choice had grown rapidly in the 1940s as well. Where the sciences of control were concerned with how the individual came to accept the rules of the game, the sciences of choice focused on how to win a given game. In wartime, of course, everyone was part of the same game. The incalculable diversity of human desires could be simplified, reduced to seeking victory (or avoiding defeat). All that remained was to find the most efficient way to do so.

The sciences of choice flourished in this rich, well-defined environment, and the 1940s saw remarkable advances in a host of fields concerned with optimizing the allocation of scarce resources toward given goals, such as game theory, operations research, utility theory, and statistical decision theory. Operations research, for example, grew out of the military's needs to allocate scarce resources (such as convoy escorts or sub-hunting aircraft) to its best advantage. Game theory similarly sought to develop methods for finding the optimum strategies to employ in various kinds of conflicts, particularly the two-player, zero-sum conflict called the cold war. Statistical decision theory combined earlier work on the theory of statistical tests with game theory, seeing scientific inquiry as a contest with Nature.[6]

These developments in the sciences of choice shared four common features. First, they were the products of high-level, but nontraditional mathematics, especially set theory and statistics. Second, they were fundamentally probabilistic rather than deterministic. Third, the models these sciences created were usually, though not necessarily, static rather than dynamic. Fourth, they all depended on the existence of an unchanging utility function for each individual; that is, they all assumed that the environment did not shape the individual's goals and preferences. The environment merely presented alternatives, among which the individual choose freely—and rationally.[7]

These features combined to make decision theory very powerful, very abstract, and very unsatisfying to persons outside of econometric circles. Statistical analysis, for example, was widely accepted as a means of assessing the reliability of data, but the idea that the world was governed by chance was more difficult to accept for many social scientists. Simon, in particular, did not like stochastic analyses of human behavior, believing that "it does not seem natural to assume that human behavior is fundamentally stochastic; its regularities showing up only with averaging (as in statistical learning theory); rather, Freud's dictum that all behavior is caused seems the natural one."[8]

The static nature of the models these sciences constructed also was troubling. While the players in a game did make a series of moves, that series typically was treated as a single strategy, chosen at the outset. Even more, in the sciences of choice the games did not change, nor did the players, and many social scientists believed that if anything characterized the world of human behavior, it was change. In addition, the dependence of the sciences of choice on the existence of a predetermined utility function, and their assumption that players attempt to maximize these functions in the course of the game, seemed unrealistic to the sociologically or psychologically trained. Simon, for example, frequently referred to such assumptions as "heroic."

Although such assumptions troubled Simon, he, like many social scientists, believed that both the sciences of choice and those of control had developed to the point where they were ready to be reintegrated. As he wrote in "Some Strategic Considerations in the Construction of Social Science Models," one of his best known essays of the 1950s: "The social sciences—weakened by a half-century of schisms among economists, political scientists, sociologists, anthropologists, and social psychologists—are undergoing at present a very rapid process of reintegration." In Simon's view, "The common diplomatic language for the scientists participating in the process is the language of sociology and social psychology, and the common core of theory—the rules of international law, if you like—is theory drawn primarily from these two fields."[9]

The cause for this movement toward reintegration, he argued, was that "in attempting to understand and analyze the large events in the political and economic scene—the wars, elections, and depressions—the social scientist has been forced to a recognition that all such events are aggregated from the interrelated behavior of human beings ... [a recognition that has] gradually and inexorably driven social science back to the atomic phenomena of human behavior in a social environment."[10] These phenomena were tremendously complex, Simon allowed; hence the historical differentiation of the social sciences into various specialized pursuits. The time had come, however, to bring these specialized pursuits back together. In his view, the product of their confluence would not be a single, unified theory—that was too much to hope for in the near term—but it would include a coherent set of methodological criteria, and it would be based on the incorporation of the sciences of choice into the sciences of control.

The Behavioral Revolution

In Simon's view, the criteria for a reintegrated social science were that it would be mathematical, that it would be behavioral and functional, and that it would be both empirically grounded and theoretically sophisticated. These views placed him in the van of the "behavioral revolution" in social science. Indeed, they defined it.

The behavioralists were a group of social scientists who were committed to remaking social science in the image of modern physical science. Generally speaking, they strove for value neutrality, objectivity, quantification and mathematization, the operational definition of terms, and a combination of careful empirical data-gathering and rigorous formal theorizing (typically via formal model-building). They saw these goals as standing in sharp contrast to those of traditional social science, which was more historical in method and moral-philosophical in style.[11]

The behavioral revolution was no silent coup; rather, it often involved a heated debate between a vigorous group of self-proclaimed "Young Turks" and an older (by-and-large) generation that defended the traditional methods and interests of social science.[12] The behavioral revolution was considerably more complex than a simple battle of the "ancients vs. the moderns," but that characterization does capture an important aspect of the struggle.

The behavioral revolution is best understood as the extension of the ideas and methods developed by the Chicago School of the 1930s for responding to the challenges of interdependence, subjectivity, and change (discussed in chapter 2). Many of the leaders of the behavioral revolution in political science, such as Simon, David Truman, V. O. Key, and Gabriel Almond, for example, had studied at the University of Chicago during the 1930s, and they had adopted their mentors' instrumentalist view of knowledge and its purposes. They continued the attempt to come to grips with interdependence through the combination of individual specialization with team research and by the adoption of a systems perspective. In response to the challenge of subjectivity, they embraced the operationalist epistemology Merriam and Lasswell only hesitantly had accepted, and they shared the awareness (but not the fear) of the limits to human reason that had haunted their teachers. Finally, they also shared the Chicago School's fascination with process as the scientific way of understanding change within systems.

In political science, still the field with which Simon most closely identified, the main events in the institutional struggle over behavioralism were the founding of the behaviorally oriented *Journal of Politics* in 1939, the establishment of the APSA's Committee on Political Behavior (headed first by Pendleton Herring and then by David Truman) in 1945, Frederick Ogg's retirement from the editorship of the *American Political Science Review* (*APSR*) in 1949, and, of course, the beginning of the Ford Foundation program in the behavioral sciences in 1951. Governmental, especially military, sponsorship of behavioral political science also played a large role, as Hugh Elsbree, the editor of the *APSR* in the mid-1950s indicated: "Many . . . [behaviorally oriented articles] are products of governmentally sponsored projects that reflect the fascination of the sponsors as well as of the authors with the cultivation of mathematical techniques."[13]

One of the major factors affecting the course and outcome of the battle over behavioralism was the enormous growth of the social sciences in the two decades after World War II. Again taking political science as an example, in 1947 the APSA had 4,598 members, a fair number of whom had not been trained in political science; in 1967 it had 14,685 members, and the percentage of members who had Ph.D.s in political science was much higher. This radical growth, shared by all the social sciences, meant that a very high percentage of all social scientists practicing in the 1950s and 1960s had been trained after the war. Hence, the persons and institutions that defined "good work" in the field during that era had an enormous impact on the shape of the field. These leaders were predominantly behavioral in orientation. The rapid growth also meant, however, that nonbehavioral social science did not die out even though it was heavily outreproduced.

The new elements the postwar behavioralists added to the Chicago paradigm were their commitments to mathematics and to formal theory, especially in the form of model building. For Simon, there was no question that a reformed social science would be mathematical. Indeed, mathematical analysis was the goal of goals for Simon and for many of his behavioralist allies. As he wrote in "Strategic Considerations," "First, I should like to rule out of bounds the question of whether mathematics has any business in the social sciences. I will simply assert, with J. Willard Gibbs, that mathematics is a language; it is a language that sometimes makes things clearer to me than do other languages, and that sometimes helps me discover things that I have been unable to discover with the use of other languages."[14]

In a similar vein, Simon opened his book *Models of Man,* a collection of "Mathematical Essays on Rational Behavior in a Social Setting," by quoting Fourier's hymn to mathematics: "Mathematical analysis is as extensive as nature itself; it defines all perceptible relations, measures time, spaces, forces, temperatures . . . Its chief attribute is clearness; it has no marks to express confused notions. It brings together phenomena the most diverse, and discovers the hidden analogies which unite them . . . It seems to be a faculty of the human mind destined to supplement the shortness of life and the imperfection of the senses."[15]

To Simon, mathematics was a way of thinking as well as of expressing ideas. It was a means of seeing the world and of communicating thoughts about it.

In particular, mathematics was essential to the integration of "islands of theory" into broader, more general theories, for the possession of a similar mathematical form by several "narrow gauge" theories often revealed the existence of similar mechanisms. The best example of this in Simon's work is his paper on the Yule distribution, in which he developed a statistical model that described "(A) distributions of words in prose samples by their frequency of occurrence, (B) distributions of scientists by number of papers published, (C) distributions of cities by population, (D) distributions of income by size, and (E) distributions of biological genera by number of species."[16] In this case, the "mechanism" was a probabilistic one in which the chance of something recurring was related positively to the number of times it had already occurred. The similarities between these phenomena existed on the level of the mathematical models that described them, and no flash of nonmathematical insight was likely to perceive a similarity between them.

Mathematics had a moral meaning for Simon as well, as the following passage makes clear:

> Of all the possible dynamic systems we might feel impelled to write down to represent particular sets of mechanisms, only a very small proportion have known mathematical solutions. Nonmathematicians can derive no comfort from this fact, however. For if it is impossible to determine mathematically the path of a dynamic system whose differential equations have been written down explicitly, we wonder how verbal reasoning, starting with a vague and indefinite word description of the mechanisms of the system, can reach any answers to this same question. The answer is, of course, that it cannot—except by a legerdemain that consists in introducing a host of implicit and unacknowledged assumptions at each step of the verbal argument. *The poverty of mathematics is an honest poverty that does not parade imaginary riches before the world.*[17]

Because of the great importance of mathematics in science, Simon believed that intensive training in mathematics for social scientists was vital. It was not enough to "buy" one's mathematical expertise by collaborating with a mathematician or statistician; one had to learn the math in order to think scientifically about the world; hence Simon's avid support for the SSRC's Summer Training Institutes in mathematics described in chapter 7.[18]

Nevertheless, Simon also held that "Mathematical tools, like Humpty Dumpty's words, must be servants, not masters."[19] In particular, "The strategy of mathematical theorizing must come primarily from the field about which the theorizing is to be done. The aim of a language is to say some-thing—and not merely to say something about the language itself."[20] As he wrote to his old friend from the Bureau of Public Administration, Kenneth May, "Mathematics is a language. We want scientists to be able to read it, speak it, and write it. But we are not training them to be grammarians."[21]

A key step on the road to mathematization was the adoption of a *behavioral-functional* approach. Although these two modes are separable, dur-ing the 1940s–1960s they went together almost invariably. This tight coupling was not accidental.

Perhaps the most important link between the two was their common grounding in a systems-based approach to understanding human behavior. Here the link between behavioralism and functionalism was both practical and epistemological. In Simon's view, an individual only could be known by his behaviors, and his behaviors could be known and identified only by their effects on the other elements of the system to which the individual belonged. Hence the significance of his earlier work on the operational definition of mass: even such a seemingly "natural" and "individual" quality of an object as its mass actually belonged to the scientist and the system, not the object itself.[22] The same was true *a fortiori* of human behavior.

Behavioralism and functionalism also were products of a systems-based perspective in that both enabled the radical simplification of the analysis of phenomena by concentrating on the formal rather than the intrinsic proper-ties of the system components. An individual performing a function is far eas-ier to understand than an individual with a unique history and nature, and individuals can be analyzed in terms of their functions only if they are parts of systems.

Similarly, what counts in such functional analyses is the performance of the function. The means by which it is performed is far less important, so long as the output fits the necessary operational parameters. From this perspective, it matters not whether one uses a hammer or a piece of plywood to drive a nail so long as the nail is driven when and where it needs to be driven. The same is true of a behavioral approach; the specifics of the internal mechanisms that generate a given behavior are only important insofar as they affect that behav-ior. These internal mechanisms can be "black boxed," and one can concen-trate, like the engineer, on the functions that box must perform (by whatever means) in order to transform a certain input into a certain output.

Behavioral-functional analysis did not lead inevitably to mathematical analysis (witness Talcott Parsons), but it made the development of mathematical social science more feasible and more desirable. It did so in several ways. First, the construction of a mathematical model requires simplification. The myriad details of individuals and localities must be stripped away. The analysis of individuals as components of systems enabled such simplification, for it restricted the scientist's attention to a finite set of functions. Similarly, a behavioral-functional approach allowed researchers to define theoretical terms in terms of systematic, measurable effects on a finite set of phenomena, making it possible not only to quantify relationships but also to describe behaviors as functions of each other in the mathematical sense. *Function* is, after all, a mathematical, as well as a mechanical, term, used to describe a relationship between two (or more) variables, and Simon employed the term in precisely this sense. In his "Formal Theory of Interaction in Social Groups," for example, "friendliness" is a function of "interaction" (among other things).[23]

The third key element of Simon's vision of a reintegrated social science was empirical research. Simon says less about the need for empirical work in "Strategic Considerations" than he does elsewhere, but it was a vital part of his intellectual agenda nonetheless. (Recall, for example, the discussion in chapter 4 of the importance of experimentation to Simon and his commitment to the radical empiricism of the Vienna Circle, discussed in chapters 2 and 5.)

The ability to conduct experimental investigations of social behavior was limited at midcentury, but it was growing. The war had given experimental analysis of behavior in small groups and of "human factors" in technical systems a marked boost, and these would continue to find extensive military support during the late 1940s and 1950s. Similarly, the surveys, opinion polls, and various large-scale data gathering and analysis projects that had begun in the 1920s and 1930s expanded dramatically in the years after the war, as social scientists surveyed everything from presidential to sexual preferences. (The analysis of presidential sexual preferences, however, would have to wait until the 1990s.)

Still, the most widely used experimental system in the social sciences in the early 1950s was the rat in a maze (or some variant of this basic model). The potential of the electronic digital computer as a tool for social, especially psychological, research was as yet unforeseen, though it would soon come to have enormous importance for Simon and for the social sciences generally. Since such uses of the computer lay in the future, and since Simon was no pollster, his empiricism, as of 1950, was more of the rough-and-ready sort. Concordance with common experience was the usual test he applied in his theory-building, with more detailed analyses being prescribed, but not performed, for the most part. Hence his colleague James March's uncharitable, but not entirely inaccurate, description of Simon's oeuvre as a "collection of prolegomena."[24]

The final element in Simon's behavioralist agenda was the development of formal theory and formal models. Simon and his fellow behavioralists sought to unite the empirical data-gathering characteristic of American social science with the philosophically rigorous theorizing associated with European social thought. The objective was to create a rigorous empirical theory that would organize and so give meaning to brute facts while avoiding the enticements of the descent into the mists of metaphysics. In the strongly empiricist context of American social science, this goal largely translated into a concerted effort to legitimize theory in the social sciences. To the behavioralists, science was the product of the organization of facts into conceptual schemes, and the progress of science was due primarily to the development of more sophisticated, elegant, and parsimonious theoretical systems, not simply the discovery of new facts.[25] Simon's experiences with computers and simulation techniques—and the criticism those techniques provoked—later would lead him to a more strongly inductivist position. In the early 1950s, however, he was interested in formal theory and formal models.

In keeping with this theory-centered view, Simon believed that the proper products of a reformed social science would be formal theoretical models. The importance of modeling in his science can be seen from the titles of his numerous volumes of collected works: *Models of Man, Models of Discovery, Models of Thought,* and *Models of Bounded Rationality.* Simon even titled his autobiography *Models of My Life.*

The subject of models and their relationship to theories is worth exploring in some detail, for Simon came to see modeling as basic to all science. Indeed, over the course of the 1950s, Simon increasingly came to see model-building as the central feature of all cognition, lay or scientific. All thought involved the creation of simplified models of the world. Science simply made this process rigorous and systematic via the use of formal languages, such as mathematics, and through the empirical testing of hypotheses.

To Simon, a scientific model was necessarily a simplified description of the real world. That was its nature and purpose. To criticize a model as being a simplified depiction of the world, therefore, was to miss the point. A model that was as complicated as what it represented would *be* what it represented and, therefore, would be useless.

If a formal theoretical model was science's destination, then its starting point was the empirical observation of a consistent relationship among a certain set of variables. A theory's purpose was to provide an explanation of that relationship. A model's purpose was to describe a concrete (physically realizable) system in terms of a theory. A theory, therefore, was not what one tested in an experiment, except indirectly: rather, one tested one's model, which might be one of many ways of expressing that theory. (Indeed, as Simon later would argue, a model might not be drawn down from theory but built up from empirical generalizations.) Simon used the term *microtheories* to describe his individual concrete system models.

The tricky thing in Simon's use of these terms is that, since a model describes a concrete, realizable system and since the world is a complex place, full of many interacting systems and subsystems, there often is more than one set of significant variables at play in the situation one wants to model. As a result, a model that strives to depict a situation accurately almost inevitably will make use of more than one theory. Thus, in one sense a model can be "bigger" than a theory, for it may incorporate elements of many theories in order to capture the complexity of a concrete system. At the same time, a theory can be "bigger" than a model, for a single theory may be a part of a great many models.

Simon, like others, also used the term *model* in a second way, thinking of models as heuristics, as opposed to representations of concrete systems. While concrete system models are intended to represent the essential features of a system in a manner that is simultaneously abstract yet sufficiently detailed so as to be empirically testable, heuristic models serve more as intellectual guides. They inspire theory construction rather than being built from theories. Heuristic models are thus much more like root metaphors or sets of basic assumptions than they are like concrete system models.[26] While he did not specify the difference between these two types of models, it is clear that Simon (and many others scientists) consistently used the word *model* to refer to both of these distinct types of constructs.

Some examples may help to illustrate these terms. Let us take an example that would come to have great importance for Simon: evolutionary theory. The basic empirical generalization at the core of Darwin's theory of evolution by natural selection was that while all species have distinct traits, the organic world is organized into clusters of species that share a number of traits, as is evidenced by our ability to group species into kingdoms, phyla, classes, orders, and so forth. Evolution by natural selection was Darwin's primary theory—his basic explanation—for this hierarchical clustering of species.[27] His theory of evolution, as elaborated over the years, has been built into a host of concrete system models, from models of specific ecosystems (the Chesapeake Bay watershed, the tallgrass prairie) to models of how individual species have evolved. Evolution by natural selection has proven to be such a fruitful theory, productive of so many powerful concrete system models, that it has become the root of a number of heuristic models, of which the most basic is that the economy is like an ecosystem (with firms and individuals being like species).[28]

To move a bit closer to home, Simon's theories of decision-making in organizations, as expressed in *Administrative Behavior* and *Public Administration*, rested on a number of empirical generalizations, among the most important being that individuals in organizations usually obey their superiors, but not always. One of Simon's theories in this regard was that individuals obey their superiors so long as the orders they are given fall within their "zone of indifference"—that is, so long as the orders do not violate their expectations regarding the kinds of orders their superior may give. The size and shape of this zone, in Simon's theory, depends on how thoroughly the individual identifies with the organization's basic goals, the specifics of an individual's value system (largely inherited from the broader society), and the organization's structure and function in society.

Thus, while his theory of authority was intended to apply to all organizations, the concrete system models Simon constructed that made use of this theory were models of decision-making in specific types of organizations. Although they shared many common features, his models of decision-making in the army were different than his models of decision-making in a civilian agency or political party or business firm. These concrete system models of decision-making in specific organization types by necessity also comprised other theories regarding the different structures and missions of these different types of organizations, the internalization of social value systems, patterns of communication within organizations, and so on.[29] As a result, his concrete system models of decision-making were both more and less general, both bigger and smaller, than his theory of authority.

The basic heuristic model related to Simon's theory of authority, circa 1950, comprised three basic ideas: first, that humans are plastic; second, that they intend to be rational; and third, that minds are like bureaucratic organizations in that both have hierarchical structures that limit choices as one descends the ranks of the hierarchy. As the next two chapters show, during the 1950s Simon's basic heuristic model changed subtly, but significantly, as he added a third term to the equation of mind and bureaucratic organization: by the mid-1950s, to Simon minds were like bureaucratic organizations, which were like programmed digital computers.

So much for theories, concrete system models, and heuristic models. But what is a *formal* theory or *formal* model? To create a formal model or formal theory is to express that model or theory in a formal language, such as mathematics. A formal language is a language that has rules that enforce internal consistency on its users: in other words, there are strict rules regarding the *forms* of statements. For example, one does not need to know whether a system of equations is empirically valid in order to judge whether it is internally consistent because algebra and the calculus are formal languages.

A formal model, to Simon, was the ultimate objective of any true social science, but far too few social scientists understood that goal, in his view. Too many social scientists developed "islands of theory" based on observation or experimentation but did not move on to the next, critical stage of formalizing these verbal theories.[30] Hence, Simon believed that the "translation" of such verbal theories into mathematics was "itself a substantive contribution to the theory."[31] Eventually, after his encounter with stored program digital computers at RAND, Simon would come to believe that the computer program was the ideal formal language for constructing theories of human behavior.

Two Islands

In "Strategic Considerations," Simon had argued that it was important to develop multiple, partial theories of human behavior as a preface to their eventual integration into larger structures. In 1950–51, Simon put this strategy into action himself, constructing his own "islands of theory." Two of these islands were his theories of the "employment relationship" and of "interaction in social groups."[32] In keeping with his step-by-step approach to synthesis, neither theory attempted a full synthesis of choice and control. The former drew primarily on econometric decision theory and was undertaken under the auspices of the Cowles Commission, while the latter drew on sociology and grew out of the controllership study.[33] Nevertheless, both were undertaken as steps toward an eventual synthesis.

In "A Formal Theory of the Employment Relation," Simon views organizations from the point of view of the individual trying to decide whether to join a firm and from that of the employer trying to decide whether to hire him or her. In both cases, the organization is "purely instrumental" in that neither the employer nor the employee views membership in the organization as an end in itself.[34] In keeping with the traditions of the sciences of choice, the situation is treated as a "two person nonzero-sum game, in the sense of von Neumann and Morgenstern."[35]

The article begins by noting that traditional economic theory regarding the employment contract "involves a very high order of abstraction—such a high order, in fact, as to leave out of account the most striking empirical facts of the situation as we observe it in the real world." Simon takes as his aim, therefore, the limited reintroduction of "some of the more important empirical realities into the economic model."[36]

The bridge that allows for this reintroduction of empirical realities is the concept of authority, a concept familiar to us from *Administrative Behavior*. The employment contract, Simon observes, differs from a sales contract in that a specific good is not exchanged for a specific price. Rather, the employee agrees to accept the authority of the employer over his behavior (within a certain range) in exchange for a wage.[37]

The employer and the employee are competitors in this game, for each has his own utility function he is trying to maximize at some cost to the other. Still, this is a nonzero-sum game, so both the employer and the employee gain something from playing (else they would choose not to play). What do they gain? They gain a certain "liquidity" of choice in the face of uncertainty, postponing the need to choose specific courses of action until more is known. Thus, the employment contract is fundamentally a tool for use in solving the problems of "planning under uncertainty."[38]

Simon admits that this model, though more realistic than traditional models, still leaves out "numerous important aspects of the real situation." In particular, "it is a model of rational behavior in an area where institutional history and other nonrational elements are notoriously important." There are ways, he argues, to extend his model to incorporate some such concerns, but it is still based on strong "assumptions of rational utility-maximizing behavior."[39]

The same cannot be said of Simon's other major project of mathematical formalization of 1950, his "translation" of George Homans's *The Human Group* into a "Formal Theory of Interaction in Social Groups." Instead of using game-theoretic formulations, Simon's "Homans model" is constructed as a system of interdependent variables whose relations are described by a set of differential equations. Indeed, Simon chose *The Human Group* as the model to mathematize precisely because the systems Homans described could be characterized by a system of differential equations, thus providing an opportunity to employ the sophisticated tools mathematicians had developed for analyzing the behavior of such systems.[40]

Simon begins the article by stating "to a person addicted to applied mathematics, any statement in a non-mathematical work that contains words like 'increase,' 'greater than,' 'tends to,' constitutes a challenge." Simon is more than willing to meet that challenge, taking as the purpose of the paper the demonstration "by means of a concrete example, how mathematization of a body of theory can help in the clarification of concepts, in the examination of the independence or non-independence of postulates, and in the derivation of new propositions that suggest additional ways of subjecting the theory to empirical testing."[41]

After this introduction, Simon goes through a brief description of the social group as a system, defining it in terms of four variables, all of which are functions of time. These variables are the intensity of interaction among the members, the level of friendliness among them, the amount of activity carried on by the group's members, and the amount of activity imposed on the group by the environment (the "external system").[42]

In contrast to most econometric models, Simon's Homans model posits three sets of dynamic relations among the variables. Thus, in this model, the system itself changes over time. The intensity of interaction, for example, "depends upon, and increases with, the level of friendliness and the amount of activity carried on within the group."[43] As a result, changes in friendliness, say, are fed back into the system, altering other variables that affect the level of friendliness once again.

Such a model enabled Simon to apply the mathematical tools that had been developed to analyze the "time path" and equilibrium positions of other systems characterized by feedback mechanisms. Analysis of the time path of the system would require significant empirical work in order to determine the values of the coefficients in any specific system, but equilibrium analysis could be conducted even in advance of such investigations. For example, the "conditions for stability of the equilibrium might be examined," and "starting from the assumptions of equilibrium and stability" one might attempt to predict "what will happen if the independent variables or the constants of the systems are altered." This method, the method of "comparative statics," Simon argued, was a powerful tool for "deriving properties of a gross qualitative character that might be testable even with relatively crude data."[44]

Diverse Projects, Diverse Patrons

In the early 1950s Simon pursued a variety of approaches to the same basic problem of understanding human behavior in social situations. He found support for these diverse projects by appealing to the interests of a diverse set of patrons. To the Ford Foundation and the other patrons of the sciences of control, Simon was a leader of the behavioral revolution, one who would make social science behavioral science and thus lead the field into a new understanding of how the group affected the individual. To the Air Force, the Cowles Commission and the other patrons of the sciences of choice, he was an applied mathematician who sought to apply his tools to new areas. Although he perceived himself as an outsider, he was able to present himself as an insider to both camps without deception. His strayings from the mainstreams of each, therefore, were understood to be the products of creativity rather than ignorance. In his own mind, Simon beat both at their own games because they did not realize that they were playing the same game.

Simon's reliance on a diverse set of patrons had several effects on his work during the 1950s. First, it encouraged his pursuit of multiple, partial syntheses rather than a single, grand vision. Second, it encouraged the selection of concepts, theories, and techniques from one set of fields that could be translated easily into the other, particularly mathematical modeling techniques and behavioral-functional analysis. Third, it encouraged abstraction in order to facilitate such translation.

The specific nature of these patrons and the specific context of the late 1940s and early 1950s also played a role in enabling Simon to pursue his diverse researches, for these patrons all were convinced of the value of "basic research" and were disposed to grant scientists substantial freedom to follow their interests. The foundations and the military research agencies were interested in building capacities as well as advancing their specific missions and so supported methodological advances, the development of a body of basic knowledge, and the creation of a group of competent behavioral scientists.

Although Simon's patrons wound up supporting his efforts to synthesize the sciences of choice and control, they did so only indirectly, at least in the early-to-mid-1950s. His pursuit of synthesis, though respected, was not required or even expected by his patrons. The drive to synthesize came from within Simon, and it could find only limited expression so long as the sciences of choice and the sciences of control kept their distance. Thus, in 1950, Simon was a decision theorist and an organization theorist, though he sought to be something more than both.

Something did link these disparate fields, however, in Simon's view: they were both based on a systems approach to the world and to human behavior. What was needed for a synthesis, then, was a new science of systems that could encompass both choice and control, and a new context in which that interdisciplinary science could take shape. In servomechanism theory, Simon would find the seeds of such a science, and in RAND's Systems Research Laboratory he would find an environment in which those seeds could grow and flourish.

CHAPTER NINE

A New Model of Mind and Machine

Herbert Simon sought a synthesis in social science that would take place on many levels, bringing together not just choice and control but also theory and practice, research and reform. Institutionally, he fought to create a set of interdisciplinary research centers and a network of patrons that would conduct and support mathematical, behavioral-functional, and problem-centered social science. Intellectually, he hoped to unite the sciences of choice and control by focusing on the interdisciplinary "area of convergence" he had identified: the study of decision-making in social systems.

The next, crucial step toward that synthesis was the development of a new model of man to replace both the perfectly rational, perfectly free *homo economicus* of the sciences of choice and the perfectly malleable, perfectly docile *homo administrativus* of the sciences of control. This new model of man is perhaps best called *homo adaptivus*, for in it the human is a bounded but rational, limited but capable, problem-solver, with problem-solving being understood as the central process in adaptation.

The story of the development of homo adaptivus is in some ways a very complicated one, for Simon built his new model from many pieces, some of which he found and adapted to his own use and some of which he himself made. In other ways it is a very simple, very familiar story: Simon took a number of ideas, the majority of which had occurred—separately—to other people as well, and put them together in a new way, giving all of them new meaning.

The chief ideas that Simon assembled in this new model of the human came from three sources: (1) his own previous work on administrative decision-making; (2) contemporary cybernetics and servomechanism theory, especially the work of W. Ross Ashby; and (3) Gestalt psychology, particularly work on problem-solving and "productive thinking." Simon's previous work on administrative decision-making had revealed to him the importance of the principle of bounded rationality and the centrality of the problem of choice, and it had taught him to see both the individual and the organization as decision-making machines. From cybernetics and servo theory he now added the ideas that organisms, organizations, and adaptive machines were functionally equivalent, not merely similar; that feedback was an essential component of all adaptive systems, organic and mechanical; and that adaptive systems could evolve enormously complex behaviors by "nesting" rather than chaining simple behavior mechanisms. Here the term *evolve* is particularly important, for, in Simon's view, a process akin to natural selection produced both adaptive behaviors and their organization into a hierarchical system of behavior. Finally, from Gestalt theory Simon learned to think of learning and problem-solving as processes of cognitive adaptation: creatures adapt to their environment by learning to construct simplified mental models of that environment, models that serve as the reference points not only for decisions as to how to achieve the organism's goals but also as the basis for defining the goals themselves.

All of these ideas were grounded in two essential beliefs: the principle of bounded rationality and the idea that the world is a system. The overall product, the model of homo adaptivus, reflected those ideas throughout. The human in this model is a simple, error-controlled creature that, in the course of otherwise random encounters with an environment so complex as to be incomprehensible in its entirety, learns to construct simplified models of its environment, models that enable it to satisfy its goals, though not to achieve them in optimal form.

These ideas all came together for Simon in 1952 at RAND's Systems Research Laboratory (SRL). Indeed, while 1956 is typically seen as the starting point of both AI and the cognitive revolution, 1952 was Simon's *annus mirabilis*. The SRL was the nexus, and Simon the interface, in which all these ideas met and merged. At the SRL, Simon first encountered the digital computer and the concepts of the program and of simulation, not to mention his intellectual soul mate, Allen Newell. Together, Simon and Newell would develop the heuristic model at the heart of postwar psychology—that of homo adaptivus, the human as a finite problem-solver—as well as the concrete models that would serve as the exemplars for both cognitive psychology and AI during the 1950s–1970s.

Inputs: Adaptive Machines and Gestalt Scenes

Simon began to investigate servomechanism theory shortly after his arrival at Carnegie Tech, seeking in it a way to construct models of dynamic systems. Servo theory would resonate with Simon's scientific philosophy and his immediate intellectual goals, for around 1950 it was in the process of developing into both a powerful, practical tool and an all-encompassing science. This science was a science of machines—all machines, be they physical, social, or symbolic.

The science of machines, like the sciences of choice and control, had developed rapidly during the 1940s. As with all the sudden scientific advances of the World War II era, a long history of development preceded the explosion of wartime research on machines. The early decades of the twentieth century, for example, had seen the evolution of increasingly complex mechanical and electromechanical systems and the development of powerful conceptual and mathematical tools for the analysis of the behavior of such systems.[1]

The creators of the complex power-generation and telephone communication systems built during the first third of the century encountered many of the problems that later system designers had to solve—problems of interdependence, complexity, and scale. It is thus no surprise that their work gave us many of the concepts that postwar "cyberneticists" employed. The most fundamental were the ideas that the machine was a system and that systems were to be defined functionally. That is, a machine was any system that transformed a certain input into a certain output, with a certain error, according to certain rules. So defined, a machine could be made of metal, or flesh, or symbols. This broad framework underpinned all the sciences of machines, from control theory to servomechanical theory; as a result, a growing number of historians of science and technology have begun to label "systems sciences" the various sciences that emerged in the 1930s–1940s to analyze the properties of complex dynamic systems, especially human-machine systems.[2]

The great electric power and telecommunications systems of the early twentieth century also produced two other things of signal importance for the postwar science of machines: electrical engineers and electronic communications. Electrical engineers, of course, were trained to serve the needs of the power generation, telephone, and, later, radio broadcasting systems. They were trained, therefore, to think in terms of systems, particularly in terms of closed systems. They understood systems in functional terms, caring less about what went on inside the proverbial "black box" than that the box transformed a certain input into a certain output in a regular way. In addition, while electrical engineers were highly trained in sophisticated mathematics, they still were engineers, which meant that they were interested in building tangible, working machines, not just algorithms.[3]

At the same time, the electrification of machinery made possible a new understanding of machines. Electrified machinery could be controlled by feedback loops that operated in "real time." That is, electronically controlled machines could adjust their performance at least as rapidly as (and often more precisely than) their human operators could. Such capabilities were not unheard of in purely mechanical systems, and automatic systems with mechanical governors that operate in real time (such as clock-controlled automata and Watt's steam engine) long have been sources of human-machine analogies. The advent of electronic machinery, however, made such qualities far more common and far more impressive, fostering the creation of a whole new set of analogies between machines and their human operators.[4]

The rapidity of internal communication within electrical systems made it possible to think of ever-larger conglomerations of machinery as being single, unified systems. As a result, the boundaries of systems became more malleable conceptually. Anything that participated in real time in the transformation of a certain input into a certain output could be thought of as being part of the same system. In the 1940s, "anything" came to include human components, and the invisible stuff that linked all the components of systems came to be called "information."[5]

As Peter Galison has argued, by the end of World War II, the human operators of many types of complex machinery commonly were seen as components of the system.[6] Pilots, for example, were not distinguished from their aircraft, and antiaircraft gunners were seen as parts of their gunnery systems. This incorporation of the human into the machine depended on two related developments: the humanizing of the machine through the ascription of purpose and the mechanizing of the human by defining people in terms of their functions.

"Behavior, Purpose, and Teleology," the famous 1943 paper by Norbert Wiener, Arturo Rosenblueth, and Julian Bigelow, illustrates this confluence of man and machine.[7] This paper, significantly, was the joint product of a mathematician (Wiener), a scientist used to studying human systems (the physiologist Rosenblueth), and an electrical engineer (Bigelow). The fact that such different disciplinary perspectives could come together reveals how far the redefinition of the human body as a physiological machine and of the electromechanical device as a purposeful, adaptive system had come.

In their article, the authors argued that it is perfectly acceptable to describe machines as behaving purposefully if those machines are controlled by feedback. In their generalized formulation, a machine was anything that acted to transform an input into an output. A special class of machines called servomechanisms was able to compare this output to a set of predetermined goals. The difference between the output and the goal was the error, and this error was fed back into the system (as a kind of secondary input), affecting the system's performance. Servomechanisms, then, were error-controlled devices that adjusted their performance to achieve their goals. Hence they were both purposeful and dynamic. Wiener later elaborated on the ideas advanced in this article in his influential book *Cybernetics,* which he defined as the "science of communication and control" in all systems, mechanical and human.[8]

The possibility of creating a science of dynamic systems had enormous appeal to social scientists, as Steven Heims has shown in *The Cybernetics Group.*[9] In the early postwar years, many practitioners of the sciences of control—sociologists like Talcott Parsons, anthropologists like Gregory Bateson, and psychologists like Kurt Lewin—read Wiener's book, attended the Macy conferences on "Feedback Mechanisms," and adopted much of the cybernetic vocabulary. Parsons, for instance, used an increasingly abstract systems-based language to describe social processes, analyzing social institutions in terms of the roles they played in adaptation, goal-attainment, integration, and pattern maintenance.[10] Though his work on *The Social System* is often criticized as being too static, Parsons's fascination with the mechanisms of adaptation in systems led him back to the once and future creed of social science, evolution.

Simon, however, was not enthralled by Wiener's formulation of cybernetic theory, nor was he impressed with Parsons's interpretation of it. Simon always described his ideas about adaptive systems as having their roots in "classical servo theory" from the years "b.c." (before cybernetics), and he found the work of W. Ross Ashby more stimulating regarding the application of servo-mechanism theory to living systems.[11] Simon was such a fan of Ashby's first book, *Design for a Brain,* that he wrote Ashby in 1953 to say it was "the most exciting book I have read in a decade."[12]

Why Simon preferred Ashby's formulation is not precisely clear. Perhaps the bumptious brilliance of Wiener put him off; perhaps he liked that Ashby had a better feel for the physical properties of the systems from which he generalized; or perhaps Wiener had made his ideas too popular for Simon to cite. In any event, Simon's repeated professions of admiration for Ashby and his relative silence regarding Wiener indicate that it is Ashby's work and not Wiener's that we should explore in order to see what Simon took away from his study of the new science of machines.

Ashby, a biochemist and neurophysiologist, was the director of research at Barnwood House, a center for psychiatric research and treatment in Britain.[13] Ashby's primary research areas were the endocrine system and the biochemistry of brain functions. Like all physiologists influenced by L. J. Henderson and Walter Cannon, Ashby was fascinated by the ability of organisms to maintain their internal equilibriums in changing environments.[14] He understood the central nervous system to be the key agent in producing the adaptive behavior necessary to the maintenance of equilibrium, so in *Design* he took as his purpose the discovery of "the origins of the nervous system's unique ability to produce adaptive behavior."[15] The work had as its "basis the fact that the nervous system behaves adaptively and the hypothesis that it is essentially mechanistic; it proceeds on the assumption that these two data are not irreconcilable."[16] Hence, Ashby soon rephrased his question as "what sort of machine can be 'self-coordinating?' "[17]

Ashby deduced the properties that any such mechanistic but adaptive system must have, finding that it must be error-controlled and that it must be able to switch from one "way of behaving" to another in order to respond to changes in its environment. Such a system, he argued, could be both "wholly automatic" and "yet actively and complexly goal-seeking." Indeed, "Once it is appreciated that feedback can be used to correct any deviation we like, it is easy to understand that there is no limit to the complexity of goal-seeking behavior which may occur in machines quite devoid of any 'vital' or 'intelligent' factor. Thus, an automatic anti-aircraft gun may be controlled by the radar-pulses reflected back both from the target aeroplane and from its own bursting shells." "Such a system, wholly automatic, cannot be distinguished by its behavior from a humanly operated gun."[18]

In these passages we see many of the crucial aspects of the new theory of machines that emerged during the 1940s: the introduction of electronic communication (via "radar-pulses") enabled real-time adjustment of the system, allowing one to speak of the machine as purposeful and thus blurring the line between human and machine. This blurring was no accidental elision. In a letter to Simon, Ashby writes that "It is my firm belief that the principles of 'organisation' are fundamentally the same, whether the organisation be of nerve cells in a brain, of persons in a society, of parts in a machine, or of workers in a factory." Ashby continues, "I have long been of the opinion that the problem of how ten billion nerve cells work harmoniously together in the brain is the same problem of how two billion people can work harmoniously together in a society." *Design,* he concludes, shows the way such harmonious organizations are created biologically.[19]

The way in which harmonious, homeostatic organizations were created, Ashby argues, was through the "principle of ultrastability." As he explains, feedback systems are usually actively stable or actively unstable, depending on whether the feedback corrects or reinforces the error. Simple feedback is the mechanism by which the various states of the system are selected in an unchanging environment—the processes of the system are sped up, slowed down, or otherwise altered in order to maintain the system's "critical variables" within tolerable limits.

In a changing environment, however, stability is not enough. The system must be able to change its whole "way of behaving" and to "act selectively" toward the various possible "ways of behaving" it might choose, rejecting those that lead to instability and retaining those that maintain stability. In short, the system must be able to respond to a change in its parameters as well as in the values of its variables. Ashby calls such a system "ultrastable."[20]

Imagine an automatic thermostat-controlled furnace system. It compares the output (the desired air temperature) to the input (the current air temperature) and raises or lowers the flow of hot air produced by the furnace accordingly. Such a system is stable within certain limits—limits set by the outside air temperature, the furnace's heating capacity, the thickness of insulation, and other similar parameters. The system will reach equilibrium (or hover near it) and remain there, even in the face of minor disturbances such as a door being opened and shut. Let us now change the parameters of the system, however, opening all the windows and leaving them open, say, or blocking a major vent. The heating system then might overheat, produce large temperature swings, or show other signs of instability. Unless the thermostat system were ultrastable—that is, able to select a new mode of operation and thus adapt to the changed circumstances—it would not be able to produce equilibrium.

In Ashby's formulation, ultrastable systems are systems that are able to adapt to their environment through a kind of natural selection of behaviors. Various "ways of behaving" are generated randomly and then tested against the environment. If the time path of the system is toward stability, then the way of behaving is kept until the environmental parameters change again. If the path is toward instability, then the system shifts to another randomly generated way of behaving. Ultrastability is thus the natural selection of behaviors, and it is the fundamental characteristic of all adaptive systems, even complex organisms and social organizations.[21]

Such attempts to apply the principle of natural selection to behavior were not new. Rather, they lay at the heart of behaviorist psychology, from Pavlov's creation of conditional reflexes to Skinner's association of stimuli with responses through "reinforcement."[22] Such psychological systems were powerful, but they always had foundered on the shoals of complexity. Complex human behaviors, such as language, required the composition of so many smaller behaviors into larger actions that it was hard to see how humans could ever learn through selective reinforcement all the things people commonly do learn. If the stimulus-response-reinforcement model were applied to language acquisition, for example, it might be able to explain how children learned words, for their number is finite, but it never could explain how children learned sentences, for their number is infinite, as the psychologist George A. Miller pointed out in the 1950s.[23]

Ashby recognized this problem, and he offered a new solution. The key to understanding complex behavior, he argued, was to remember that systems, including ultrastable systems, were composed of subsystems. If a single organism were composed of a great many ultrastable systems, each acting largely independently and each adapting to its environment (which included the higher-level systems that encompassed them), then the overall system would be capable of effecting fantastically complex adaptations. Complex behaviors, therefore, were not produced by *chaining* simple behaviors together but by *nesting* them inside each other. The result: "By the time a human being has developed an adult's skill and knowledge, he has been subjected to the action of ultrastability repetitively to a degree which may be comparable with that to which an established species has been subjected to natural selection. If this is so, it is not impossible that ultrastability can account fully for the development of adaptive behavior, even when the adaptation is as complex as that of Man."[24]

Thus, in *Design for a Brain*, Ashby produced much more than merely a solution to a "specific problem." He created a generalized science of adaptive machines that held all things, from learning to evolution to gunnery control within its sights. According to Ashby: "We start by assuming that we have before us some dynamic system, i.e. something that may change with time. We wish to study it. It will be referred to as the 'machine,' but the word must be understood in the widest possible sense, for no restriction is implied other than that it should be objective." Hence, by the third chapter of *Design* we learn that the "animal . . . is a machine."[25]

What Simon took away from servomechanism theory, generally, was this expansive, functionalist definition of machines, coupled with the idea of an adaptive machine as being error-controlled and thus constantly testing its current state against a goal state. What Simon derived from Ashby, in particular, was the idea of behavior as the product of a process akin to natural selection: behavior alternatives were generated randomly and then selected on the basis of their viability, not their optimality. Evolution taught that the *fitter* survived, not the *fittest*. As Simon later wrote, "Bracketing satisficing with Darwinian [evolution] may appear contradictory, for evolutionists sometimes talk about survival of the fittest. But, in fact, natural selection only predicts that survivors will be fit enough, that is, fitter than their losing competitors; it postulates satisficing, not optimizing."[26]

While this view of behaviors as being selected through a process of organized trial and error fit in well with the psychology Simon had learned from E. C. Tolman, this much more direct connection to evolutionary theory was new to him. He, like almost every student of the life and social sciences of his generation, was broadly familiar with evolutionary theory, of course. Unlike many of his colleagues, however, natural selection had not played a major role in his thinking before the early 1950s. Like Tolman, Parsons, and other functionalists, he had written of means being adapted to ends, but this use of the term *adaptation* was fairly loose. Now, Simon began to believe that evolutionary adaptation, understood in terms of ultrastability, might provide the heuristic model that could unite the sciences of choice and control.

Some concepts related to feedback and adaptive behavior in dynamic systems, drawn from "classical servo theory" had appeared, indirectly, in Simon's "Homans Model." Simon's first attempt to employ such ideas directly, however, did not come until a year later, when he wrote an article titled "Application of Servomechanism Theory to Production Control."[27] This article linked the sciences of choice and those of control in a new way, with the linking mechanism being the concept of adaptive behavior. In a telling indication of how important this concept would come to be in Simon's later work, he included this article as a direct part of the ancestry of his most prized creations, the computer programs Logic Theorist and General Problem Solver.

Simon begins the article by stating that "Powerful, and extremely general, techniques have been developed in the past decade for the analysis of electrical and mechanical control systems and servomechanisms. There are obvious analogies between such systems and the human systems, usually called production control systems, that are used to plan and schedule production in business concerns." He admits "the notion of a servomechanism incorporating human links is by no means novel," since "many gun-sighting servos involve such a link." In his view, however, "the idea of social, as distinguished from purely physiological, links is relatively new."[28]

From this starting point, Simon then goes on to define the terms input, output, load, and error, noting that servomechanical systems are defined by the presence of a control loop (feedback loop) in which the output is compared with the input and the difference is fed back into the system. He also observes, however, that both the input and the load affect the system but are unaffected by it; they are "unilaterally coupled" to the other elements of the system. Thus, his definition of a servomechanism differs somewhat from that of the other cyberneticists. To Simon, "a servomechanism, then, is a system (1) unilaterally coupled to an input and a load, (2) with one or more feedback loops whereby the output is compared with the input, and (3) with a source of energy controlled by the error that tends to bring the output in line with the input."[29] Simon explains that the ordinary household thermostat is a common example of such a system, and he notes that such systems are usually described by sets of linear differential equations with constant coefficients.

Using this broad, functional definition of servomechanisms, Simon then argues that the production control system he describes in the article (one that he believes is typical) is, quite literally, also a servomechanism. In this system customer orders are the load, and optimum inventory size is the input. The system's operations are controlled by a feedback loop wherein the difference between production output and optimum inventory size (the error) influences planned production. Planned production, in turn, affects actual production, which influences inventory, which is compared with optimum inventory to produce a new error measure, beginning the loop once again.[30]

Such a system is both adaptive and optimizing. The theory permits "actual numbers to be inserted for the construction of specific decision rules [for] actual situations," and these rules guide the system in its evolution toward a previously calculated optimal state (the optimum inventory). Later, as Simon turned ever more to the study of the psychology of the individual decision-making organism, he redefined the optimal in terms of stability and survival; real creatures do not calculate optimal states, they (or their environment) select viable or satisfactory ones. These developments still lay in the future, however, and Simon was pleased with his enlistment of adaptation in the service of optimization.

Simon's next application of the science of machines was to the study of political power, a central issue in the sciences of control. As Simon had noted in his article on the employment relationship, one of the salient features of human behavior in organized groups is the existence of authority. Authority had been an important concept in Simon's work ever since *Administrative Behavior*. Indeed, the concept was of sufficient importance that he grouped all the sociologically based sciences (the sciences of control) together because they all dealt with the mechanisms of influence, power, and authority.[31]

Simon was not alone in his concern with the problems of power and authority; many political scientists in the postwar years struggled to understand the hold totalitarian states had on their people and the power interest groups had to define (or to defy) the "public interest," even in free societies.[32] Many of these works were none too cheery about the future (or even the reality) of democracy: David Truman's *The Governmental Process*, for instance, treated democratic politics purely as a contest between organized interest groups, and Harold Lasswell and Abraham Kaplan's *Power and Society* emphasized that hierarchies of power and authority were inescapable; democracy was simply one method of choosing those higher up in the hierarchy. These analyses of power and authority, however, did not satisfy Simon, who found them too loose and informal—and too pessimistic. At the same time, Kenneth Arrow and his student Anthony Downs both tried to develop formal mathematical models of the political process, starting from the vantage point of the sciences of choice. Simon applauded the formal rigor of Arrow's *Social Choice and Individual Values* and Downs's *An Economic Theory of Democracy*, but he believed that they suffered from a lack of realism.[33]

Every path to formalization seemed to lead through the thickets of causal analysis, however, which was forbidden territory for any good positivist. Hume had taught all empiricists not to speak of causes and effects but of correlations, and positivists like Simon took Hume's injunction quite seriously. In addition, the study of power and influence led into the even more shadowy realm of the mind. Political power may grow out of the barrel of a gun, as Mao once observed, but the remarkable thing is that the gun rarely is present. Power often is as invisible as it is inescapable. How could one ever observe the mechanisms of influence in action? How could one measure power? And if it could not be measured, how could it be studied scientifically?

These concerns may seem quite removed from servomechanism theory, but Simon saw in the servomechanism a model for understanding influence as well as change. In his usual fashion, he addressed the broadest philosophical questions first: how was one to understand causation, and could causation be defined in such a way that it could be testable and measurable?

Simon tackled this question directly in a series of articles, beginning with "On the Definition of the Causal Relation" and "Causal Ordering and Identifiability."[34] In these articles, written during 1951–52, Simon argues that the customary avoidance of the notion of causation is unnecessary, if one defines the concept properly. Causation, in his view, is best understood not in terms of the relations between events in the real world but rather in terms of the relationships between variables in the scientist's model of the world. Quite simply, "the concepts to be defined all refer to a model—a system of equations—and not to the 'real' world the model purports to describe."[35]

If one adopted this restricted notion of causality, he argued, then it was possible to define causation in a philosophically valid and operationally meaningful way. One could define causation as an "asymmetrical relationship" between variables in a system of equations, making "both Hume's critique and the determinism-indeterminism controversy . . . irrelevant."[36]

But are such asymmetrical relationships between variables found in any working scientific models of the world? Certainly, Simon says. One sees such relationships in Cannon's and Henderson's models of homeostatic physiological systems and in the operation of servomechanical systems. Indeed, "our concept of causal ordering is essentially identical with the concept of unilateral coupling, employed in connection with [such] dynamical systems."[37] Here the analysis becomes difficult to render without mathematics, but its details are important, for they bear quite directly upon Simon's work in computing, as well as his emerging ideas about models and experimentation in science.

Simon's procedure in discovering the causal relationships in a system of equations ran as follows: a self-contained system of linear equations is posited. (A self-contained system is one in which there are exactly as many variables as equations.) This system then is broken down into a hierarchy of sets and subsets of equations through repeated factoring. The process stops when none of the remaining sets of equations can be broken down any further without becoming un-self-contained. Each of these subsets can be regarded either as a part of the larger system of equations or as a complete system of equations in its own right. In the latter case, the variables of the higher-level system that have been eliminated from the subsystem by substitution become, in effect, parameters, not variables. As such, they are "unilaterally coupled" to the subsystem, affecting the variables within the subsystem without being affected by them.[38]

The result of this analysis is a system of equations in which those equations are organized hierarchically into systems, subsystems, and sub-subsystems, with the "givens" for each level in the hierarchy being established by the higher levels of the system. It is not too difficult to translate this model into the language of Simon's organization theory, with each system or subsystem becoming a department of an agency and the "givens" being "decision premises" rather than coefficients. In these articles Simon already anticipates a connection between such hierarchies of causal relations and the eventual development of "programs" that could "set the sequence of computation" of the equations' solutions, programs that would be "easy" to implement on "electrical computing devices."[39]

How to give this abstract model of causation meaning in the real world? Simon was quite interested in answering this question, though his first attempts at application were not to the "electrical computing devices" he saw on the horizon. Rather, the attempts to operationalize these concepts had to do with experimentation and with political power. As he wrote in "Causal Ordering and Identifiability," "The causal relationships have operational meaning, then, to the extent that particular alterations or 'interventions' in the structure can be associated with specific complete subsets of equations. We can picture the situation, perhaps somewhat metaphorically, as follows. We suppose a group of persons whom we shall call 'experimenters,' . . . we may say that they control directly the values of the nonzero coefficients . . . [while they] control indirectly the values of these variables."[40] Thus, Simon operationalizes his idea of causality through the concept of experimental intervention in the world, seeing both the scientist and the administrator as managers of complex, hierarchical systems.

In "Notes on the Observation and Measurement of Political Power," written shortly after the completion of "On the Definition of the Causal Relation," Simon applied his ideas about causal relationships to the real world by examining the problem of authority.[41] Although the article could be "regarded as a series of footnotes on the analysis of influence and power by Lasswell and Kaplan," its rigorous operationalism and mathematical style of formalizing their views was quite radical.[42] The reviewers for the *American Political Science Review* (*APSR*) were strongly divided regarding its publication, with one reviewer writing that "most of the observations are elementary, but are expressed in abstract language which irritates and obstructs the willing reader's attention . . . All he has done is to restate the obvious in an abstract way."[43] Even a more sympathetic reviewer admitted that this article might not be suitable for the *APSR* since it was "actually part of a long philosophical debate that has been going on in a variety of other journals . . . it may be that the article would have its maximum impact if the normal readers of this kind of material found it in one of the journals which they habitually read."[44]

Such responses almost made Simon despair of ever being able to reach political scientists, who increasingly seemed to him to be a foreign, "primitive" people. As he wrote in a long letter to Dwight Waldo, with whom he had recently had quite a "go-round":

> What I was angry at—if "angry" is the word—is perhaps best expressed toward the end of my philippic: "the standard of unrigor that is tolerated in political theory would not receive a passing grade in the elementary course in logic." I think I can defend that statement, and if I applied it to your writing (in spite of the fact that you are well above par among political theorists in the matter of

clarity and rigor) it was because I have hopes for you, while I have very few hopes for most political theorists ... When I say "hopes," I do not mean hopes for your "conversion" to positivism. I mean hopes that you will impose on yourself a discipline of thought and expression that, unfortunately, is not imposed on you by your profession.[45]

Simon had no sympathy for those who could not follow his arguments because they were too difficult or abstruse, like the *APSR* reviewers: "So long as I do my best to express myself in lucid English, I cannot accept the additional constraint that what I write should be readable by those who are unwilling to master what must be mastered in order to follow the problem." Ending his letter on a somewhat bitter note, he wrote, "Perhaps I will only communicate to a few, but then they are the few to whom it is important to communicate. On rereading what I have just written, I find that it sounds condescending. I don't know how to say what I feel about the state of the political science profession today without sounding condescending, and I do not always feel like remaining silent."[46]

Despite such tribulations, "Notes on the Observation and Measurement of Political Power" eventually did find its way into print in the pages of the *Journal of Politics,* which had been created in 1939 specifically to give the more behaviorally inclined practitioners of political science a publication outlet.

Simon begins "Notes" by observing that because power is one of the "central phenomena" of political science, the "sentences of political science" will contain phrases like "the power of A is greater than the power of B," or there was an "increase (or decrease) in the power of A." Simon holds that the presence of such statements indicates that power and influence are properties of dynamic systems, just as "friendliness" and "interaction" were in his "Homans Model." The proper definition of influence, then, necessarily involved a relationship between variables over time. In addition, since the definition of power was the ability to "affect the policies of others than the self," there was necessarily an "asymmetrical relation between influencer and influencee."[47]

At this point Simon observes, "We are wary, in the social sciences, of asymmetrical relations. They remind us of pre-Humeian and pre-Newtonian notions of causality. We have been converted to the doctrine that there is no causation, only functional interrelation, and that functional relations are perfectly symmetrical." Drawing on his work on causation, he counters by arguing that causation does exist, as is evidenced by the phenomenon of unilateral coupling in "reassuring" and "respectable" electrical and mechanical systems.[48]

Simon proceeds to describe influence mechanisms as those involving feedback, with political structures appearing to be directly analogous to linear structures of differential equations. Both are divided into a hierarchy of asymmetrical relationships, for example, and in both cases the scientific observer is interested in how the regime sustains its equilibrium (survives) rather than how the components assert their individuality.

Because of this parallel, the mathematical analysis of systems of power is both possible and useful, Simon argues. The relations in such systems might even be quantified, for "most of the arguments against 'quantitizing' or 'measuring' the 'qualitative' variables encountered in the social sciences stem from ignorance of how flexible the concept 'quantity' is, and how indefinite the line between quantity and quality."[49] In the case of political power, the proper kind of quantity to use is not a numerical count, represented by a cardinal number, but a "partial ordering," represented by ordinal numbers. The concept of a partial ordering had received renewed attention with the rise of game theory in the 1940s and early 1950s but was still unfamiliar territory to most social scientists and even to many mathematicians. Cardinal numbers (1, 2, 3, etc.) are fully ordered on a uniform scale, but ordinal numbers (first, second, third, etc.) are only partially ordered because there is no uniform scale to which they can be related. Ordinal numbers are useful in dealing with the relationships between sets and in describing relations of more or less, but they are not appropriate for measuring absolute quantities or for performing precise counts.

In "Notes," Simon had begun to bring together elements of the sciences of choice and the sciences of control, using concepts from game theory (partial ordering), from his work on causality (asymmetrical relations in systems of variables), and from his studies of servomechanical theory (unilateral coupling, feedback, and adaptation), all in an attempt to create a dynamic model of influence. This theory of influence, however, was still only a partial synthesis; it was a theory of how the group adapted the individual to suit its needs. The group was the actor, not the individual. Simon, himself no passive subject, wanted to extend and expand this model of control so that it could incorporate an active, if not independent, individual chooser.

How and why did the individual adapt himself to the group? How and why did she accept the limits the group imposed upon the alternatives she considered? As was his wont, Simon sought to find the mechanisms by which this adaptation occurred, and in late 1951 he began to look for them in psychology, rather than in economics or sociology. Specifically, he studied learning theory, seeing learning as a process of adaptation linked to the bounding of one's choices.

Simon's perspective on learning was novel, but it was kin to that of experimental psychologists, such as Simon's old friend and new colleague Harold Guetzkow. Guetzkow and Simon had first met many years before on a train platform in Milwaukee, both on the way to school at the University of Chicago. They had become close friends in college, but graduation had taken them down different paths. After a stint as a high school teacher, Guetzkow had gone on to graduate school in psychology, taking his Ph.D. at Michigan under Norman Maier, a Gestalt psychologist interested in problem-solving. Gestalt psychology was a particularly strong source of ideas for Guetzkow, and Simon soon adopted many of the Gestaltists' ideas about problem-solving. Indeed, Simon once wrote that if the Gestalt psychologists had had access to digital computers, then they might well have been the first to write the kinds of problem-solving programs he became famous for writing.[50]

Guetzkow's research focused on the analysis of patterns in problem-solving behavior, infusing concepts from Gestalt psychology into that most traditional of topics in experimental psychology, learning.[51] In his work, learning was learning to solve problems, and learning to solve problems was a process of developing a mental "set." Such mental sets directed one's attention to certain elements of the problem situation rather than others, allowing people to develop efficient, routine responses to problems they encountered frequently.

Guetzkow was a psychologist in the mold of E. C. Tolman, who was influenced by the Gestaltists himself. His psychology was behavioral, but not behaviorist, and his focus on human learning rather than animal conditioning linked him to other students of human cognition and communication, such as George Miller and Jerome Bruner, as well as to Gestalt psychologists such as Max Wertheimer and Karl Duncker.[52] Such behaviorally oriented but cognitively concerned psychologists were willing to accept the existence of mental processes that intervened between stimuli and responses, provided that those processes were defined operationally. Guetzkow thus provided Simon with a complementary new perspective on cognition, and their collaboration would be significant for both of them.

Guetzkow joined Simon at CIT in the fall of 1951. Their first project together was the creation of a laboratory for social science research (later known as the Organizational Behavior Laboratory).[53] In this laboratory, they would explore the ways in which individuals learned to adapt their behavior to the needs of the group. In keeping with this mission, their first studies were replications of the social psychologist Alex Bavelas's experiments on the effects of communications networks on group problem-solving and on the implications of the theory of games for understanding small group behavior.[54]

Simon and Guetzkow soon expanded their studies to the analysis of group pressures toward uniformity.[55] Becoming a member of a group or organization, as they understood it, involved learning the mental "set" appropriate to that group.[56] This process of learning to behave properly, in their view, was guided by various selection mechanisms. It was an adaptive, dynamic process similar in many ways to the processes by which servomechanical systems maintained their equilibrium. Learning theory thus blended with the ideas Simon had drawn from the science of machines. Their confluence taught him many things, but probably the most important was that learning to solve problems was not an optimizing process. Like other adaptive processes, it was about finding a viable, not an optimal, solution. There was no test beyond survival.

Interface: The Systems Research Laboratory

Although Simon undertook a serious study of the psychological literature on learning in 1951–52, his real introduction to the field came at RAND's Systems Research Laboratory (SRL). Because of his experience directing large-scale studies of organizational behavior (the BPA studies), the creators of the SRL—Robert Chapman, John L. Kennedy, William Biel, and Allen Newell—asked Simon in early 1952 to be a consultant on its design.[57] Simon agreed, beginning an association that would change the trajectory of his career, leading him down a new path to the synthesis of choice and control.

The SRL was the product of the Air Force's "concern for the performance of today's complex military systems."[58] The specific military system it analyzed was SAGE (Semi-Automatic Ground Environment), the network of command centers that was to be the heart of the nation's air defense system. SAGE was a vast project whose consequences for American science and society are still being explored.[59] It was the key to IBM's dominance of the computer industry, and it led to the development of many new technologies on which the computer revolution would depend, such as modems, CRT display screens, analog-to-digital conversion techniques, and the use of digital computers as real-time control devices. SAGE cost billions and taught thousands. It ensured that the Whirlwind model of the computer (electronic, digital, serial, centralized information processing) would predominate and that the United States would take the lead in computing from our allies and rivals, the British.

The Air Force funded an enormous amount of research on the human, as well as the technical, aspects of SAGE. The Air Force's experience with flight simulators, combat command centers, and gunnery control systems had taught its leaders that the human components of such systems were vital. As a result, in 1951 they launched a major program of research into "human factors."[60]

Personnel with varied backgrounds carried out this research at a number of sites. The most important of these were the Human Resources Research Laboratory (HRRL) at MIT (part of MIT's Lincoln Labs, the vast enterprise the Air Force established to build the SAGE system), and RAND's Systems Research Laboratory. RAND was responsible for programming the SAGE computers, and the SRL was designed to analyze the performance of SAGE's control centers.[61]

Both of these large research centers brought together interdisciplinary teams of physicists, mathematicians, electrical engineers, and psychologists, all of whom were mathematically sophisticated. At the HRRL, these teams focused on the performance of the human individual, studying the perceptions, reaction times, and learning abilities of the operators of the SAGE machinery.[62] The SRL team was more group- and process-oriented. It studied the structure of the whole "task environment," including the organization of personnel and interpersonal communication. In keeping with these different focuses, the psychologists on the SRL's staff tended to be social psychologists while those at the HRRL, by and large, were experimental psychologists with backgrounds in psychophysics.[63]

Despite these differences, the staffs of the two projects had much in common. First, though they were *behavioralists,* they were not Skinnerian behaviorists: they believed that the mechanisms that intervened between stimulus and response were the proper study of psychologists. The staffs of both centers also shared a common belief that a "convergence" of great importance was in progress.[64] They believed that their studies were going to show the way to link individual and group behavior, and the key was understanding the individual as a component of a "man-machine" system.

Chapman and Kennedy described the SRL project as follows: "The Systems Research Laboratory will be studying particular kinds of models—models made of metal, flesh, and blood. Many of the messy and illusive variables of human and hardware interactions will be put into the laboratory." "Why put them in a laboratory?" they asked rhetorically—because "the laboratory [is] a specialized computer which grinds out the consequences of humans' interaction with hardware and with each other as only they know how."[65]

These "specialized computers" made up of men and machines were called "Information Processing Centers." The IPCs were designed to simulate the operations of a SAGE control center. There were to be twenty-three such centers in the real world, each of them a giant, windowless block of concrete connected to the outside world solely through a network of electronic communications lines (and a door for the staff, presumably).

That these centers were such close approximations to closed systems made the task of simulating them easier. A closer relationship between the world of the laboratory and the real world could hardly be hoped for. The experimenters at the SRL, however, never tied themselves to the particular. They understood the IPCs to be models, abstractions of organizations in general. To them, the IPCs were "human-machine systems" designed to enable experimental investigations into the nature of communications and decision-making in all organizations, not just those housed in windowless cubes. The SRL practiced, as Allen Newell put it, "organization theory in miniature."[66]

For Chapman and Kennedy, it was the Information Processing Center—the entire complex of men, machines, tasks, and experimenters—that made up the laboratory. The whole laboratory, the complete experimental system, was an "organism." As they defined it, "An organism is a highly complex structure with parts so integrated that their relation to one another is governed by their relation to the whole." Significantly, this meant that "Although adaptive behavior in the wider sense may yet remain the province of the human, the moral of the story is the same. There is an assemblage of components united by some form of regular interaction or interdependence whose performance can be studied only as a unit."[67] Functional interdependence an organism makes, not flesh or bone or steel. So defined, the organism was equivalent to a machine or to an organization of men and machines. A model of one could, therefore, be a model of the others.

Why choose one organism instead of another, then? By choosing an "organism made up of a number of men and machines rather than an individual, the experimenter has, in effect, magnified the operation he wishes to study. The flow of information is now over a set of communications channels rather than in an individual's nerve net. A good deal of the interaction is brought out into the open where it can be observed."[68]

Chapman and Kennedy strove to bring everything out where it could be observed. Indeed, "Observation facilities [were] an integral part of the layout." The IPC was designed with an observation bay that ran the entire length of one wall, several feet above floor level, allowing the experimenters a view of the entire IPC. The actions of the forty-odd human components of the IPC were filmed, and they were required to speak into microphones so that their conversations could be recorded. In addition, "A mike pickup system enable[d] the experimenters to catch the conversations not directed into the telephone net," and "each of the telephone circuits and mike outputs is brought to the monitor station so that any conversation can be tapped at will." The result was that the Information Processing Center was truly "an ideal laboratory" in which "fact is the great leveler."[69]

Of course, not even the IPC was totally subject to the experimenter's control. Some compromises had to be made. For example, Chapman and Kennedy decided not to vary either the kind or amount of equipment or the operation policies in the Information Processing Center during a "run," and they made a "concentrated effort" to choose the tasks "so that their accomplishment did not depend upon individual's social history, social perception, or set of personal values."[70] Thus the behavior they studied in the laboratory was behavior in a very restricted domain.

In order to achieve such controlled conditions, the people involved had to be chosen as carefully as the equipment. An initial run, for example, was termed "an instructive failure" because the college students who formed the experimental group found their tasks too easy and did not respond to the experimenter's exhortations to act as though the safety of the free world was in their hands. (The students apparently preferred to catch up on their homework or to doze so long as no crisis was at hand.) Later experiments used real military personnel, though the experimenters did find it necessary to "issue partly unintelligible" and "occasionally contradictory" instructions in order to simulate military life accurately for the soldiers.[71]

Simon was fascinated by what he saw at the SRL, and he returned to RAND each summer during the 1950s to work with the SRL staff and with RAND's computer scientists. Simon gained three things of great importance to his later work from his association with the SRL. The first was an acquaintance with Allen Newell, who would collaborate with Simon on his most significant work in computer science and psychology—and who would teach Simon to "think big about money." The second was experience with and continuing access to the most advanced computers available. Such access was vital to Simon's work, so much so that he later called it his "secret weapon" in his psychological research.[72] The third was experience with the use of computers as simulation devices.

This last experience may have been the most important. Although it took him some time to work out its full implications, Simon quickly became convinced that the computers he saw being used to construct simulations at the SRL were more than simply big, fast calculators. They were something much greater: general-purpose symbol processors. In addition, he became convinced that simulation was a powerful technique for experimentation and that something profoundly interesting was involved in the human operators' processing of the symbolic information that appeared on the screens before them. These ideas formed the seeds of a new path to the synthesis of choice and control, a path that would take Simon across the bridge and up "to the mountaintop."[73]

Outputs

The first major milestones along this path were a pair of articles in which he outlined a new approach to rational choice. These articles were "A Behavioral Model of Rational Choice," and "Rational Choice and the Structure of the Environment."[74] Although these were separate pieces, published a year apart in 1955 and 1956, they were written to go together, and the seeds of both grew out of a "series of discussions . . . at RAND in the summer of 1952."[75] These articles began in conversations at RAND related to work on a contract for the Cowles Commission that was part of a project funded by the ONR and completed with the aid of a grant from the Ford Foundation. Hence, in his work on modeling rational choice, Simon found a way to link the interests of all his major patrons in one project.

These seeds did not sprout immediately. Rather, Simon tended to them carefully over the next three years, crossing them with other strains, particularly W. Ross Ashby's ideas about adaptive, ultrastable machines. The result was a sturdy hybrid that soon assumed a dominant position in Simon's mental ecosystem. In many ways, this hybrid was a translation of Simon's earlier work in organization theory into a more precise, more formal, mathematical format. As with his "translation" of the Homans Model, however, this was no mere transliteration: the new language of rational choice and problem-solving subtly altered both his thinking and his daily practice. The cumulative effects were striking.

In his collection *Models of Man,* issued shortly after these articles were published (an indication that Simon believed he had reached the end of one period in his career and the beginning of another), Simon introduced these two articles by noting that "the publication in 1945 of von Neumann and Morgenstern's *Theory of Games and Economic Behavior* has attracted enormous attention to the theory of rational choice. This has been reinforced and amplified by parallel developments in mathematical statistics ... which have reinterpreted the theory of statistical tests as a theory of rational decision." While these developments in the sciences of choice were of "the greatest importance," Simon argued that "the approach taken in the theory of games and in statistical decision theory to the problem of rational choice is fundamentally wrongheaded." In his mind, it was wrong in "precisely the same way that classical economic theory is wrong—in assuming that rational choice is choice among objectively given alternatives with objectively given consequences that reflect accurately all the complexities of the real world." It was mistaken, in short, in "ignoring the principle of bounded rationality, in seeking to erect a theory of human choice on the unrealistic assumptions of virtual omniscience and unlimited computing power."[76]

It was time, therefore, for a "fundamental change in our approach," to "take account—and not merely as a residual category—of the empirical limits on human rationality, of its finiteness in comparison with the complexities of the world with which it must cope." In Simon's mind, the key to making this fundamental change was to study the psychology of rational behavior; that is, to take into account the psychological properties and limitations of the individual chooser. "The alternative approach employed in these papers is based on what I shall call the *principle of bounded rationality:* The capacity of the human mind for formulating and solving complex problems is very small compared with the size of the problems whose solution is required for objectively rational behavior in the real world—or even for a reasonable approximation to such objective rationality."[77]

In keeping with this outlook, in "A Behavioral Model" the task was to "replace the global rationality of economic man with a kind of rational behavior that is compatible with the access to information and the computational capacities that are actually possessed by organisms, including man, in the kinds of environments in which such organisms exist." Simon admitted that the "kinds of empirical knowledge ... required for a definitive theory" of rational choice were lacking. He argued, however, that "none of us is completely innocent of acquaintance with the gross characteristics of human choice, or the broad features of the environment in which this choice takes place." Hence, Simon felt free to "call on this common experience as a source of the hypotheses needed for the theory about the nature of man and his world."[78]

In typical fashion, Simon begins "A Behavioral Model" with a discussion of rational choice that draws heavily upon game theory, moving from this "unrealistic" model of human behavior to one resting on less "heroic" assumptions. The result is a model of great simplicity and generality. He describes the "choosing organism," for example, in terms that could apply to any organism in any situation, analyzing it in terms of the set of behavior alternatives available to it, the subset of those behaviors it actually considers, the possible future state of affairs, and the "pay-offs" for the organism attendant on these various possible future states.

Simon then sketches the basic game theoretic and probabilistic rules that have been developed to guide the organism's choices. He does so only to point out their limitations, however, concluding the discussion by noting that these "classical" concepts of rationality make "severe demands" on the organism. The organism, for example, must be able to "attach definite payoffs (or at least a definite range of payoffs) to each possible outcome."[79] There is thus no room in such models for "unanticipated consequences" or for incomplete orderings of preferences. To Simon, this narrowness invalidates these theories. Indeed, he argues, "there is a complete lack of evidence that, in actual human choice situations of any complexity, these computations can be, or are in fact, performed."[80] As he wrote to the psychologist Ward Edwards, "About the kindest thing I can say about the maximization model is that: humans being the compliant creatures they are, you can sometimes induce them (by defining the game that way in real life or the laboratory) to behave more or less like the model of rational man (if you make the choice-situation sufficiently simple-minded)."[81]

In his view, the choosing organism is limited not only by the external constraints on its choices that the game theorists recognized but also by internal constraints: "Some of the constraints that must be taken as givens in an optimization problem may be physiological and psychological limitations of the organism."[82] In particular, the "limits on computational capacity" of an organism are powerful constraints, requiring that the organism simplify the calculations behind its choices dramatically: "For the first consequence of the principle of bounded rationality is that the intended rationality of an actor requires him to construct a simplified model of the real situation in order to deal with it. He behaves rationally with respect to this model, and such behavior is not even approximately optimal with respect to the real world. To predict his behavior, we must understand the way in which this simplified model is constructed, and its construction will certainly be related to his psychological properties as a perceiving, thinking, and learning animal."[83]

One of the most significant simplifications that the organism makes in constructing its model of the world, in Simon's view, is the adoption of an extremely simple pay-off function. Instead of creating a vast, precise ordering of all possible outcomes, listing each one as marginally better or worse than the next, organisms tend to judge outcomes as either satisfactory or unsatisfactory. The organism either survives or it dies; the game is either won or lost. The chess player, to use Simon's favorite example, does not have to defeat his opponent in the best, most efficient way. He simply has to win. He does not have to choose the best move (or strategy); he just has to choose one of the many potentially winning ones. This simplification drastically reduces the computation necessary for the organism, making it possible for it to conduct simple tests of possible actions and to choose the first acceptable option.

Learning more efficient, more nearly optimal strategies, in this model, could be accounted for by a simple mechanism that raised or lowered the aspiration level of the organism, redefining what was satisfactory according to the situation. Such a higher-level aspiration-adjusting mechanism could be understood as a kind of feedback or control mechanism. In combination with a simple selection mechanism, the resulting system would have the properties of an Ashby-style "ultrastable" system.

Simon believed that this simple model of rational choice was the key to a synthesis of choice and control. "The paradox vanishes, and the outlines of theory begin to emerge when we substitute for 'economic man' or 'administrative man' a choosing organism of limited knowledge and ability."[84] In "Rational Choice and the Structure of the Environment," Simon moved to a description of the nature of the choice mechanisms—and the nature of the environment—that would enable such an organism to survive.[85]

Simon begins "Rational Choice" by asking the reader to consider an extremely simple organism with only a single need (food) and only three modes of activity (resting, exploring, and food-getting). The organism can travel over the "bare surface" dotted with "little heaps of food" that constitutes its "life space." (Or, to recall Simon's story "The Apple," discussed in chapter 1: Hugo can travel through a castle with dinners laid out for him in some rooms but not others.) This organism has a certain limited perceptual capacity that allows it to see a certain number of "moves" (rooms) ahead along the "branching system of paths, like a maze" that make up its mental world. In addition, this organism's "needs are not insatiable, and hence it does not need to balance marginal increments of satisfaction," a quality that distinguishes it further from "economic man." Given such a simple creature, the "problem of rational choice" is correspondingly simple: how to "choose its path in a way so that it will not starve."[86]

Simon builds complexity into this simple model step by step, first adding choice mechanisms for allocating time toward the achievement of multiple goals. He then adds the ability to detect "clues" in the environment that lead the organism to food more rapidly and aspiration-adjusting mechanisms that allow it to adapt to the richness or barrenness of the environment. Even with these additions, the result is still a remarkably simple model of rational choice, one more appropriate to "rat psychology" than to human psychology. Nevertheless, Simon believed that "Economics and administrative theory both need models of rational choice that provide a less God-like and more rat-like picture of the chooser. The assumptions of omniscience and intelligence implicit in the utility function model are truly fantastic."[87]

Simon believed that the ability of even this simple concatenation of mechanisms to produce adaptive behavior was a sign of the model's power. Just as Ashby had argued that the adaptive powers of an ultrastable system could be magnified almost infinitely if that system were composed of a hierarchy of subsystems, so too did Simon believe that complex behavior could be constructed from a set of nested systems. Each level of the hierarchy might consist only of a small, simple set of mechanisms for selecting alternatives, testing them against a standard, learning from these experiences where to look for more alternatives, and discovering how high to set the bar for future tests. The product of such a system of systems, however, could be extremely complex, adaptive behavior.

The influence of Guetzkow and Newell on Simon's thinking is clear in these articles, for in them he has redefined the problem of choice from being one of making decisions to one of solving problems. The goals of choice have shifted from making the best decision to finding a feasible solution. "Since the organism . . . has neither the senses nor the wits to discover an 'optimal path,' . . . we are concerned only with finding a choice mechanism that will lead it to pursue a 'satisficing' path."[88]

This redefinition of choice also entailed a redefinition of rationality and of the scope of a theory of rational choice. As he wrote in *Models of Man,* "The central task of these essays, then, is not to substitute the irrational for the rational in the explanation of human behavior but to reconstruct the theory of the rational." If this task were accomplished, Simon believed, then a "return swing of the pendulum will begin," and "we will begin to interpret as rational and reasonable many facets of human behavior that we now explain in terms of affect."[89] Simon later would expand the sphere of rationality to encompass (nearly) all mental activity, including the seemingly ineffable acts of inspiration associated with the highest flights of creativity and the deepest springs of emotion (see chapters 11 and 12).

Despite his belief that his papers on rational choice had pointed out the one true path to synthesis, Simon continued to work on parallel tracks during the mid-1950s. For example, in 1954, while he and Newell were meeting for day-long bull sessions in which they discussed how to model choosing organisms, Simon, Guetzkow, and their colleague James March continued their "inventory" of organization theory for the Ford Foundation.[90] Simon also continued his work with Jack Muth, Charles Holt, and Franco Modigliani on linear and "dynamic" programming techniques for the Cowles Commission and the ONR.[91]

While Simon's collaborators on each of these projects rarely crossed project boundaries, Simon did. Though he assumed a different role in each project area, maintaining his identities as organization theorist and econometrician even as he created a new identity, he was still the same person. His mind, unlike the systems he studied, was not "decomposable" or fully compartmentalized. Thus, even as he and Newell developed their models of simple problem-solving organisms, Simon strove to integrate these models of rational choice into his models of complex organizations.

His most significant attempt at such integration came in a series of working papers written for the Ford project team: one on the "theory of departmentalization," and two on "Functional Analysis and Organization Theory."[92] In these papers, which formed the conceptual structure for the book *Organizations,* Simon reinterpreted the organization as being an organism itself, forging a link of functional equivalence between bureaucratic organizations, evolving organisms, and adaptive machines.[93] In the end, the computer became a model organism for Simon because it was to him a model bureaucracy.

Simon begins "Functional Analysis and Organization Theory" by acknowledging his debts to Parsonian social theory: "The conceptual framework of structural-functional analysis underlies a large part of existing organization theory. This is evident in the case of 'survival' theories like the Barnard-Simon inducement-contributions schema; in their use of terms like 'purpose' and 'process' in the description of departmentalization; and generally, in the treatment of organizations as adaptive, self-maintaining systems." Structural-functional analysis allowed Simon to make connections between organizations and organisms and thus between sociology and physiology. As he writes, "Since functional analysis is well developed in biology—specifically physiology—and since it is easier to avoid objectionable metaphysics in this field than in sociology, I shall resort freely to physiological examples." Significantly, Simon believed that "at the level of abstraction of our analysis these will be genuine examples, not analogies."[94]

This foundation established, Simon then proceeds to define two key terms: *unit* and *function.* "By a unit is meant a system of variables in which certain invariant relations over time make possible the identification of an 'individual' with temporal, and usually spatial, continuity. In organization theory, we are generally concerned with at least two levels of units: (1) the human individual, and (2) the organization. In physiology, we are generally also concerned with two levels of units: (1) the cell, and (2) the organism."[95]

Defining the individual unit in such abstract, functional terms made it possible to understand units at each level of the organizational hierarchy as entities distinct from their components. "The cell represents a stable configuration of molecules (that is, certain relationships among them are stable), although the individual molecules that enter into the configuration may change. Similarly, an organization (to take the other extreme) is recognizable through a stable system of interpersonal relationships among persons, although the individuals who make up the organization may change over time."[96]

After defining these concepts, Simon then moves to an analysis of "Process and Structure." Under this heading, he analyzes "organs and tissues," as well as the "systems" they compose. "An organ," to Simon, "is a localized subsystem," and a tissue is "generally a sub-organ characterized by homogeneity of process and function among its units." Although these terms are not employed commonly in the analysis of organizations, Simon argues that they do apply: "the scheduling department of a factory is an 'organ' in terms of the above definition ... the problems of analysis of the two kinds of systems—organizational and organismic—are identical (and not merely analogous) at this level of abstraction." Similarly, to him "a system is a collection of organs connected by a flow such that the output of one organ in the chain becomes the input of the next. The circulatory system is an obvious biological example; an assembly line, an organizational example."[97]

Continuing this analysis of system functions, Simon notes that the same means (or "way a function is performed") may be employed toward a number of ends. Organs, for example, are "generalized means," for they provide the sites and the mechanisms for processes that enter into successive means-ends chains meeting the functional requisites of the organism. They "constitute ready-made elements out of which such means-end chains can be constructed and serve as building blocks for large structures of goal-oriented behavior." Thus, "A factory is a generalized means for 'making things'; a sales department a generalized means for 'selling things.'"[98]

Harking back to Ashby's notion of ultrastable systems, Simon argues that "this capacity of organs to serve as generalized means is based on two characteristics: (1) the organ's output varies in a determinate way with the input of signals ... [and] (2) the signaling system contains switching elements that permit an organ to be fitted into a variety of relations with the rest of the system." Since generalized means "enter into a multiplicity of means-end chains, survival of the system requires that competing demands upon the means, arising from the several functional requisites, be handled appropriately by the programming mechanism." Learning, in this model, becomes the process by which an organism develops efficient switching mechanisms. That is, "learning consists in considerable part in the development of preprogrammed responses, or elements of responses, for various situations."[99]

Thus, a person's mental "set," formerly his or her set of values and beliefs, has now become a "program," according to which "specific activities are performed at definite times in response to signals and stimuli of one sort or another."[100] In other documents from this period, Simon increasingly begins to describe all human behavior in such functional-mechanical terms, using phrases like "load balancing," "error control," "switching rules," and information "storage and filtering" to describe the activities of human individuals and organizations.[101]

But how are such programmed, conditional activities learned? How are new programs developed? The development of new programs Simon now called problem-solving. He distinguished problem-solving from "routine" behavior by its dependence on "inductive rather than deductive logical processes," reflecting that problem-solving was oriented toward finding a "feasible, not an optimal course of action."[102]

Simon believed that he—and others—had a firm grasp on deductive, programmed responses. The sciences of choice, especially operations research, had developed powerful algorithms for finding optimal solutions to such "routine" optimization problems. The inductive, adaptive processes by which such programs were developed, however, were still largely unknown. Simon knew the functions the organism must perform in order to survive, but that was not enough. A vast number of mechanisms could perform such functions, but only some of these mechanisms actually were employed by human choosers. How could one discover which mechanisms in truth lay at the roots of human behavior? To him, until one could devise a "program" that could learn, adapt, and solve human problems in human ways, the mind would remain a mystery.

Decisions, Values, and Problems

As a part of creating "organization theory in miniature," of bringing decision-making into the laboratory, the subject of inquiry had been transformed. Decision-making had become problem-solving and the individual had become a behavioral system. Simon had written in *Administrative Behavior* that "In a given situation, and with a given system of values, there is only one course of action which an individual can rationally pursue." In the rich, chaotic context of human organizations, the situation never was wholly given and values never were truly separable from facts, as Simon well knew.[103] But in the laboratory, the situation and the system of values could be given, almost completely, leaving choice to the designer alone. The new science of problem-solving systems, as a result, would be markedly different from either the sciences of choice or of control.

While problem-solving in a "given" situation was the more fundamental activity in Simon's view, it was less inclusive than decision-making. The problem-solver assumed the goals that the decision-maker had to choose. The decisions that all of us have to face as members of multiple organizations with differing goals, including that crucial decision whether to accept the organization's authority, have been simplified to a less contentious, less political, process of allocating "processor time" to different tasks. Choices were now less decisions about which set of values to accept and more decisions about what set of data to process.

One consequence of this shift from decision-making to problem-solving was that Simon's research increasingly began to focus on situations that could be described unambiguously. Simon's first studies had been large field experiments in the very real bureaucracies of the California State Relief Administration and the controllers' offices of large corporations: the city had been his laboratory, as had the firm. His experiments now simulated human behavior in chess games or in the solution of mathematical proofs rather than describing the behavior of caseworkers trying to work around the rules in order to do their jobs as they saw them: "A Behavioral Model" originally contained an appendix sketching the outlines of a program that would play chess, not a description of an administrator making decisions.[104] In keeping with this narrowed focus, his later studies also were ever more restricted in place. At first, the laboratory stood in for the city. Then, the bounds of his research tightened even more, with the experimental subject and the experimental situation becoming embodied in the computer. This most adaptable machine soon became both his means of simulating homo adaptivus and his model of the human machine.

CHAPTER TEN

The Program *Is* the Theory

In his work at the Systems Research Laboratory, Herbert Simon attempted to bring together the sciences of choice and control via the science of machines. The product of their union was a new science of behavioral systems, abstractly defined, and the goal of this new science was the construction of formal models of the psychology of rational choice.

The crucial development in Simon's mature efforts to construct such models was his adoption of the digital computer as his primary research instrument. The computer was the vehicle for the redefinition of Simon's science, and in it he traveled far. For example, by the late 1950s, Simon's theories took the form of computer programs, and his experiments compared runs of these programs to individuals engaged in solving specific problems, a distinct change in practice from his earlier field experiments in public administration.

By redefining and reorienting the science of machines, Simon did not lose his general perspective, for he understood the computer to be a universal machine capable of simulating the behavior of any set of mechanisms. It was thus the machine of machines, the system of systems. Those, like Simon, who developed the science of computers would redefine our understanding of all machines, physical, biological, and social. For Simon the universal science of machines became the science of a universal machine. The irony behind this perspective was that the computer's ability to simulate any machine made it unique, not typical.

Because adopting the computer as his primary research tool had such important consequences for Simon's work, and because the computer became a model technology as well as tool for modeling, the study of the role of the computer in Simon's science promises to shed light on the role of instruments in modern science more generally. Indeed, the recent surge of interest in the role of instruments in science has been generated, in significant part, by the awareness of the powerful role the computer plays in present science—and in our daily lives.[1]

Instruments, particularly powerful and expensive ones, can be important parts of the paradigms that shape fields. They can serve functions similar to what George Lakoff and Mark Johnson have called "generative metaphors," and which I have called heuristic models, for they organize experience.[2] Instruments, models, and metaphors can be vehicles for bringing disparate fields together, serving as the conceptual, linguistic, and experiential common ground that enables communication in interdisciplinary "trading zones."[3] The power of certain instruments and metaphors is perhaps best illustrated by the fact that their use often becomes an end in itself, as researchers seek to test the limits of the machine or the bounds of an analogy.

Instruments also can shape the social and practical aspects of science, for scientific paradigms include models of both technical and intellectual practice. That is, disciplinary paradigms are organized around certain technical and instrumental capabilities and certain social relations, as well as around specific concepts. These models of technical and intellectual practice are interrelated, though not in strictly deterministic ways. Each sets bounds on the other without determining fully what lies within those bounds, and both are grounded in a common set of basic assumptions about the nature of the world and our ability to know things about it.

Simon reoriented his intellectual and technical practice around the computer. On the technical level, digital computers became a focal point of his work as he attempted to program them to simulate human behaviors. On the intellectual level, his research agenda became the elaboration of the information-processing model of mind. His basic questions—and answers—

all were translated into the language of information processing. The adoption of the computer as his primary research instrument also involved the creation of a new set of professional relations, including a new professional identity and a reconfigured network of patrons for his work. These changes in Simon's technical and intellectual practice are apparent in the origins of Simon's first computer model of mind, the Logic Theorist program, as well as in one of his most significant claims: the idea that "the program is the theory" in cognitive psychology.

Logic Theorist

The place to begin any discussion of LT, as Logic Theorist was known, is RAND in February 1952, when Herbert Simon first met Allen Newell. The two immediately recognized each other as kindred spirits. As Simon recalls, "In our first five minutes of conversation, Al and I discovered our ideological affinity. We launched at once into an animated discussion, recognizing that though our vocabularies were different, we both viewed the human mind as a symbol-manipulating (my terms) or information-processing (his term) system."[4] In addition to a shared outlook, Simon also found in Newell a colleague of great intellectual power: to Simon, Newell was the "only person I know to whom I can apply term 'genius' non-metaphorically," which was saying something, coming from a man who could count a dozen future Nobel laureates among his friends.[5]

Thus began an intellectual partnership that lasted more than twenty years and a friendship that lasted until Newell's death in 1992. This partnership was immensely productive and intensely frustrating for anyone interested in apportioning credit (or blame) to one or another member of the team. When Simon read an early draft of the manuscript for this book, one of his three main criticisms was that it did not give enough credit to Newell for their joint work. (The other two were that I made Simon appear "larger than life" and rather more calculating than he thought he had been, especially in his institution-building campaigns.) He suggested that I discuss their joint work as the product of "Newell-Simon, or Simon-Newell," rather than either man individually. This was difficult advice for a biographer to take: shifting from "Simon" to "Simon-Newell" sounded as if my subject suddenly had grown a second head.

Despite the similarities in their intellectual interests and approaches, however, Newell and Simon did differ. Generally speaking, Newell was more interested than Simon in creating formal languages and notation systems for the description of behavior, while Simon usually was more concerned than Newell with ensuring that their models corresponded to observed human behavior. Hence, though the Information Processing Languages were joint products (with Clifford Shaw's assistance as well), they were a bit more Newell than Simon. Similarly, though both conducted experiments with humans in order to test their simulations, it was Simon who wrote on the use of "Verbal Reports as Data."[6] In addition, Simon was more interested in the implications of their work for other fields, from economics to political science to biology.

In some ways, Newell and Simon did act as if they were one person. After Newell came to Carnegie Tech's Graduate School in Industrial Administration in the spring of 1955 to get his Ph.D. with Simon, it begins to be quite difficult to determine where Newell left off and Simon began. Accordingly, some background on Newell is in order.

Newell did his undergraduate work in physics at Stanford, where he encountered the mathematician George Polya, whose book *How to Solve It* introduced him to the concept of heuristics in problem-solving. Newell then went to Princeton in 1949 to study mathematics with John von Neumann and Alonzo Church. He absorbed his instructors' common interest in formal mathematical analysis, especially in logic and game theory, though he rapidly came to feel that the Princeton atmosphere was too rarified, too remote from application. He left Princeton after one year and went to RAND to work in the mathematics section under John Williams, who recruited a number of other Princeton graduates to staff his section.

At RAND, Newell's first project was an attempt to formalize organization theory along game-theoretic lines. In the course of this work, Newell became interested in experiments on decision-making in small-groups, seeing the work of R. Freed Bales (of Harvard's Department of Social Relations) and Alex Bavelas (of MIT's Center for Group Dynamics) as a way of grounding decision theory in empirical reality. Hoping to extend such experimental investigations of decision-making, he sought out the psychologists in the social science division at RAND, working with Robert Chapman, John L. Kennedy, and William Biel to create the Systems Research Laboratory (SRL).

Newell's role at the SRL was to help create a "thick, rich" simulated environment so that the decision-making they observed would be as true to life as possible. The key to creating such a simulated environment was to control the information presented to the subjects: the simulation consisted of presenting the subjects with different, but mutually consistent, sets of data on printouts and, later, CRT screens. Newell was familiar with Shannon's information theory and, thanks to his training in logic, was predisposed to conceive of decision-making as a logical process of drawing conclusions based on certain premises. The combination of these ideas in the environment of the SRL led him to think of decision-making in terms of the communication and processing of information.

Newell thus shared Simon's interests in decision-making, information processing, and formal theory, adding to them a fascination with simulation and early experience with computers as simulation devices. For Newell, these interests coalesced rather suddenly in September 1954, when he had what he later called his "conversion experience." Oliver Selfridge, an electrical engineer turned computer scientist, gave a talk about a computer program he had designed that could recognize letters and other patterns. While Newell was aware that computers were more than big, fast number-crunchers, the implications of the computer's ability to process symbols in general had not yet dawned upon him. Although Selfridge's program was crude, the light went on for Newell: suddenly he saw that "intelligent adaptive systems could be built that were far more complex than anything yet done."[7] He decided to join Simon at Carnegie Tech the following spring in order to build just such a system.

When Allen Newell joined Simon at CIT in the spring of 1955, the two took as their mission the creation of a computer program that could solve problems. They decided to focus on problems that were too complex and ill defined to be solvable by deductive algorithms, but not so ill defined that success or failure in their solution was not clear-cut. In keeping with these goals (and their own hobbies), they chose to focus on playing chess and constructing geometry proofs: both were complex tasks that were unassailable by deductive logic, yet they had definite outcomes.

How to implement such a program? How to test it? These questions troubled Simon and Newell. Algorithms for finding solutions to well-defined problems had a familiar format, the formal logical proof, and they could be written in a standard language—set theoretic or probabilistic mathematics.

But what language could one use to describe complex, ill-defined problems? What techniques were appropriate for experimentation with hypothetical organisms?

The SRL provided the best available model for the experimental analysis of such model organisms, and ever since his first summer at RAND in 1952, Simon had sought to shape the GSIA's Organizational Behavior Laboratory in the SRL's image.[8] Significantly, though Simon and his colleagues were interested in studying organizational behavior in the laboratory, the construction of simulated environments suitable for long-term, large-scale experiments using human subjects was prohibitively expensive. Though Simon was wise in the ways of getting grants, his lab at Carnegie Tech was simply never going to have the thousand-man support staff that the Systems Research Laboratory employed to construct their simulations. This limitation of resources meant that where the SRL's experiments "ran" over a period of several weeks, those at Carnegie Tech often took about ninety minutes—a change which markedly reduced the importance of certain kinds of social interaction and limited the kind of tasks that could be assigned to the group.[9] In such an environment, using the computer for simulating both the actors and the environment seemed quite attractive, for it would make effective use of existing resources without sacrificing anything that had not been sacrificed already.

Before the computer could be used in this way, however, it first had to be understood as something more than a superfast adding machine. Thinking of computers as general-purpose information or symbol processors comes so naturally to us today that it is hard to remember that this understanding of computers had to be constructed. The first electronic computers, like Howard Aiken's Mark I or Eckert and Mauchly's ENIAC, were merely big calculators used to compute firing tables and the like. Even Jay Forrester, the chief designer of Whirlwind, thought of the computer as a control device, not a general information processor, and Alan Turing's "universal machine" was an algorithmic fantasy (though a suggestive one) because it required an "infinite tape" for its operations.[10]

At the SRL, however, computers had been enlisted to help generate the simulated environment of the air defense controllers. Simon "became fascinated by the method that Al [Newell] and J. C. (Cliff) Shaw . . . had devised for using a card-programmed calculator to produce imitations of radar maps for air-defense simulation": "The computer was generating not numbers but locations, points on a two-dimensional map."[11] If computers could be used to generate—and transform—locations in two dimensions, why not three? And if three, why not vectors of multiple dimensions?

Simon's studies of symbolic logic had taught him that numbers were symbols, just like any other kind of symbol. In addition, he also knew that single symbols (like a location on a map) could be used to represent multidimensional vectors and, conversely, that symbols representing such vectors could be broken down into sets of simpler symbols representing certain aspects of the whole. As George Boole had shown in the nineteenth century, and Alan Turing in the twentieth, there was no real limit to how simple the elements, or how complex the composites could be. The implication, as Claude Shannon had proven for electronic circuits and Warren McCullough and Walter Pitts had shown for neurons, was that a machine that merely counted, added, and subtracted ones and zeros (ons and offs) could use strings of such binary digits to represent any number and, by extension, any symbol.[12] An electronic calculator could be a logic machine (Shannon), and a logic machine could be a general symbol processor (Turing). The end of this logical chain, as John von Neumann argued in 1946, was that the digital computer, formerly seen as merely a big, fast electronic calculator, was truly a generalized symbol processor.

Simon became aware of von Neumann's work on computing in the late 1940s, and he was well prepared mentally to make a connection between humans and computers by his belief in the functional equivalence of organizations, organisms, and machines. Understanding both humans and computers as general symbol processors, however, required sustained experience with programming a computer capable of storing programs in its memory. Simon was not alone in needing this material prompt: no one thought of computers as general symbol processors until they encountered digital computers capable of storing programs in an internal memory.

What was so important about the ability to store programs? The first computers, such as ENIAC, EDSAC or the Harvard Mark I, had required direct human intervention every time they needed to change their program.[13] The program, or "way of behaving," was the sequence of operations the machine would perform on a set of inputs. Machines like the ENIAC were "programmed" physically—the jungle of wires that connected the various components of this hundred-foot long set of vacuum tubes, capacitors, and resistors had to be rearranged every time a new "program" had to be run.[14] The very use of the word "program" to describe this operation is an anachronism; the persons who planned and executed this rearranging typically called it "setting up" the machine.[15]

In mid-1945, however, John von Neumann began working with J. Presper Eckert and John Mauchly on a design for the EDVAC, a successor to Eckert and Mauchly's first computer, the ENIAC. In his "First Draft of a Report on EDVAC," von Neumann advanced the idea that one could treat instructions as sets of symbols just like data.[16] The implication of this was that one could store these instructions in electronic form instead of storing them in the form of a certain arrangement of wires and switch positions. This meant that computers could be programmed and that these programs could be stored in the computer's memory.[17]

Storing a program in the computer's memory was a crucial step in making the link between computers and humans both reasonable and fruitful as a spur to inquiry. There are three main reasons why the concept of the stored program was so important: first, if you can store one program, you can store two, or three, or more. By adding a higher-level mechanism (yet another program) for shifting from one program to another, you will have created W. Ross Ashby's dream—a truly "ultrastable" system capable of making startling adaptations in real time. Second, this real-time adaptation would occur within the computer, without direct human intervention, making the machine more of an entity and less of an object.[18] Third, the fact that one was storing instructions in memory, not just numerical data, meant that one was storing nonnumerical symbols in the computer's memory. (What was actually stored, of course, was a pattern of ones and zeros, but these ones and zeros represented nonnumerical symbols, not just other numbers.) The concept of the stored program, therefore, contained within it the implication that the computer was a general processor of symbols, not simply a calculating machine.

Note that the idea of the stored program assumed that the computer in question was a digital machine, not an analogue device. A machine that makes calculations by manipulating physical analogs to the system under consideration is necessarily a specialized device; to shift such a machine to another "program" would require physical alteration of the machine, typically via human intervention. Data and instructions cannot be stored in the same memory unless they are stored in the same format, and digitization provides such a format. This concept also played an important role in structuring thinking about the computer, for digitalization is inherently an abstracting, generalizing technique: it was the digital computer, not the analog, that was the "universal machine."

At RAND in the summer of 1954, Simon learned to program the IBM 701, IBM's first stored-program computer. The experience taught him a great deal, enabling him to see the significance of the ability of the computers to plot locations, not just one-dimensional numbers. By the end of that summer, he, like Newell, had begun to think of both humans and computers as behavioral systems that manipulated symbols in order to adapt to their environment. From there it was no great leap to thinking that one such system might serve as a model of the other.

The thought of modeling human problem-solving on a computer was exciting, almost intoxicating to Simon. It was exciting to use the best and latest technology, of course, and even more so to use it in a new way. In addition, vast resources were being poured into the development of computers and computing, far greater than ever would be available to traditional social science. Tying social science to computer science, therefore, made it possible to siphon some of that flood of dollars and talents into study of human behavior.

But there was something more behind the excitement Simon felt. Simulating human behavior with a computer equated humans with machines, and such an equation was deliberately provocative. It was a thumbing of the nose at every humanist believer in the ineffable and the unobservable, and there were very few things Simon enjoyed more than deflating the pretensions of others. Part of the added thrill thus came from the feeling that he was unmasking the greatest of impostors, undercutting the most irrational of faiths—the belief that humans were somehow different, special, unlike the rest of creation. As he and Newell wrote in 1959, the idea that "a computer can only do what you tell it to do ... seems to give great comfort to people who are worried about the computer's challenge to the uniqueness of man—the same challenge raised, a century ago, by Darwinism, and four centuries ago, by the Copernican picture of the universe."[19]

The computer simulation of human behavior also was an implicit attack on those, like radical behaviorists or believers in "economic man," who had allowed practical difficulties to become epistemological impossibilities. Simon believed that the successful use of computers to model human problem-solving processes would validate not only his specific theories about human cognition but also his whole approach to science and his basic understanding of humanity's place in nature. LT was much more than a program precisely because it was only a program.

A Virtual Machine

The first successful use of a computer to model human problem-solving came in late 1956, when Simon, Newell, and Shaw succeeded in getting their program Logic Theorist to run on RAND's JOHNNIAC.[20] To Simon, however, this first successful run on a computer was not the first successful test of Logic Theorist: that had come on December 15, 1955.

Simon viewed this first run of LT as a defining moment in his career. In his autobiography he writes that 1955 and 1956 were the "most important years of my life as a scientist," for they brought a marked change in his intellectual goals, theories, and practices:

> During the preceding twenty years, my principal research had dealt with organizations and how the people who manage them make decisions. My empirical work had carried me into real-world organizations to observe them and occasionally to carry out experiments on them. My theorizing used ordinary language or the sorts of mathematics then commonly employed in economics ... All of this changed radically in the last months of 1955 ... the focus of my attention and efforts turned sharply to the psychology of human problem-solving, specifically, to discovering the symbolic processes that people use in thinking. Henceforth, I studied these processes in the psychological laboratory and wrote my theories in the peculiar formal languages that are used to program computers. Soon I was transformed professionally into a cognitive psychologist and computer scientist, almost abandoning my earlier professional identity.[21]

As Simon tells the tale, he and Newell were in New York for a meeting of the Institute of Management Sciences in October 1955, when he (out for a noon stroll along the Hudson in Morningside Heights) had a "sudden conviction" that they could program a computer to do proofs in geometry. This would be a far easier task, he believed, than programming a computer to play chess, Newell's project at the time. Simon soon tracked down Newell, meeting him in Merrill Flood's hotel room, where they began to work out a rough plan for a program that would solve problems (by which they meant construct proofs) in geometry.[22]

Geometry, like chess, however, posed difficult problems of visual representation. Symbolic logic, on the other hand, did not pose such problems, and it eliminated a big step in the programming process, since the programmer did not have to reduce chess moves or geometric figures into symbolic logic. By December 15 of that year, Simon had developed the heuristics that would guide such a program to the point where he was able to "hand-simulate" its operation, producing a complete proof of one of Russell and Whitehead's theorems from *Principia Mathematica.*

By January, the program was ready for a sterner test:

> While awaiting completion of the computer implementation of LT, Al [Newell] and I wrote out the rules for the components of the program (subroutines) in English on index cards, and also made up cards for the contents of the memories (the axioms of logic). At the GSIA building on a dark winter evening in January 1956, we assembled my wife and three children together with some graduate students. To each member of the group, we gave one of the cards, so that each person became, in effect, a component of the LT computer program ... So we were able to simulate the behavior of LT with a computer constructed of human components. Here was nature imitating art imitating nature.[23]

There was a very close parallel between the bureaucratic organization and the programmable computer in Simon's mind. As he argued a year later in his (in)famous talk to the Operations Research Society, "As you will see ... physicists and electrical engineers had little to do with the invention of the digital computer ... the real inventor was the economist Adam Smith, whose idea was translated into hardware through successive stages of development by two mathematicians, Prony and Babbage."[24] The idea to which Simon refers was Adam Smith's concept of the division of labor, as exemplified by the pin factory, which Prony applied to the calculations of logarithms (and which Babbage sought to mechanize). The digital computer was powerful because it extended the division of labor to its ultimate end: the worker (logic gate) that was either on or off, open or closed.

By the beginning of the spring term, Simon was confident enough of his team's accomplishments that he announced to his class at Carnegie Tech, "Over the winter holiday, Al Newell and I invented a thinking machine."[25] Soon, Logic Theorist went on to prove three quarters of the theorems of the *Principia,* including proving one in a "more elegant" fashion than had Russell and Whitehead. With these accomplishments as proof, in 1957 Simon proclaimed to a startled audience of operations researchers that "Intuition, insight, and learning are no longer exclusive possessions of humans: any large high-speed computer can be programmed to exhibit them also."[26]

Although Logic Theorist clearly was a remarkable machine, Simon's conclusion was curious. What had he and his colleagues actually done, after all? First, a human had simulated the problem-solving processes of another human. Then, a group of humans had simulated those same processes in the same manner, with several people combining to do the work that Simon had done alone. It was not until nearly a year later that a nonhuman machine conducted the same simulation. The point here is that when Simon referred to a "machine that thinks," he was referring to the *program* he and Newell and Shaw had developed, not to any physical machine.[27]

How did this nonphysical machine work? Logic Theorist was a servomechanism in symbolic form. It was an error-controlled machine that tested its current state against a goal state, measured the difference between the two, and applied one of a set of operations in order to reduce the difference.[28] It was adaptive on two levels: the results of a given operation were fed back into the machine, generating a new error, which could trigger a different operation (still aimed at reducing the difference between present state and goal), and the successful completion of a proof meant that the newly proven theorem was added to the stock of axioms against which a new theorem was tested.

Logic Theorist proceeded by trial and error, like one of Ashby's ultrastable systems, with that trial and error being guided by "rules of thumb" that Simon called "heuristics." The concept of the heuristic was important enough to Simon that he called the class of programs that Logic Theorist exemplified "heuristic problem solvers."[29]

The pursuit of heuristic problem-solving differentiated Simon and Newell's work in (what came to be known as) artificial intelligence from that of other pioneers in the field. Other researchers such as John McCarthy and Marvin Minsky initially had sought to create rigorously algorithmic programs. A heuristic, however, is not an algorithm. It is a rule of thumb, not a rule. Its purpose is not to search through all possible answers but to narrow the "search space" of alternatives to examine. Within this search space, the alternatives are selected for testing and, typically, the first satisfactory alternative is chosen. One heuristic can be applied in conjunction with another to further limit the search space, putting off as long as possible the final stage of generating and testing alternatives.

For example, take a common problem, such as finding a mechanic who will repair your car at a price you can afford. One heuristic might be to call your friends and ask them for recommendations; another might be to look in the yellow pages. The alternatives generated by each method would then be tested in random order until one met your price criterion. These heuristics might be further refined, of course, perhaps by limiting the set of friends you call to those you know have had their cars repaired recently or by calling only those mechanics in the yellow pages that advertise their ability to service cars of your make. One heuristic may be better or worse than another for a certain purpose, but heuristics cannot be right or wrong in absolute terms. Indeed, it is a fundamental characteristic of a heuristic that it may lead one down a blind alley in some situations, while an algorithm always will find a solution, if one exists, since it searches all possibilities. The problem with algorithms, of course, is that the search space is usually far too large to investigate exhaustively: one algorithmic solution to the problem above would be to call every number in the phone book.

Simon and Newell's concept of heuristic programming was a clear descendant of Simon's principle of bounded rationality. As noted earlier, to Simon one way in which organisms compensated for the limits to their knowledge was to use radically simplified tests of the alternatives under consideration, judging options as satisfactory or unsatisfactory, as leading to survival or death. While this tactic simplified matters enormously, it was not sufficient by itself. Humans also adopted heuristics to guide them in narrowing the field of alternatives to test, with successful heuristics being incorporated into "mental sets."

The gains in an organism's ability to solve problems that heuristics enabled were nothing short of spectacular. These gains far exceeded those that would come from any conceivable increase in processing speed. In a classic example, Simon observed that chess has something on the order of 10^{120} possible outcomes. This is an absurdly large number; one cannot possibly calculate the optimal path through the maze of decisions one must make in a chess match. Yet people do play chess, and, what is more, some people play much better than others. Chess grandmasters almost invariably beat newcomers to the game. How can this be? Simon's reply is that chess masters have better heuristics. At any given step in the search for a good move, they are able to focus on a much smaller set of alternatives to analyze, enabling them to carry the computation many more moves ahead.

In another analogy of which Simon was fond, he asked his readers to imagine the task of a safecracker trying to open a safe with ten dials each having one hundred numbers. The possible number of combinations, then, is 100^{10}, or 10^{20}. All the time in the world would not suffice to open this safe if one had to try them all (or even half of them). If it were possible, however, to employ the heuristic of "factoring" the problem (breaking a big problem down into sub-problems), and if it were possible to solve these sub-problems independently, then the situation would improve dramatically. For example, suppose that the dials made faint clicks when turned to their proper number; then the safecracker could "factor" his big problem into ten smaller ones because he could be sure that he had gotten each one right as he went along. The impossible has now become trivial: instead of one-half of 10^{20} trials, on average he will have to try but 500 combinations (fifty for each of ten). In addition, each test in this case is a simple yes-no test, further reducing the complexity of the problem-solving task.

Logic Theorist employed a few basic heuristics, the most important of which were factoring and means-ends analysis (comparison of goal and present states and selection of steps that will reduce the difference between the two). By its success, LT demonstrated that remarkably complex behavior could be simulated by a very small set of simple operations (seven in total) applied in sequences ordered by definable heuristics. As Simon later wrote, LT demonstrated "that a computer could discover problem solutions in a complex nonnumerical domain by a heuristic search that used humanoid heuristics."[30] To him, that meant it was a "thinking machine."

Simon and Newell took Logic Theorist to be a vindication of their understanding of both men and machines. They had reached "the mountaintop," and they saw a vast field unfolding before them: "We are now poised for a great advance that will bring the digital computer and the tools of mathematics and the behavioral sciences to bear on the very core of managerial activity—on the exercise of judgment and intuition; on the processes of making complex decisions."[31] Indeed, Simon was so emboldened by Logic Theorist's successes that he was willing to predict:

1. That within ten years a digital computer will be the world's chess champion, unless the rules bar it from competition.
2. That within ten years a digital computer will discover and prove an important new mathematical theorem.

3. That within ten years a digital computer will write music that will be accepted by critics as possessing considerable aesthetic value.

4. That within ten years most theories in psychology will take the form of computer programs, or of qualitative statements about the characteristics of computer programs.[32]

In short, Simon held that "there are now in the world machines that think, that learn, and that create. Moreover, their ability to do these things is going to increase rapidly until—in a visible future—the range of problems they can handle will be coextensive with the range to which the human mind has been applied."[33]

The Program Is the Theory

Though it attracted less attention than their startling pronouncements regarding the potential powers of intelligent machines, perhaps the most remarkable claim Simon and Newell made for their creation was that "the program is the theory." They believed that in the program they had discovered the formalism appropriate to the description and explanation of the behavior of complex, adaptive systems. To them, the program was to the explanation of adaptive behavior as the system of differential equations was to the explanation of the behavior of nonadaptive systems, such as bridges, billiard balls, or planets. Simply put, "At this level of theorizing, *an explanation of an observed behavior of the organism is provided by a program of primitive information processes that generates this behavior.*"[34] To give an indication of the outsize implications of this argument, one should note that it equated their development of formal computer languages—the Information Processing Languages—with Newton's creation of calculus, a claim they wisely left implicit rather than explicit in their publications.

In order to understand what it meant to say that "the program is the theory," Simon and Newell's understanding of explanation needs some explaining: to them, one explained something by writing down a set of statements that described the mechanisms that produced that something. The falling of a heavy body is explained by the equations describing the actions of the forces that govern its fall. Such mechanisms had to be described in a formal language appropriate to the phenomena being studied, for formal languages had limits and no one such language, not even calculus, could be used in the description of all phenomena. Formal languages were limited in the phenomena to which they applied because all such languages were based on certain assumptions about the world, assumptions that held in some circumstances but not others.

For example, the formal languages of Euclidean geometry and Riemannian geometry are based on different assumptions about the world—most notably, whether or not parallel lines ever can meet—assumptions that have marked implications for how one understands phenomena. Einstein's theory of relativity could not be expressed, or even imagined, in the language of Euclidean geometry. Similarly, calculus is based on a number of fundamental assumptions, among them being the assumption that the continuous world of our experience is the product of the assembly of nearly infinite arrays of discrete units (what one might call the digital faith) and the assumption that temporal sequence can be ignored.

This latter claim may seem strange, for calculus commonly is used to describe dynamic systems—that is, systems whose parts are in motion relative to one another and whose arrangements thus change over time. It is not particularly well suited to the description of the time path of systems that are subject to contingent events, however. For all calculus cares, time could run backward, even though time in the physical universe does have a direction. Calculus also is not well suited to the description of the sequence of steps necessary to solve a problem, even though each of those steps may in itself use calculus: when asked to calculate the volume of a sphere, for instance, a student first will try to discover the radius of that sphere, then plug that result into an equation, and then solve for the volume. The student uses calculus in this case in that he or she uses (or perhaps even derives) an equation relating the volume of the sphere to its radius, but the sequence of steps in the solution of this problem is not describable in the language of calculus. Some other formalism is required to describe sequences of operations, even when each of those individual operations is describable in the language of calculus.

While the digital faith at the heart of calculus largely went unchallenged in the twentieth century—indeed, the development of electronic computers has strengthened this faith and carried it to new realms—the difficulties of describing both path-dependent (contingent and irreversible) phenomena and sequences of operations with calculus led a number of mathematicians, physical scientists, and engineers to attempt to develop new mathematical formalisms. Some, in the tradition of Bertrand Russell, Kurt Gödel, and Alan Turing, investigated the question of "decidability," attempting to discover sequences of operations that would enable one to decide always and in every case whether a given statement was valid. While Gödel proved that such a universal method did not exist, Turing and others inverted the question of decidability, attempting to find an elemental set of operations that could be used to reproduce any other operation.[35]

A number of similar attempts to create formalisms for describing path-dependent systems and sequences of operations followed in the 1930s and 1940s. The two most notable of these are the mathematical analysis of servomechanisms and other feedback systems, epitomized by the work of Claude Shannon, Norbert Wiener, and W. Ross Ashby, and the theory of games developed by John von Neumann and Oskar Morgenstern.[36] Social scientists—mathematically skilled ones, at any rate—displayed an immediate interest in these two new mathematical formalisms in the late 1940s and 1950s. Psychologists such as George Miller, Wendell Garner, and David Grant attempted to mix Shannon's information theory with game theory in order to understand sequential actions, as did decision theorists, such as Abraham Wald and Merrill Flood. At the same time, operations researchers invented a variety of scheduling and queuing techniques, such as George Dantzig's technique of linear programming, using a combination of servomechanism theory, Shannon's information theory, Wiener's analysis of the use of time series in predicting future events, and von Neumann's game theory.[37]

All these innovations shared a common origin in the need to develop new formalisms in which one could describe contingent, sequential actions, and many of them shared an institutional and temporal locus: the organized effort by the National Defense Research Council and the military research agencies to apply mathematics to the problems of modern warfare, primarily gunnery control and logistics.[38] These innovations also shared an interesting combination of assumptions about the world: they analyzed phenomena in terms of systems, which allied them with the sciences of control, but they assumed that the actors in these systems acted rationally in the pursuit of optimal results, which allied them with the science of choice.

To Simon and Newell, these all were steps in the right direction. Nevertheless, they believed that information theory, game theory, and servo theory all fell short when it came to describing human behavior. The program was the only appropriate formalism for the description of the behavior of complex, adaptive systems, for only the program permitted scientists to describe contingent, sequential events in a precise, demanding language.[39]

Only the program enabled the reduction of complex adaptive behavior to a set of hierarchically organized elementary processes. Only the program allowed one to generate unpredictable complexity from determinate simplicity.

The Computer in Simon's Science

The programming languages Simon and Newell developed—the Information Processing Languages (IPLs)—were based on a particular set of assumptions about the nature of the human mind. In particular, they assumed that information is stored in the form of symbols, that the basic structure of the human memory is associative and hierarchical (meaning that information is stored in lists whose elements may refer to other elements or other lists), and that the human information processing system is fundamentally a serial processor. Thus, Simon and Newell's assumptions about the mind structured their understanding of the computer and the program.

At the same time, Simon's experience with computers began to feed back into his understanding of the mind. At the most mundane level, his experience with computers affected the time he devoted to different research activities: programming computers is a difficult, time-consuming, and fascinating activity, one that threatened to demand all of Simon's time and energy. At a different (though equally prosaic) level, adopting the computer as his basic research instrument allowed Simon to find support for his work from a new set of patrons whose goals and expectations were different his previous patrons'. At the methodological level, Simon's advocacy of computer simulation and his claim that the program is the theory, both of which were controversial positions, led him to undertake a number of research projects intended primarily to justify those positions, research that led him to modify his understanding of human problem-solving behavior, especially as it related to creativity and scientific discovery. Finally, at a yet more basic level, Simon's focus on the computer simulation of human cognitive processes led him to privilege those aspects of human thought that could be simulated most readily—or that he saw as the most important to simulate in order to defend his claims. To use his own language, the idea that the computer can be programmed to simulate human problem-solving was his heuristic; it was an enormously fruitful heuristic, but it was fruitful because, like all heuristics, it limited the alternatives under consideration.

CHAPTER ELEVEN

The Cognitive Revolution

Although it was a machine (of sorts), the Logic Theorist was created as a theory of *human* problem-solving. It was, therefore, a theory that belonged to the realm of psychology, not engineering. It was a new kind of psychological theory, in form and in content, but it would win surprisingly many adherents in a remarkably short period of time. It was able to do so because it was created at an opportune moment. Many psychologists, especially those involved in military research on communication, simulation, and man-machine systems (such as George A. Miller, Ulric Neisser, J. C. R. Licklider, Jerome Bruner, Carl Hovland, and Bert Green), were chafing at the restraints imposed by behaviorist orthodoxy in the 1950s. To them, LT was a source of inspiration.

Such psychologists understood the analogy between programs and minds to have strong antibehaviorist implications. As I have argued elsewhere, they did so for three main reasons.[1] First, the computer metaphor was connected to an approach to language—Noam Chomsky's mathematical linguistics—that saw mind as necessary to explain human language. Second, the computer metaphor was associated with an interdisciplinary approach to psychology, an approach that stood in sharp contrast to the behaviorist effort to establish psychology's independence. Third, the adoption of the computer metaphor of mind entailed the adoption of a new research program for experimental psychology, one that sought to replace the lab rat with the human-computer system as the chief source of psychological data.

Those three factors explain why the analogy between programs and minds was understood to have "revolutionary" antibehaviorist implications. But why was this new approach to psychology so successful? Why were Simon and LT welcomed into a discipline to which they were strangers? What role did Simon and LT play in the cognitive revolution? To answer these questions, it is necessary to examine the context of postwar experimental psychology, the intellectual and institutional environment Simon entered.

Background to Revolution: Behaviorism and Operationalism

What was the behaviorist psychology against which Simon, Miller, Bruner and the other cognitivists rebelled? Behaviorism was a movement in early twentieth century psychology that emerged in response to the challenges of subjectivity, interdependence, and social change, all of which had produced deep concerns regarding modern society's ability to maintain control over an increasingly diverse populace (see chapter 2).[2]

Although John Broadus Watson did not create the behaviorist movement out of whole cloth, his 1913 article "Psychology as the Behaviorist Views It," became a touchstone for psychologists dissatisfied with both traditional "brass instrument" experimental psychology and the "mystical" new psychology of Sigmund Freud. Watson argued that for psychology to become a true science it must be put on the firm, positive ground of empirical observation. Traditional psychology was not sufficiently empirical because it attempted to study unobservable phenomena, such as consciousness. In contrast, Watson's scientific psychology would start from a very different basis: "Psychology as the behaviorist views it is a purely objective experimental branch of natural science. Its theoretical goal is the prediction and control of behavior. Introspection forms no essential part of its methods, nor is the scientific value of its data dependent upon the readiness with which they lend themselves to interpretation in terms of consciousness."[3]

Watson declared the traditional Wundtian method of introspection invalid because it dealt with unobservable entities like mind, will, purpose, and consciousness.[4] In his view, such mentalistic terms had to be banished from psychology for it to become truly scientific. Teleological language in particular was ruled unscientific, since purpose is as unobservable as the mind. Because only actions, not intentions, can be observed, Watson argued, the proper arena for scientific psychology was the observable behavior of animals.

Humans could be considered part of the animal realm, so behaviorists could, in principle, study human behavior as well as that of rats or pigeons. Nevertheless, behaviors that are uniquely human—complex spoken and written language, for example—did not suit the behaviorist program as well as did behaviors that humans and other animals share, such as "learning," understood as adaptation to environmental stimuli.[5]

Watson's programmatic statements did much to define the behaviorist movement, giving form and focus to a widespread but diffuse dissatisfaction with the "the old 'new' psychology" (in E. G. Boring's phrase) of the previous generation.[6] Watson and his allies set new goals and outlined a new research program for the discipline. Behaviorist psychology would study the conditioning processes by which animals "learn" various behaviors. Experiments would be conducted to determine what kinds and what patterns of reinforcement were most effective, to discover how complex behaviors were created out of sequences of simple stimulus-response reactions, and to reveal the mechanisms by which responses became associated with specific stimuli.

This redefinition of the field was part of a deliberate effort to establish psychology's independence as a scientific discipline. Watson and his behaviorist allies took great pains to distinguish their field from other sciences, particularly from biology, psychology's closest kin. They feared the perception that "the science of behavior must at bottom be a study of physiology."[7] In order to defend psychology's autonomy, many behaviorists followed E. C. Tolman's lead in distinguishing between "molecular" and "molar" actions. Biology, they argued, should study the molecules of nerve physiology, while psychology should focus on the larger molar units of behavior. In similar fashion, these terms were often used within psychology to distinguish between various levels of action: the "molecular" actions of muscles or sets of muscles versus the "molar" actions of the organism as a whole. Such a distinction was difficult to apply in practice, however, as molecular and molar soon revealed themselves to be quite relative terms.[8]

At the same time they defended psychology's independence from other sciences, behaviorists sought to justify its status as a science. To do so, many believed that they needed to identify the essential features common to all true sciences and remake psychology according to those principles. Watson's banishment of mentalistic terminology had been vital as a first step, but for the philosophically minded it was not enough. Watson had shown what psychology should *not* look like; neobehaviorists like Clark Hull, on the other hand, sought a systematic philosophy—one based on a model science—that could provide a positive picture of proper psychology.

In the 1930s, Hull and a number of other behaviorist psychologists found in logical positivism or its close kin, operationalism, the systematic philosophy of science they sought, and in experimental physics, the blueprint for the reconstruction of psychology.[9] Positivism, whether logical or operational, gave behaviorism a coherent, systematic philosophical foundation. The positivist principles of mechanico-physical causation and parsimony were applied vigorously, and the attempt was made to purge psychological discourse of metaphorical concepts and language.[10] Any concept that was not defined in the strictest operational terms was suspect, and Occam's razor was wielded with a will.

For example, Hull's classic text, *The Principles of Behavior* (1943), begins with a chapter on the philosophy of scientific knowledge that mirrors Percy Bridgman's *Logic of Modern Physics*.[11] From this positivist-operationalist base, Hull then derives his methodology for psychology. Applying his "hypothetico-deductive" method to a limited set of postulates, he then derives numerous theorems about behavior. The result is a comprehensive "mathematico-deductive" system of psychology.[12]

In keeping with his positivist roots, Hull suggested a "Prophylaxis against Anthropomorphism" to guard against the tendency to introduce mentalistic terms into psychological analysis. This prophylaxis was to regard "the behaving organism as a completely self-maintaining robot, constructed of materials as unlike ourselves as may be."[13] Hull took his own advice to heart, constructing various "learning machines," which he used to demonstrate that mind is not necessary for learning.[14] Thus, to Hull, the statement that the human is a machine meant that human behavior could be explained without reference to an internal, conscious agent (or to heredity). Ironically, only a few years after Hull's death, cognitivists like Herbert Simon and George Miller turned Hull's argument upside down, using the analogy to a more powerful, more purposeful machine—the programmed computer in general and Logic Theorist in particular—to argue that mind does exist and that it can be studied scientifically.

Although operationalism held wide sway, Hull's mathematico-deductive system won few converts in a psychological community weaned on a fierce distrust of theory.[15] For Edwin Boring and many other experimentalists, steady, solid experimental data gathering, not deductive system building, was the key to establishing psychology as a science. But what kind of data from what kind of experiment? Some, such as S. S. Stevens, advocated the study of psycho-physics as the foundation for a rigorous, empirical experimental psychology.[16] Others, such as Karl Lashley and his student Donald Hebb, found their data in the neurology of brain function and used surgical techniques as the foundation of their experimental methods.[17] The third avenue for American experimental psychologists was the more radical behaviorism of B. F. Skinner, who looked to the lab rat (or the pigeon) in the Skinner box for his data.[18] Skinner's approach appealed to psychologists interested in preserving both their discipline's autonomy and their own reputations as "tough-minded" scientists, but his extreme views troubled those who were uncomfortable making inferences about human behavior from experiments on rats, pigeons, and other lower animals.[19]

There was a fourth avenue, but it was a road less traveled. This was the path of Gestalt theory. Gestalt psychology, as Mitchell Ash has argued, was no woolly-minded rejection of "objective" behaviorist psychology.[20] Rather, it was driven by the same quest for objectivity that spurred Watson, Skinner, and the behaviorists. The differences between Gestalt and behaviorist psychology instead stemmed from the different social locations of psychology in Germany and America. In Germany during the first half of the twentieth century, psychology still was considered a part of philosophy, and its central questions emerged from a complex contest between experimentally oriented psychologists and their philosopher colleagues, most of whom were neo-Kantians at the time. In America, by contrast, behaviorist psychology was far closer to biology than to philosophy. As a result of their different locations, Gestaltists and behaviorists focused on different problems: Gestaltists were deeply concerned with the problem of knowledge, with sensation, thought, and action, while behaviorists primarily were concerned with the problems arising from the plasticity of animal behavior.

A number of prominent German psychologists, such as Harold Guetzkow's teacher Norman Maier, came to the United States during the 1930s and 1940s, and the translated works of Gestalt psychologists Max Wertheimer, Kurt Koffka, and Wolfgang Kohler became available in the United States after the war. As a result, the ideas of the Gestaltists were well known in American psychology by the late 1940s, though such work lay outside the mainstream of the discipline.[21]

Gestalt psychology was one alternative for experimentalists interested in straying from the behaviorist path, but it was the military's interest in communication, simulation, and man-machine systems that had the strongest influence on the reorientation of the discipline in the years after World War II. Defense research in these areas was a particularly fertile breeding ground for psychologists willing to try something new because psychologists working in these areas often were not socialized into the profession in the same way their mentors had been. Simon, for example, was never formally trained as a psychologist, and George Miller and his colleagues at Harvard's Psycho-Acoustic Lab, from whence came many early cognitive psychologists, "became professionalized as military research workers, more engineers than psychologists."[22] Such psychologists were used to working on interdisciplinary teams, and they were accustomed to using humans and machines as their experimental subjects, not rats in T-mazes.

The lure of military funding exerted a powerful influence on psychology, and even radical behaviorists like Skinner caught the communications bug.[23] Language was a complex phenomenon, however, and it resisted all attempts to annex it to the stimulus-response regime. As a result, in the early 1950s, psychologists interested in communication, simulation, and man-machine systems began to look outside behaviorist psychology for ideas, being convinced by their military research experience that "grass grows in the cracks between disciplines."[24]

The first place they looked was information theory, a powerful mathematical perspective on communications—and on choices.[25] In Shannon's novel definition of information and his sophisticated mathematical analysis of messages, proto-cognitivists like George Miller and Frederick Frick saw a chance to unify psychology through the systematic analysis of language and communication. Though they pursued Shannon's information theory with the goal of reforming behaviorism, it soon took them down a very different path. Information theory was a theory of messages, and to develop a complete theory of messages, Miller and others would find they needed a theory of the message generator—that is, a theory of mind.

Information theory spread like wildfire throughout experimental psychology in the 1950s. One reason why it was accepted so eagerly was that behaviorist psychologists were used to seeing the body as a communication system. To a behaviorist, psychology was the study of correlations between stimuli and responses. Stimuli were readily equated with sources of messages, the brain with a switchboard, and nerves with transmitters and receivers.[26] The step to information theory therefore seemed a small one.

While fully compatible theoretically with even the most radical behaviorism, information theory-based psychology outlined a different research program from standard behaviorism. As they had since the time of Watson, behaviorists in the late 1940s and early 1950s focused on the study of animal behavior. Rats in T-mazes or pigeons in Skinner boxes were the paradigmatic experiments. Behaviorists did not wholly neglect the study of human behavior, of course, but they generally focused on the "animal" aspects of humanity, rather than the "higher qualities," such as language.[27] As Watson once wrote, "I have devoted nearly twelve years of my life to experimentation on animals. It is natural that one should drift into a theoretical position which is in harmony with his experimental work."[28]

In contrast, the research program set by information theory was aimed at establishing the specifications of the human communication system. The thresholds of sensation for various types of stimuli, the amount of information that could be perceived at one exposure, the channel capacity of the human nervous system, the amount of "noise" or interference present in various situations, the rate of transmission of signals and the length of communication routes in the human body—these were the important subjects of experimental study for those interested in information theory.[29] This experimental program set by information theory eventually would lead psychology away from not only the animal behavior lab but also the behaviorist philosophy of psychology.

In his work at MIT's Human Resources Research Laboratory, the sister psychology laboratory to the Systems Research Laboratory, George Miller studied the ways the human operators of the SAGE computers related to their charges. He observed how they perceived and processed the information the machines presented to them (in the form of arrays of lights or switches or symbols), and he attempted to discover the limits of the human organism's communications system.

In his groundbreaking paper, "The Magical Number Seven," Miller synthesized the results of these studies of "absolute judgment" in a way that was to have great importance both for him and for Simon. Absolute judgment refers to the ability to distinguish among one-dimensional sensory stimuli, for instance, how someone can identify a sound via differences in pitch or volume, but not both. In the experiments on absolute judgment that Miller summarized, "the observer is considered to be a communication channel." The object of experiments on absolute judgment was to determine the "channel capacity" of the observer; that is, "the upper limit on the extent to which the observer can match his responses to the stimuli we give him."[30]

Miller reviewed the results of various such experiments and came to the conclusion that "there seems to some limitation built into us by learning or by the design of our nervous systems." This limit turned out to be surprisingly small: around three bits of information. A channel capacity of three bits of information means that we can identify roughly seven items, plus or minus two, if they differ only in one dimension (and are presented in isolation). The repeated discovery of limits on absolute judgment in the neighborhood of seven items caused Miller to remark, "I have been persecuted by an integer. For seven years this number has followed me around, has intruded upon my most private data, and has assaulted me from the pages of our most public journals ... There is, to quote a famous senator, a design behind it."[31]

Miller does not stop with the discovery of the magic number seven. He notes the coincidence that the number of items that can be held in the immediate memory was also known to be around seven. At first, one might suspect that absolute judgment and the span of immediate memory might therefore be one and the same. Miller, however, quickly shows that this Pythagorean insight is misleading. He describes the distinction as follows: "Absolute judgment is limited by the amount of information [measured in bits]. Immediate memory is limited by the number of items. In order to capture this distinction in somewhat picturesque terms, I have fallen into the custom of distinguishing between bits of information and chunks of information."[32]

Miller then reveals his discovery that "the span of immediate memory seems to be almost independent of the number of bits per chunk."[33] (In another work, he described the immediate memory as being like a purse that can only hold seven coins but does not care whether they are seven pennies or seven silver dollars.)[34] This curious feature of immediate memory is crucial, for it allows us to "increase the number of bits of information that [our memory] contains simply by building larger and larger chunks, each chunk containing more information than before." To give this successive "chunking" of information a name, Miller borrowed from communications theory the term *recoding*.[35]

Though the concept of recoding was taken from information theory, Miller's use of the term was strongly influenced by his early experience with computers. For example, one of the key experiments on which he based his idea of recoding was the "chunking" of sequences of binary digits into larger units, which was exactly what the computers (and their operators) he was working with did every day. Miller has stated that the idea of recoding came from watching human operators in the control rooms observing complex arrays of lights and switches and "recoding" these arrays in their minds into easily recognizable patterns of light and switch arrangements.[36]

Though recoding marked a convergence of ideas, it also heralded a transition, for recoding dealt with how information is processed, not how it is transmitted. For the study of information transmission, the telephone system had seemed a good model. To understand information processing, however, would require a new model. Miller—and psychology—would find two: Noam Chomsky's linguistics and Simon and Newell's Logic Theorist.[37]

Miller was a close friend of Chomsky's at MIT and Harvard, and he saw in his colleague's ideas the basis for a new psychology. The two began to work closely in 1955, and they coauthored a number of influential articles over the next several years. Miller's talk on Chomsky's linguistics became famous on the university lecture circuit, and, apocryphal as it may sound, Miller saved the only draft of Chomsky's manuscript of *The Logical Structure of Linguistic Theory* from a fire at Harvard's Memorial Hall. The two were so closely identified with that B. F. Skinner apparently thought Miller was behind Chomsky's famously savage review of Skinner's book, *Verbal Behavior*.[38]

Chomsky argued that the statistical analysis of communication was insufficient to account for the complexity and regularity of natural language.[39] He held that when people hear or read something, they remember only a "kernel" sentence. A kernel sentence is the simplest declarative form in which the idea of the sentence can be expressed. This kernel sentence is then transformed according to various rules to suit the situation at hand. Chomsky drew from this analysis the conclusion that grammar, not wording, was the key to meaning. Syntax governed semantics. In addition, he concluded that syntax was rule-governed and analyzable through symbolic logic.

Chomsky then tried to discover what the rules for sentence generation must be. In order to do so, he defined a language as a certain set of possible sentences. The set of rules that specify that particular set of sentences—all and only the syntactically correct sentences in a language—Chomsky called a grammar. Further, Chomsky argued that each language's grammar was derived from a universal "transformational grammar" (the existence of which is proven, in his view, by our ability to translate between languages).[40]

Chomsky's definition of a grammar made it almost identical to a computer program, for both involved the manipulation of symbols according to the sequential application of a set of logical rules. Recoding clearly was a similar process. The chunks of information produced by recoding were symbolic representations formed, organized, and related to one another according to a set of rules akin to a grammar—or a program. If each language's grammar was like a program, then the transformational grammar must be analogous to the operating system of the human mind. This link between programs, grammar, and mind was made explicit by Chomsky's later proof that a transformational grammar is formally equivalent to a Turing machine and thus, to many cognitive psychologists, to a mind.[41]

The particular program that Miller and other experimental psychologists had in mind when they talked of mind was the Logic Theorist. In a telling convergence of this branch of experimental psychology, Chomskyian linguistics, and computer science, Simon and Newell presented their first paper on LT, Chomsky delivered a paper outlining his new approach to language generation, and Miller gave his paper, "The Magical Number Seven" all at the same conference.[42] To Simon and the rest of the audience, proto-cognitivists and proto-AI researchers all, it was clear that a revolution was in the making.

Spreading the Gospel of Information Processing

Despite many promising signs, the community of true believers in the new cognitivist approach to psychology still was quite small in 1956–57. Very few people anywhere had had much exposure to computers and computing at that time, and psychologists were no exception to this rule. There were only forty computers installed and operating on university campuses nationwide in 1956.[43] This was a severe limitation, for the research program Simon and his fellow cognitivists embraced depended on access to computers. To spread the information processing model of mind would require the expansion of access to and experience with computers for psychologists.

In addition, the research program implicit in the information processing model of mind was quite alien to that of most psychologists, since the group interested in the psychology of communications and man-machine interaction that formed the core of cognitivism was a definite minority in the field. It was a peculiarly well-funded and increasingly prominent minority to be sure, but it was a minority nonetheless. Neither Simon nor Newell, for example, was recognized as an experimental psychologist at this time, and their language of computers and programming branded them as foreigners to the discipline. Spreading the gospel of information processing, therefore, would depend on Simon's ability to define his work as psychology, to psychologists.

The first two major successes in Simon's struggle to convert the psychologists came in 1958. First, he and Allen Newell organized a summer seminar at RAND that introduced thirty leading behavioral scientists (primarily psychologists) to advanced computers and to nonnumerical programming techniques. Second, Simon, Newell, and Shaw published an article in the *Psychological Review,* the journal of record for mainstream experimental psychology, in which they set forth their model, their program, and their programme for psychology. Titled "Elements of a Theory of Human Problem-solving," this article translated their work on the Logic Theorist into a language familiar to psychologists, even creating an ancestry for their work in psychology.[44]

As noted in chapter 7, the 1958 summer seminar on simulation techniques was a classic example of Simon's ability to bring together diverse resources in the service of a synthetic vision. Simon and Newell organized it under the SSRC's aegis with funding provided by the Ford Foundation and the RAND Corporation, and it quickly attained an almost legendary status in the fields of cognitive psychology and AI.

The summer seminar was an intensive training program, eight or more hours a day for eight weeks, and the lessons Simon and Newell taught there took hold in many minds. For example, this seminar had a marked impact on George Miller, who stayed out in California at the Center for Advanced Study of Behavioral Sciences for the 1958–59 academic year digesting what he had learned the previous summer. The product of his meditations on computers, information processing, and cognition was the book *Plans and the Structure of Behavior,* written in 1959 in collaboration with Eugene Galanter and Karl Pribram (two fellow fellows at the center that year).[45] *Plans,* though no slavish imitation of Simon and Newell's work, was heavily indebted to it. Even more, it was immensely popular and soon became the standard introduction for psychologists to the information-processing model of mind. The summer seminar had borne even richer fruit than hoped, and a vanguard of eager cognitivists, all carrying their copies of *Plans,* soon went forth to spread the good news.

The second major event of 1958 was the publication of Simon, Newell, and Shaw's article, "Elements of a Theory of Problem-solving." This article was Simon's first epistle to the psychologists, a crucial—and highly successful— effort to introduce new terms and new concepts into psychological discourse.

Simon and Newell begin by stating that their goal is to create a theory that explains how human problem-solving takes place. This means discovering "what processes are used and what mechanisms perform these processes." In their view, "an explanation of an observed behavior of the organism is provided by a program of primitive information processes that generates this behavior." Hence, the goal of their research is to discover programs that generate behavior that matches the behavior of human problem-solvers in similar situations. They are "well aware of the standard argument that 'similarity of function does not imply similarity of process' " but hold that "the specification of a set of mechanisms sufficient to produce observed behavior is strong confirmatory evidence for the theory embodying these mechanisms," especially since all competing theories have been unable to establish their sufficiency.[46]

Simon, Newell, and Shaw follow this introduction with some comments on the relation of their theory of human problem-solving to digital computers. Observing that "the ability to specify programs precisely, and to infer accurately the behavior they will produce, derives from the use of high-speed digital computers," they take pains to "emphasize that we are not using the computer as a crude analogy to human behavior." They "are not comparing computer structures with brains, nor electrical relays with synapses"; to them, "a program is no more, and no less, an analogy to the behavior of an organism than is a differential equation to the behavior of the electrical circuit it describes."[47] Thus, they argue, while their work was inspired by computers, its intellectual roots derive from an attempt to synthesize behaviorist and Gestaltist psychology rather than an effort to conquer psychology for engineers.

Their argument, therefore, is methodological, not metaphorical. "The appropriate way to describe a piece of problem-solving behavior is in terms of a program ... This assertion has nothing to do—directly—with computers." In fact, "Such programs could be written (now that we know how to do it) if computers had never existed." Computers "come into the picture only because they can, by appropriate programming, be induced to execute the same sequences of information processes that humans execute when they are solving problems."[48]

After this discussion, Simon, Newell, and Shaw turn to an example of their theory of human problem-solving—the Logic Theorist—testing its ability to solve problems in symbolic logic under different circumstances (principally, with theorems presented to it in different order). In keeping with their view that a program describes an individual's behavior, each different configuration of the program is referred to as being a new experimental subject. (New "subjects" are obtained by erasing some parts of the memory of the computer and by entering different data in different sequences.) The sum of their experimental data, they argue, demonstrates that "the program of LT is qualitatively like that of a human faced with the same task."[49]

Using the data provided by LT regarding the methods it employed in attempting to solve problems under different circumstances, Simon, Newell, and Shaw then ask whether LT "shows any resemblance to the human problem-solving process as it has been described in psychological literature." Not surprisingly, they find many such resemblances, arguing that their theory of human problem-solving has close kin in psychological theories about "set," "insight," "concept formation," and the "structure of the problem-subproblem hierarchy."[50]

The connections they drew to these established issues in psychology sometimes were stretched and typically depended on a significant redefinition of the issue (e.g., insight is displayed by LT if one redefines insight as being selective search for problem solutions). Similarly, though they admit LT "shows no concept formation," they argue that it "uses concepts" because it recognizes when two equations "look alike."[51] Nevertheless, in this article Simon, Newell, and Shaw were able to make a persuasive case for their work as psychology, and for themselves as a new breed of psychologists.

In part, the success of these two evangelical efforts depended on the intellectual appeal of the information-processing model of mind to a community of psychologists eager to blend the rigor of behaviorism with the appreciation of the complexity of human behavior that characterized Gestalt theory. As Simon was well aware, the growing dissatisfaction with behaviorist theory coexisted uneasily alongside a persistent belief that the behaviorist methodology still was the only acceptable one. He was careful, therefore, to present his theory as a way of addressing issues of longstanding interest to psychologists, especially Gestalt psychologists, while still satisfying behaviorist methodological criteria (e.g., strict operationalism). Thus, he argued that, far from being foreign to psychology, the concept of the program was the key to uniting its two strongest traditions.

Patrons of a Second Revolution

The success of these efforts did not depend on intellectual appeal alone, however. Powerful patrons were interested in supporting work on information processing in man and in machine, and Simon's programme for psychology would allow cognitivists to win support from both. RAND, the Ford Foundation, the Office of Naval Research, and the Air Force supported work on behavioral models throughout the 1950s and 1960s, concentrating their social science resources more and more on models of individual psychology, a departure from the statistical analyses of groups that had dominated the psychological literature since the 1940s.[52] With the exception of the Ford Foundation, these agencies also were extremely interested in advancing computer technology as far and as fast as possible. So too was the Department of Defense's new umbrella agency for defense research, the Advanced Research Projects Agency (ARPA), which became a key player in the development of both the behavioral sciences and computer science.

ARPA was created in 1958 in the post-Sputnik drive to prepare America for scientific and technological competition with the Soviet Union. It funded work in many fields, but its directors had a special fondness for computing and communications technologies. As Paul Edwards and Arthur Norberg both have shown, by the late 1950s the military had become fascinated with the potential improved computing and communications technologies held for developing effective centralized command and control.[53] This fascination was in part a response to the threats posed by nuclear warfare, for in the new world of missile-borne nuclear weapons there "would be no time." Decision-making needed to take place rapidly, and orders had to be executed at a moment's notice.

These new needs suggested to military leaders that decision-making need-ed to be systematized and centralized, while the physical resources for execut-ing decisions needed to be decentralized so as to be less vulnerable to attack. Computing and electronic communications technologies seemed heaven-sent, for they enabled both physical decentralization by eliminating distance as an obstacle to rapid communication and decisional centralization by enabling information gathered at a variety of sites to be collected and processed centrally.

In order to speed development of computing technologies, ARPA estab-lished IPTO, the Information Processing Techniques Office, in 1962. IPTO's first head was J. C. R. Licklider, a psychologist, computer scientist, and vet-eran of both Harvard's Psychoacoustics Laboratory and MIT's Human Re-sources Research Laboratory.[54] As his background indicates, Licklider was intimately connected to the budding field of cognitive psychology, as well as to computer science proper, and he was interested in advancing the cause of "Man-Computer Symbiosis."[55]

While Licklider's tenure in office was relatively short (1962–64 and 1974–75), his decisions—and the relationships he developed between certain research groups and IPTO—rapidly became institutionalized. As a result, IPTO has taken an active interest in sponsoring work in the computer simulation of human behavior since its inception. For example, in addition to funding a marked expansion of campus computing facilities, particularly at eight "cen-ters of excellence" in computing (including Carnegie Tech), Licklider and his successors were interested in developing high-level computing languages.[56] Such high-level languages, particularly ones designed for nonnumerical infor-mation processing, were vital to advancing the cause of human-computer interaction. Newell, Simon, and Shaw's work on their Information Processing Languages (IPLs) was a major beneficiary of IPTO's interest in this area.

A second source of support for cognitive psychology was the National Science Foundation (NSF). As World War II drew to a close and planning for the postwar era began, the younger generation of social scientists—like their counterparts in the natural sciences—sought ways to continue their associa-tion with the government and to expand governmental support for social sci-ence research.[57] The prospect of a national foundation that would support science, including social science, exerted a powerful pull on the imaginations of this younger generation, as Talcott Parsons's 1945 proposal for a new Department and Laboratory of Social Relations at Harvard reveals:

There seems ... to be a strong probability of the passage by Congress within a few months, of a bill which will in some way provide federal support for research in the social sciences as well as the natural sciences and medicine.

As at present organized, however, the social sciences at Harvard are exceedingly poorly prepared to live up to the challenge which seems certain to come ... Almost any major field of research which is likely to be sponsored by a federal agency is in the nature of the case bound to cut across the at present dominant divisions into departments and faculties. A situation in which the University was, with its present organization, asked to undertake a program of research would be likely immediately to unloose a process of competition between the various units of university organization within whose competence it might conceivably fall, with the very possible result either that none of them might get it, or that the one which did would not be in a position to handle it competently.

In our opinion, this situation calls for an organization for social science research which is broad enough in scope to include all the major agencies of the University which might have an important contribution to make in that area.[58]

This new organization for social science research, a "nucleus for receiving funds from the Government and private sources," eventually became the Laboratory of Social Relations, an interdisciplinary research center that played an important role in setting the agenda for social science during the 1950s and 1960s.[59]

The elders of the social sciences greeted the prospect of an NSF with mixed feelings. They knew that government patronage would allow for a marked increase in social science research and that it might provide a new path to greater influence on policy, but they were wary of the government's influence on such research, knowing full well that he who pays the piper calls the tune. They were particularly fearful that the lower status and greater political salience of social research would tempt politicians to meddle with their work.

The hesitations of the elders in the social sciences were mirrored by the doubts of the natural scientists who led the campaign for the NSF. As a result, when the long battle to create the NSF finally ended in 1950, the social sciences found themselves on the outside looking in. This defeat, however, was not nearly so momentous as it might have been. Because the groups supporting the NSF did not unite behind any single proposal until 1950, the opportunity to use the foundation to chart a new course for science-government relations passed. The patterns and procedures established during the war were continued and expanded, and the NSF, once established, was relegated to an insignificant role until well after Sputnik.[60] If the NSF had been what Senator Kilgore had hoped it would be, then it might have helped restore the balance between natural and social science, as well as between military and civilian science.[61] The omission of the social sciences from the NSF that actually existed, however, was primarily of symbolic importance. The Ford Foundation's disbursements for 1952–60, for example, exceeded those of the NSF (in total) for that period, and its aggressive program in the behavioral sciences meant that it played the role for social science that the NSF was intended to play for science generally.[62]

In addition, though the social sciences were not explicitly included in the NSF's mandate, neither were they excluded from its purview. It allocated some funds for social science research almost from the beginning, and in 1954, sociologist Harry Alpert was hired to supervise a grants program for the social sciences. Although it took him many years to build support for the social sciences at the NSF, it also took years for the NSF to achieve a significant role in *any* area of science. Only after Sputnik did the NSF acquire the funds to accompany its prestige. The social sciences thus rose to acceptability, though not to equality, within the NSF at precisely the time that it rose to importance.

Mindful of the continuing need to justify social science as truly scientific, Alpert and his successor, Henry Riecken, chose to support research projects that promised mathematical or methodological advance. While neither man had a brief for or against cognitive psychology per se, the fields of psychology that were most likely to use powerful mathematics and sophisticated instruments were those that fed cognitivism, especially psychophysics, psycholinguistics, and other forms of psychology influenced by communications theory.[63]

The NSF also was interested in promoting access to computers for scientists generally. In the 1960s, this interest was extended to improving the use of computers in the behavioral sciences, with the primary beneficiaries of such support being psychological researchers interested in developing computer simulations of human behavior. Simon never was entirely happy with the NSF programs in the social sciences or in computing; in 1976 he chaired a committee appointed to review its social science program, and the "Simon report" that the committee produced was quite critical of its efforts.[64] Despite its problems, however, the NSF's support for both social science and computer science helped cognitive psychologists win both money and prestige.

The most important single patron of the cognitive revolution was, like ARPA and the NSF, another new node in the institutional web supporting behavioral science: the National Institute of Mental Health (NIMH). Congress had created NIMH in 1946 to be a "National Psychiatric Institute," and its proponents intended it to be a center of support for training and research (in that order) in psychiatry and clinical psychology, as well as a sponsor of improved clinical psychiatric services. NIMH's first director, Robert Felix, and all of the initial members of the institute's National Advisory Mental Health Council (NAMHC) were members of the Group to Advance Psychiatry (GAP), an association of psychiatrists who sought to promote psychodynamic approaches to psychiatric treatment.[65] Thus, NIMH had a strong initial orientation toward improving clinical psychiatric services, which was unusual in the basic research-oriented NIH.

Felix did have a strong interest in research, however, and he had an expansive vision of the kind of research that could improve clinical practice. A former colleague of Adolf Meyer at the Johns Hopkins University, Felix had absorbed Meyer's belief that many mental health problems should be treated as public health problems.[66] As a result, Felix supported social research related to mental illness, especially studies of what might best be called the epidemiology of mental illness. He convened an external panel of social science consultants to advise him in this area, and its members—sociologists H. Warren Dunham and Robin Williams, anthropologist Margaret Mead, social psychologist Ronald Lippitt, and sociologist/social psychologist Lawrence Frank—encouraged Felix to broaden his support of social science research. All of these social science consultants shared a broadly functionalist approach, and all were interested in the culture-and-personality approach to understanding human behavior. They were thus charter members of the postwar sciences of control.

Despite this encouragement, in NIMH's first nine years its training and research grants were heavily oriented towards medical-psychiatric research: a 1958 review of the NIMH training program that categorizes the training grants by discipline of recipient does not even have a category for psychology, let alone the other social sciences.[67] Felix did create an intramural Laboratory for Psychology in 1954, but up to 1958, both NIMH's budget and the percentage of that budget devoted to social science remained small.

In 1958, however, Felix decided to expand the training program to include support for graduate training in psychology, including experimental psychology.[68] This tentative broadening of NIMH's program met with great enthusiasm among academic psychologists, and it proved to be the proverbial camel's nose inside the tent. The NAMHC gave this new program a ringing endorsement in 1959, even going so far as to encourage Felix to make NIMH the leading source of support for "basic behavioral science."[69] Felix heeded this advice, and by 1964, 55 percent of NIMH principal investigators were psychologists while only 12 percent were psychiatrists.[70]

At the same time, NIMH's budget began to grow with great speed, thanks to the post-Sputnik wave of support for scientific research. Over its first decade, NIMH's budget had grown at a respectable rate, rising to $50 million in 1959. Between 1959 and 1964, however, NIMH's budget more than tripled, reaching $189 million. In part, this rapid growth was simply a product of the growth of the NIH more generally, with Felix and NIMH riding on James Shannon's coattails. NIMH's growth also was due to its embrace of the research ethos of the NIH, however, an embrace manifested through support of what it termed "basic behavioral science."

The basic behavioral science Felix and the NIMH wished to support was exactly the kind of science Herbert Simon and other cognitive psychologists wanted to conduct: behavioral, not behaviorist, psychology. Felix and the NAMHC members were psychodynamic psychiatrists, which meant that they believed in the mind, in consciousness, and in the ability of trained observers to gain insight into a person's mental processes by analyzing that person's verbal reports. Behaviorist strictures against all of the above seemed counterproductive to Felix and his staff, and the traditional behaviorist research program centering on animal psychology was far less appealing to them than was a focus on how humans think and learn.

Hence, Simon and other early cognitivists, such as George Miller, Bert Green, Carl Hovland, Jerome Bruner, and Lee Gregg, found in NIMH a generous patron with a strong interest in research on human thought processes. In addition, NIMH actively supported the use of computers in psychology, creating a special funding program in the early 1960s specifically to expand access to computers for behavioral scientists. NIMH's generous patronage—which involved sums larger than the NSF and the private foundations put together—thus played a major role in promoting the cognitivist approach to experimental psychology. NIMH today remains the largest single agency supporting behavioral science research. The other institutes of the NIH also have embraced the behavioral sciences since the 1960s, so much so that the NIH now sponsors more behavioral science research than the rest of the federal government combined.[71]

Not only did the old (RAND, Air Force, ONR) and the new (ARPA, NSF, NIMH) patrons of behavioral science support cognitivist approaches to psychology, but they also were key patrons of computing. Thus, even when Simon shifted his focus from administrative decision-making to human problem-solving, it remained at the center of a convergence of multiple interests. All of these patrons were interested in spurring the advance of computing technologies and in making better, faster computers available on a wider scale to academic researchers; they all wanted to encourage researchers to take up projects that would advance both their specific science and that of computing; and they all were interested in sponsoring efforts to use computers for the simulation of human behavior. This last point is of particular importance, for it is often assumed that these patrons thought of computers exclusively in terms of their application to natural science. To these patrons, however, the whole arena of the behavior of complex systems was ripe for the application of computers, and the importance of computers for improving decision-making was a prominent concern.[72]

Because of these factors, resources for the use of computers in psychology expanded rapidly from 1958 on. This increased availability of dollars and equipment in one sector of psychology occurred at a time when the field as a whole was growing with great rapidity.[73] Just as in the concurrent "behavioral revolution" in political science, a vastly larger new generation of psychologists was being trained at precisely the time when Simon's was the "hottest scent in the woods."[74]

Revolution or Coup d'État?

It has not been quite so clear to historians, however, that the cognitive revolution was a true revolution in science. Thomas Leahey, for example, has called the cognitive revolution a "mythical" revolution: in his view, it was shift within the broader behaviorist tradition, not a break from it, and, what is more, there never was a unified behaviorist paradigm to rebel against. Similarly, John Mills has argued that the essential feature of American behaviorism was its orientation toward developing technologies of control, a trait he believes characterizes cognitivist psychology as well.[75]

Such criticisms of the idea of a cognitive revolution do have a point: cognitive psychologists such as Simon and Miller were thoroughgoing operationalists; they believed that psychology could accept only behavioral data as evidence; and they continued the tradition, not limited to the behaviorists but certainly exemplified by them, of seeing the human as a biological machine. If using a mechanical analogy a behaviorist makes, then the cognitivists clearly were behaviorists, for they replaced the telegraph and the switchboard with the computer as the mechanical half of the human-machine analogy. Thus, although Simon and Miller reintroduced "mind" to experimental psychology, it was quite a different mind than psychology had known in the nineteenth century—and quite different from the popular understanding of the term.

The cognitivists, however, were convinced that they were leading a revolution against behaviorism and neobehaviorism. Were they deluded? Was this mere rhetoric? The evidence says no. Cognitive psychology was a substantially different enterprise than was behaviorist or neobehaviorist psychology: the shift from the switchboard to the computer as the basic mechanical analogy made an enormous difference in how psychologists understood both the mind and their science, just as the shift from the balance to the engine as the chief metaphor underlying nineteenth-century political economy made a huge difference in how political economists understood both the economy and their science.[76]

This new heuristic model of the mind as a symbol-processing, problem-solving machine entailed a focus on the mechanisms by which humans turn sense perceptions into symbolic representations and then manipulate those symbols, a focus that led to a markedly different set of research questions from the behaviorist focus on the various forms and schedules of effective conditioning. Going along with this new set of questions was a new set of experimental practices: cognitivists typically studied human perception and problem-solving through computer simulations, often using "talking aloud" protocols with human subjects to test their simulations' empirical validity, while behaviorists typically learned about humans from animal experiments.[77]

In addition, the cognitivists saw themselves as a new research community, constructing for themselves a new understanding of psychological tradition and pointing to a different set of exemplars as their guides and inspirations: LT and "The Magic Number Seven" rather than Skinner's *The Behavior of Organisms* or Hull's *Principles of Behavior.* This perception of being part of a new community is borne out by the institutional history of psychology during the period, for the cognitivists not only "took over" the *Psychological Review,* they also established a host of new journals devoted to cognitivist psychology, established new professional societies, and developed links to new patrons. In doing so, Herbert Simon organized these new ideas, new practices, and new patrons into a research machine of astounding productivity.

CHAPTER TWELVE

Homo Adaptivus, the Finite Problem Solver

Simon and Newell had caught wind of their "hottest scent" as early as 1955, but their pursuit of it did not become their exclusive focus until roughly 1958, when they were able to secure long-term funding for research on "complex information processing" itself, rather than for such work as an adjunct to other research. Simon's interests remained diverse after this time, so no interest can truly be said to be exclusive for him. Nevertheless, computer simulation of human behavior became the organizing principle behind almost all his work in the years after LT, and even his excursions into other fields usually were undertaken with an eye toward their contribution to this central project.

Simon and Newell capitalized upon their successes with LT quickly, winning another large grant ($750,000) from the Ford Foundation for the GSIA in 1957. Of this sum, $500,000 was for a rotating distinguished research professorship and $250,000 was for an additional five years of work on "organization theory and business decision-making." This heading no longer described their work with perfect accuracy (though it does reveal the continuing connections between Simon's computer science and his organization theory), but it was close enough to win the approval of the Ford Foundation's new program officer for behavioral science, Richard Sheldon. Sheldon, like many foundation officials, had become increasingly interested in the possible application of computer simulation techniques to behavioral science, seeing in them the potential to deal with systems of great complexity.[1]

RAND also continued to provide significant resources for Simon and Newell's work, particularly access to state-of-the-art computing facilities and the assistance of numerous staff members. For instance, in addition to paying Newell's salary, five members of RAND's professional staff worked full time with Simon and Newell in 1960. Within another two years, that number had risen to ten.[2] In addition, RAND sponsored Simon's summer-long visits to Santa Monica each year and paid his salary during his sabbatical there during the academic year 1960–61.

The new research program Simon envisioned, however, would require additional resources and even greater flexibility and security than these arrangements allowed. By 1960, he already felt his interests diverging from Ford's, leading him to restructure the grant to pay for a smaller portion of his group's work over a longer period. He would need a new constellation of patrons, a new network of support, in order to pursue his new interests.

New Patrons, New Programs

The first new star in this constellation was the Carnegie Corporation of New York, whom Simon and Newell petitioned successfully for a grant of $175,000 (over five years) in early 1960.[3] The Carnegie Corporation had maintained an avuncular, though not parental, relationship with Carnegie Tech since the 1940s, and under its directors Charles Dollard (1948–55) and John W. Gardner (1955–67) it had expanded its mission to include the support of the social sciences, particularly as they related to education. Gardner, in particular, was interested in supporting research into the psychology of learning and cognition, so much so that Jerome Bruner referred to him as the "Medici" of cognitive psychology.[4] Simon was aware of Gardner's support for the new Harvard Center for Cognitive Studies, founded in 1960 by Bruner and Simon's friend George Miller, so he turned to Gardner and the Carnegie Corporation in order to get his new project on "complex information processing" (CIP) up and running.

In their proposal to the Carnegie Corporation, Simon and Newell describe their new direction for psychology as part of a broader tradition in science of advance through instrumentation and technique. For example, they begin by arguing that "rapid periods of progress in the development of a science have usually been associated with new developments in analytical technique. One could, for example, write the history of classical and modern physics in terms of such instruments of observations as the telescope, the clock, the cloud chamber, and the cyclotron." "For the first time in its history as a science, psychology may now be in possession of techniques that are commensurate in power with the complexity of the phenomena we seek to understand and explain." Notable among the new instruments, of course, was the modern computer. Hence, their request for "expanded support of a research program aimed at the use of digital computers as analytic tools for the study of human cognition" was simply the pursuit of a traditional psychological subject with a newer, better instrument.[5]

Simon and Newell state that the computer is used in a twofold capacity in their work. First, "the theories of human cognition themselves are written in the form of computer programs (quite analogously to the way theories in the physical sciences are written as systems of differential equations)," and second, "the computer, with the theory as program, simulates, and hence predicts in detail the behavior of the human subject." Here they emphasize that "the computer is not used in this work as an instrument for numerical analysis, but as a general symbol-manipulating device that is equally at home with nonnumerical as with numerical symbols." Thus, "the psychological theories developed in this research are not 'mathematical' in any traditional sense; hence they are not limited to phenomena that can be quantified or mathematized."[6]

The "basic procedure" Simon and Newell employ in this work has four components: "(1) to write computer programs capturing, as far as possible, the elementary symbolic processes employed by humans in their thinking; (2) to simulate the behavior of systems so programmed on the computer; (3) to record the behavior of human subjects performing identical cognitive tasks; and (4) to compare the trace of the computer simulation with the tape recording of the human subjects."[7]

This work, they note, is team research, occupying the time of five faculty members at CIT, five at RAND, and a "similar number" of graduate students. In keeping with Simon's organizational style, "there is no single monolithic organization, but a cluster of small, interlocking research teams (typically one, two, or three persons) working on different sub-projects within the general area, and communicating through seminars."[8] Order and focus within this "cluster of small, interlocking" teams would be maintained through Simon and Newell's senior status, intellectual power, and personal influence.

The sums Simon and Newell requested in this grant, it is worth noting, were exclusively for faculty salaries, graduate student fellowships, and secretarial services.[9] The substantial sums required to pay for equipment, especially for computing facilities, were not included. These sums could be omitted from consideration because the CIT group was able to assume that "large amounts of computer time will continue to be made available free through Carnegie Tech's own computation center and through the computers at RAND."[10] Should they be "obliged to begin paying for computer time, all these budget estimates would have to be revised upward drastically," an indication of the vital role that expanding access to computers would play in the spread of Simon's approach to mind and machine throughout psychology.[11]

The Carnegie grant thus provided essential seed money for Simon's growing research program. Though the sums involved were smaller than those received under the Ford grant, that they were given specifically for work on computer simulation of information-processing in humans validated Simon's conceptual reorientation. The big money would come later, but this grant enabled others to share Simon's unblinking confidence in the course he had chosen.

In particular, Simon's success in winning the grant helped get other faculty at CIT interested in interdisciplinary ventures like his new interdepartmental program in Systems and Communications Sciences.[12] Simon had begun to think of creating a new interdepartmental program covering the whole area of the behavioral analysis of complex systems as early as 1959. This program, as he envisioned it, would teach graduate students from a number of disciplines how to analyze complex systems through computer simulation, thus bringing a unified method and common instrumental practice to a wide range of fields. It would involve staff from many departments, including mathematics, physics, electrical engineering, economics, industrial management, and psychology, and it would be funded, he hoped, by foundations interested in pure science.

Of all these varied players, the hardest group to get in the game was the psychologists. Although there were several psychologists at Carnegie Tech who were interested in cognition, broadly understood, the chair of the Psychology Department, Haller Gilmer, opposed Simon's repeated requests that he make cognition the department's central focus.[13] After Simon returned from his sabbatical year at RAND in 1960–61, he was not willing to wait any longer. He had come to see the psychological aspects of decision-making as fundamental to understanding human behavior and so viewed the participation of psychologists in the new program as essential.

When Simon won the Carnegie grant in the summer of 1961, he shifted into high gear. Gilmer did not, and that fall Simon issued an ultimatum to Gilmer and to CIT president Jake Warner: "I can fruitfully carry on my work at Carnegie only if there is on campus a strong graduate psychology program ... While we have made some progress in this direction ... we have made it only because GSIA was willing to supply the financial resources ... To reach the goal will require vigorous leadership in the Psychology Department from a chairman who is thoroughly sold on the objective. Because of what I perceive as a drift in the department over the past two or three years ... I no longer have confidence that [Gilmer] will provide that leadership."[14]

The message was clear; it was either Gilmer or Simon. Gilmer resigned the chairmanship in short order, and Simon's handpicked choice for chair, Bert Green, took over in 1962.[15] The program in Systems and Communication Sciences was launched immediately, and the psychologists were now quite firmly on board. By 1965 this program had evolved into the Department of Computer Science, one of the first—and best—in the country.[16]

Simon and Newell were not content with sums involved in the Carnegie grant, however. It was a good beginning, but it did not allow for the dramatic growth they had in mind. By 1962, Newell would write that "There is an almost unanimous felt need to get Tech's effort in this area funded in some sort of stable long term way at an appropriate level and with appropriate freedoms."[17] Like almost all researchers in fields related to computing, they found this support and these freedoms in the arms of the federal government.

The first major source of governmental support for their work came in 1962 in the form of a $400,000 grant, renewable annually, from the Advanced Research Projects Agency to Newell in support of his work on programming languages.[18] CIT also won ARPA's designation as a center of excellence in computing—in large part because of Simon and Newell's work—and a steady flow of millions of dollars came to Carnegie Tech over the course of the 1960s and 1970s, enabling a dramatic expansion of its computing facilities. Though these grants did not support Simon's work directly, they meant that a large, first-rate group of computer scientists and top-level computing facilities were available at Carnegie Tech by the mid-1960s, providing an invaluable infrastructure for Simon's work and removing any incentive for him to leave for more prestigious institutions. In the field of computing, particularly the sub-fields of programming and simulation, there were no greener pastures than those of Schenley Park.

More directly related to Simon's own work was another governmental grant, also secured in 1962. As a part of their continuing effort to secure long-term support for their work in complex information processing, Newell "had an informal talk with Dr. Bruce Waxman of NIH, and Dr. Lee Lusted [chair of the NIH Advisory Committee on Electronic Computers]." This committee was "taking a very active role in supporting (and even initiating) work in computers as they relate to the life sciences." Their response was "Why not ask us for the money?"[19] In November of that year, Simon did, asking for (and receiving) just over $1.2 million to be spent over five years. With such resources at their fingertips, Simon and his colleagues could pursue their work confident that the only "limits on our rate of progress" would by imposed "by our own capabilities to generate fruitful ideas," not by dollars or "hardware."[20]

The basic structure of the research program Simon and Newell present in their proposal to the National Institute of Mental Health is much the same as that described in their proposal to the Carnegie Corporation. The major dif-ferences are ones of scale—Simon and Newell's ambitions had grown—and of rhetorical strategy. Because NIMH was used to funding laboratory research and clinical trials, and since it was interested in the promise of utility, Simon and Newell are careful to emphasize that their research program seeks to develop models of "complex psychological systems"; that it "maintains close interaction between experimental and theoretical work," including the use of "verbal protocols of laboratory subjects"; and that it "looks to parallel research on artificial intelligence" primarily as a source of "hints" as to how the human psychological system operates. In addition, in this application Simon argues that there may be practical, clinical benefits to their research, noting that "information processing models tend to be models of an individual" and that therefore "a common bond of concern with the individual ties clinical work and simulation together."[21]

While the latter claim was rather a stretch, the proposal's emphasis on experimental, laboratory work was not pure grant-swinging rhetoric. As Simon and Newell repeatedly had found in their work, computer simulation techniques had strong implications for experimental design. For example, Simon and Newell sought to record the moment-to-moment responses of individual subjects. Gross statistical measures derived by averaging data for groups of subjects no longer sufficed.[22] Such a "microscopic study of contiguous segments of behavior" posed "novel methodological problems," for it "departed widely from the more common patterns of research in psychology." Among other things, it required the development of "measurement techniques that give an immensely detailed moment-by-moment account of what the subject is doing." Computers could be made to print a record of each step in their calculations, but how could one do the same for humans? Simon and Newell's response was to develop new techniques for gaining insight into the operations of the human machine, such as "thinking aloud protocols"— extensive records of the descriptions of the subject of his thought processes— and new devices, such as a $10,000 camera designed to track eye-movements.[23]

To say that these techniques departed from standard practice in psychology is an understatement. While the use of instruments to measure human physiological responses and to correlate them with psychological mechanisms had a venerable history in psychology, such "brass-instrument" psychology had fallen into disfavor outside the specialty of psychophysics. Similarly, the use of a subject's verbal reports of his own internal mental processes sounded very like the techniques of "introspection" that Watson and his fellow behaviorists had banished from experimental psychology, as Simon and Newell recognized: "Evidence from a subject who thinks aloud is sometimes compared with evidence obtained by asking the subject to theorize introspectively about his own thought processes. This is misleading. Thinking aloud is just as truly behavior as is circling the correct answer on a paper-and-pencil test. What we infer from it about other *processes* going on inside the subject (or the machine) is, of course, another question."[24]

Here the association with the computer proved vital in legitimating Simon's approach to psychology, for it allowed him to depict these verbal "protocols" as analogous to a computer's report on its own operations.[25] Neither was a perfect representation of the internal workings of the machine, but both were data and both were useful provided that the researchers knew how to interpret them. Also, as noted in chapter 11, the resurgence of interest in the specifications of the human communication system that had accompanied military research into human-machine systems had prepared the way for acceptance of the value of Simon's studies of memorization time, time to task completion, eye movements, and other physical manifestations of the operation of the human information processing system. In addition, as noted above, NIMH was an institute devoted to mental health and was therefore affiliated with clinical psychology and psychiatry as well as with experimental psychology. Its grant review boards, therefore, were not as skeptical of the possibility of analyzing verbal behavior as was the community of experimental psychologists.

The Mature Model: Homo Adaptivus as Information Processor

The NIMH grant, combined with the ARPA-sponsored development of a computing infrastructure at CIT enabled the reform of psychology at CIT along cognitivist lines, the launching of Simon's interdisciplinary program in Systems and Communications Sciences, and the creation of a new Department of Computer Science. The CIP (complex information processing) project, as the Carnegie-RAND group's work in computer simulation of human behavior was known, also produced intellectual as well as institutional results, which would make homo adaptivus and the information processing model of mind central to psychology.

What was Simon and Newell's mature theory of human problem-solving? Put simply, the theory "views a human as a processor of information." Such a statement might be taken to imply that this theory is based on an analogy between humans and computers. They argue, however, that the representation of both humans and computers as information processing systems "is no metaphor, but a precise symbolic model on the basis of which pertinent specific aspects of the man's problem-solving behavior can be calculated." In fact, "the various features that make the digital computer seem machinelike—its fast arithmetic, its simply ordered memory, its construction by means of binary elements—all have faded in the search for the essential" and an "abstract concept of an information processing system" has emerged.[26]

Simon and Newell's theory "posits a set of processes or mechanisms that produce the behavior of the thinking human." It is, therefore, admittedly "reductionistic," and it takes as its goal the "explanation of behavior, not its mere description."[27] In Simon's view, the reduction of observed behavior to the mechanisms that generate it was the task of all science, natural or social.

In a marvelously direct manner, Simon and Newell disposed of any potential behaviorist objections to their theory: "As far as the great debates about the empty organism, behaviorism, intervening variables, and hypothetical constructs are concerned, we take these simply as a phase in the historical development of psychology. Our theory posits internal mechanisms of great extent and complexity, and endeavors to make contact between them and the visible evidence of problem-solving. That is all there is to it."[28]

Such are the basic assumptions of the model. What of its shape and texture? In *Human Problem Solving* Simon and Newell describe their theory in terms of seven basic characteristics. First, it is "a process theory" in that it is concerned with describing the processes used by an organism to transform an input into an output. Second, it is a "theory of the individual" in that individual differences are not "tacked on to the main body of our theory." Rather, "the models describe individuals so that the hard part is to say with precision what is common to all human information processors." Third, it is a "non-statistical theory" in that they do not "assume that human behavior is fundamentally stochastic ... Rather, Freud's dictum that all behavior is caused seems the natural one." Fourth, it is a "content-oriented theory," which is "dramatized in the peculiarity that the theory performs the tasks it explains." That is, "a good information processing theory of a good human chess player can play good chess; a good theory of how humans create novels will create novels; a good theory of how children read will likewise read and understand."[29]

Fifth, Simon and Newell's theory of human information processing is a "dynamically oriented theory" in the sense that it describes change in a system over time, characterizing "each new act as a function of the immediately preceding state of the organism and of its environment." The dynamic aspect of this theory leads them to argue that "the natural formalism of the theory is the program, which plays a role directly analogous to systems of differential equations in theories with continuous state spaces (for example, classical physics)." The only difference is that "in information processing systems, the state is a collection of symbolic structures in a memory, rather than the set of values of position and momentum of a physical system in some coordinate system."[30]

Sixth, theirs is an "empirical, not experimental theory." This feature follows from the theory's dynamic nature and from the complexity of the systems under study. As they write, "Several strategies of analysis are used, more or less, in the scientific work on dynamic theory. The most basic is taking a completely specified initial state and tracing out the time course of the system by applying iteratively the given laws that say what happens in the next instant of time." This technique is called simulation, and it is "a mainstay of the present work."[31]

Seventh, and finally, Simon and Newell note that the theory proceeds via "sufficiency analysis." That is, it puts a high premium on "discovering and describing mechanisms that are sufficient to perform the cognitive task under study."[32] That Simon and Newell's arguments were based on the sufficiency of the mechanisms they describe rather than their necessity, has attracted much criticism—so much that it will be dealt with later in the following section.

All seven features mark this theory as quite distinct from other approaches to psychology. The references in *Human Problem Solving* reflect this difference; the vast majority of citations are to the work of the Carnegie-RAND group. Simon and Newell frequently cite George Miller's work, particularly his "Magic Number Seven," as well as the work of European psychologists in the Gestalt tradition (such as Adrian de Groot, who studied the cognitive processes of chess players), but there are few references to scholarship in other psychological traditions.

The conceptual keystone of the theory is, as usual for Simon, a set of definitions. The most important of these refers to an information processing system (IPS). An IPS is defined by the existence of a "set of elements, called symbols," instances of which are gathered into symbol structures where they are connected by a set of relations. These symbol structures and the rules according to which they can be manipulated are stored in the memory of the IPS or are input from outside the system. When the symbol structures are to be acted on, they are transferred to the system's short-term memory where they are transformed by a set of elementary information processes that are applied in a sequence determined by the relevant program within the IPS. These programs are themselves symbol structures, and they are stored along with other data-symbols in the system's memory.[33]

As Simon and Newell point out, their "postulates for an IPS characterize the memory, elementary information processes, symbol structures, and interpretive process abstractly. They make no assertions about how these structure and processes are realized, physically or biologically." That "systems of this kind can be realized by a determinate mechanism is amply demonstrated by the existence and behavior of digital computers," but computers are not the only systems that are IPSs. Because "a human can simulate anything a computer can, albeit somewhat slowly," the human is as apt a model for the computer as the computer is for the human. (Apparently, simulation is not a perfectly transitive operation, however, since "the behavior of a human simulating a computer that is simulating a human solving a problem is observably different from the behavior of the human solving the problem, even if the behavior of the computer is not.")[34]

After defining these terms, Simon and Newell then observe, "it is one of the major foundation stones of computer science that a relatively small set of elementary processes suffices to produce the full generality of information processing." They apply that insight to the analysis of the human IPS as well, arguing that "the entire behavior of the IPS" is compounded out of sequences of only seven such processes: the ability to discriminate (to alter behavior in accordance with its symbol structures), to make tests and comparisons, to create symbols, to write symbol structures (to create new ones or alter old ones), to read and write externally (to receive input and produce output), to designate symbol structures (to make one symbol "point to" another), and to store symbol structures in some kind of memory.[35]

The specific characteristics of the human IPS that affect its ability to generate behavior from sequences of these elementary processes are that "it is a serial system consisting of an active processor, an input (sensory) and output (motor) system, an internal LTM [long-term memory] and STM [short-term memory] and an EM [external memory, like a book or a notepad]."[36] Simon and Newell argue that the LTM has an effectively unlimited capacity to store information, although information stored therein does have a disturbing tendency to decay over time, and that the LTM is organized associatively. Its contents are symbols and structures of symbols.

The STM, on the other hand, is much more limited in capacity, only being able to hold five to seven symbols at any given time—George Miller's "Magic Number Seven." The elementary processes that operate on symbols in the STM, however, are very, very fast. Thus, despite that the STM can only perform one operation at a time on a small set of symbols, the primary constraints on our ability to process information quickly are the much slower speeds at which information is read from or written to the LTM.[37]

Finally, the human IPS possesses a class of symbol structures that Simon and Newell call *goal structures*. These goal structures organize problem-solving by providing a set of referents against which the state of the human organism is compared, making humans adaptive, error-controlled machines.

The breakdown of the human IPS into a set of basic functional components, as described in *Human Problem Solving*, follows quite closely the "von Neumann architecture" of digital computers. In his widely read "First Draft of a Report on EDVAC," John von Neumann had defined a computer as being a set of basic functional components—input, processing, memory, control, and output units—that processed information serially.[38] This conceptualization of the computer became so widespread as to be nearly universal in the world of computing by the end of the 1950s, and almost all computers ever built have been "von Neumann machines." The only exceptions to this general rule have been computers that process information in parallel rather than serially, and these have been few in number until very recently. Simon and Newell clearly modeled their generalized IPS on the "von Neumann machine," understanding our senses as input units, our motor mechanisms as output units, the STM as the primary processing unit, the LTM and EM as the memory where data and programs are stored, and the "goal structures" mentioned above as the system's control unit.

The specification of the functional capacities of these basic components of the human IPS through computer simulation became the focus of Simon's research from the late 1950s on. LT and its bigger brother, the General Problem Solver (GPS), for example, were Simon and Newell's models of the human central processing unit; the Elementary Perceiver and Memorizer (EPAM) was a model of the input system (as were EPAM's cousins, MAPP and UNDER-STAND); and the Sequence Extrapolator (SE), General Rule Inducer (GRI), and Heuristic Compiler all were models of how the processor improved its operation over time, or learned. Thus, from 1958 to the end of his career, Simon set about building a mind, piece by piece, function by function.

The idea that "programmed computer and human problem solver are both species belonging to the genus IPS" had led to the reconstruction of Simon's intellectual and technical models of practice, as well as a reorganization of his network of patrons.[39] Though his work continued to have political implications, he was no longer a political scientist; though it continued to have implications for economics and sociology, he was neither economist nor sociologist. He was now something new, a computer scientist and a psychologist, and to him these two terms denoted the same thing—he was a practitioner of the new empirical science of complex, adaptive systems.[40]

Critiques and Contentions

There were some serious limitations to this model of human problem-solving, as Simon and Newell freely admitted. Early in *Human Problem Solving*, for example, they note that their work was "not centrally concerned with perception, motor skill, or what are called personality variables." Rather, their theory was "concerned primarily with performance, only a little with learning," and, significantly, "long-term integrated activities extending over periods of days or years received no attention.[41] Simon was convinced, however, that the model could be expanded to encompass nearly every aspect of human behavior, and his relentless efforts to extend his theory, as well as his cheerful predictions of imminent success, raised both hackles and doubts.

The limitations inherent in Simon and Newell's theories of the human IPS opened them up to considerable criticism. Some of this criticism was friendly; some definitely was not. Ulric Neisser, for example, was a fellow pioneer in cognitive psychology, and the criticisms he made of their theory of human problem-solving were those of one who shared Simon and Newell's broad approach to the study of human behavior. Simon took such criticisms very seriously, even when he did not agree with them, and he used them as starting points for further research.

In 1963, Neisser wrote an article titled "The Imitation of Man by Machine," in which he argued that: "Three fundamental and interrelated characteristics of human thought ... are conspicuously absent from existing or contemplated computer programs: 1. Human thinking always takes place in, and contributes to, a cumulative process of growth and development; 2. Human thinking begins in an intimate association with emotions and feelings which is never entirely lost; 3. Almost all human activity, including thinking, serves not one but a multiplicity of motives at the same time."[42]

Simon replied to the second and third of these criticisms directly in a paper that soon became quite famous in psychological circles, "Motivational and Emotional Controls of Cognition."[43] This article was a classic example of Simon's method of argument by redefinition, as well as an exemplary statement of how broadly applicable Simon understood his theory of human problem-solving to be. It also is one of the very few times in his entire career that he addressed the issue of emotion, a topic that the cool, calculating Simon typically avoided.

Simon opens the article by noting Neisser's criticisms and agreeing that information processing theories, at that time, had not yet met his objections. He goes on to argue, however, that they can be made to do so with only minor modifications. His argument runs as follows: first, there are two basic assumptions central to most existing information processing theories, (1) that the central nervous system is basically a serial information processor and (2) that the course of behavior is regulated, or motivated, by a tightly organized hierarchy of goals. Such systems are much more single-minded than humans are, Simon admits. In particular, humans are able to interrupt their activities and shift to the pursuit of other goals in response to various cues from the environment. In addition, humans often appear to be responding to multiple goals at once.

Simon argues, however, that the solutions to these problems, and the way to eliminate such single-mindedness, are not difficult to find. All one needs to do is to posit some mechanism that allows goals to be "queued" and to expand the notion of the goal to encompass multifaceted rather than one-dimensional criteria. Both of these mechanisms have analogues in the world of computers, Simon points out: the time-sharing systems that were then becoming widespread on university campuses were able to put the different jobs submitted to them in a queue ordered by a prioritizing program, and programmers routinely made the fulfillment of a set of criteria the goal of their programs.[44]

Regarding the interruption and shifting of goals, Simon notes that "a serial processor can respond to multiple needs and goals without requiring any special mechanisms to represent affect or emotion. We can use the term motivation, in systems like those described, simply to designate that which controls attention at any given time." Environments can place severe real-time demands on systems, so survival requires that there must be an "interrupt system" that affects "motivation" by shifting attention to the solution of more immediate problems. All that is necessary to incorporate such an interrupt system into his model, Simon argues, is the postulation of a program that monitors the status of certain critical variables. This monitor program would have to be capable of noticing "sudden, intense stimuli" that potentially can affect these critical variables and of interrupting ongoing programs when such stimuli occur. A serial information processor capable of "high-frequency time-sharing" can accomplish all this quite readily, he argues.[45]

Sudden, intense stimuli also "produce large effects on the autonomic nervous system, commonly of an 'arousal' and 'energy marshaling' nature. It is to these effects that the label 'emotion' is generally attached."[46] Taking a page from A. J. Ayer's "emotivist" account of the true nature of emotion, Simon states: "In human beings sudden intense stimuli are commonly associated with reports of the subjective feelings that typically accompany emotional behavior. We will not be particularly concerned here with these reports, but will assume, perhaps not implausibly, that the feelings reported are produced, in turn, by internal stimuli resulting from the arousal of the autonomic system."[47]

Emotional behavior is thus the behavior that is produced when the interrupt system goes into action in response to sudden, intense stimuli. It is another mechanism designed to contribute to our adaptation to a challenging environment. In similar fashion, elsewhere Simon reduces that other quintessentially human quality, creativity, to "more of the same." Creativity is simply problem-solving, and it could be reduced to the same seven basic mechanisms of symbol processing. As Simon wrote to his friend George Miller, "Is creativity as mysterious and inexplicable as you make it? At the risk of sounding presumptuous, I must confess that I had just about reached the opposite conclusion—that the mystery had now been pretty well stripped away from the higher mental processes (not from the neurology, to be sure), and that creativity was just a little more of the same."[48] This reductionist project led to other, sharper-edged criticisms from those who did not accept Simon's goals or his methods.

The first of these more fundamental critiques centered on the value of simulation and the merits of sufficiency analysis. To some critics, such as the philosopher John Searle, neither simulation, however accurate, nor sufficiency analysis, however rigorous, proved anything.[49] Searle offered the following thought experiment as a way of illustrating his argument: imagine that a man sits inside a room. He has with him a marvelous English-Chinese dictionary and a thorough set of rules for translating English sentences into Chinese. Using this dictionary and these rules, the man is able to translate every English sentence he is given into fluent Chinese. To those standing outside the room, observing only the English inputs and Chinese outputs, it would appear that there must be someone inside that understands English and Chinese. But this is an unwarranted belief, Searle observes, for it is clear to those who know what goes on inside the room that the man does not. All he knows is how to look things up. With these rules and this dictionary, the man may simulate the behavior of a person who knows both Chinese and English, but this simulation tells us nothing about his real knowledge of either language. Take away the dictionary or the rulebook, and his lack of understanding would be revealed for all to see.

Similarly, to Searle, the fact that a rulebook and a dictionary may be *sufficient* to the simulation of translation does not mean that they are *necessarily* parts of how humans actually translate things. To show that something *could* be done a certain way is not to prove that any particular actor in any given situation really does it that way.

Simon's reply to these criticisms was simple and direct: simulation is no different from any of the techniques scientists have developed to build models of the world. Physicists, for example, describe the positions and momentums of particles in a physical system via a parallel symbolic system of differential equations. They do not claim that this symbolic system is the "same" as the physical system it models. Indeed, the difference between them is essential—there is no point in creating a model that is an exact replica of the thing it represents. All that is claimed is that the relations between the symbols correspond to the relations between the elements of the physical system they represent. A model is a representation; a theory is an explanation; a simulation compares a particular model (representation) of a system with that system's actual behavior and so provides evidence for or against a certain explanation of the system's behavior.[50]

Similarly, while Simon admits that sufficiency does not prove necessity, he argues that it does not have to. Demonstrating that certain mechanisms are sufficient to produce certain behavior proves that more complex mechanisms are not necessary. Such a demonstration is a valuable discovery in itself, in Simon's view, because not every set of mechanisms is sufficient to produce the behaviors we observe. For example, most behaviorist theories of linguistic behavior are insufficient to explain human abilities in these areas. Testing for sufficiency is thus no empty test—which is good, as it is often the only test available.

In addition, the simplest set of mechanisms sufficient to generate phenomena should be the ones assumed to operate until proven otherwise. This is, in true Simonian fashion, a heuristic rather than a law: he is willing to admit there may be times when nature uses more complicated mechanisms than are strictly necessary. Nevertheless, Simon does not expect this heuristic to lead him astray, for in his view it is the core principle that has guided the search processes of Western science.

Simon's last argument on behalf of sufficiency analysis attempts to turn the tables on his critics. As he never tired of repeating, he only sought to show that one does not have to assume the mystical in order to explain behavior. His various programs had done so for significant realms of human experience. The burden of proof now lay on those who argued that something more than mechanism was needed.

There were many willing to pick up Simon's gauntlet. Joseph Weizenbaum and Hubert Dreyfus, for example, both attacked the information-processing model of mind as having omitted the truly important aspects of human experience in exchange for sophisticated analyses of the trivial.[51] Still worse, in their view, was that Simon then presumed to call these trivial aspects the fundamental ones—a galling sign, to them, of both his narrow-mindedness and his hubris.

Dreyfus took direct aim at Simon in his well-known book, *What Computers Can't Do*. In it he criticized the wide gaps between the predictions and the actual accomplishments of workers in the field of artificial intelligence, arguing that they often extrapolated wildly from early successes. Newell and Simon's famed General Problem Solver, for example, was nothing of the sort to Dreyfus. It was a clever device, but it should have been called the "Local Feature-Guided Network Searcher" instead.[52]

In Dreyfus's view, such extrapolations were unwarranted because the "artificially circumscribed gamelike domains" in which programs like GPS operated were inherently ungeneralizable. Such domains were defined by the fact that they could be described formally (and completely), but there was no evidence that the real world was so describable. The "micro-worlds" of AI were specialized compartments, not microcosms. Chess was a peculiar game, not a model of life.

Similarly, "intelligence requires understanding, and understanding requires giving a computer the background of common sense that adult human beings have by virtue of having bodies, interacting skillfully with the material world, and being trained into a culture." Simply put, "since intelligence must be situated it cannot be separated from the rest of human life." AI was thus the extreme form of Plato's separation of the "rational soul" from the "body with its skills, emotions, and appetites."[53]

Simon could respect a serious methodological dissent from another cognitive psychologist (Neisser) or a philosopher familiar with the language of logic (Searle), but the critiques of Weizenbaum and Dreyfus left him angry, sad, and uncharacteristically silent. As Simon wrote in a letter to his daughter Barbara,

"In general, I have not answered these attacks ... You don't get very far arguing with a man about his religion, and these are essentially religious issues to the Dreyfuses and Weizenbaums of the world."[54]

These criticisms did not slow his work or cause him to doubt his beliefs, however. He interpreted such dissent as a knee-jerk reaction to an assault on the uniqueness of man, a reaction just like that inspired by "the challenge raised, a century ago, by Darwinism, and four centuries ago, by the Copernican picture of the universe." Such a reaction was understandable, as "most men, including ourselves, find a comfort that is hard to give up in the idea that man is unique." But, Simon argued, "most of us also find it difficult to take comfort from chimeras." "Mankind's sense of dignity need not rest at all on claims of uniqueness ... Rather than remain apart from nature, human existence depends upon being truly a part of nature. Cognitive science, computers, information processes and psychology—far from threatening our place in the scheme of things—give us further evidence of the closeness of our kinship with the whole of creation."[55]

Simon's analysis was accurate, to an extent. Weizenbaum, Dreyfus, and many who shared their views doubtless were reacting to an attack on their understanding of humanity and its place in nature—just as it was doubtless the case that Simon took particular pleasure in defining away that uniqueness. Indeed, one could describe Simon's research agenda from the 1950s onward as being a successive attack on the basic pillars of our claims to uniqueness: intelligence, by the demonstration that machines can solve problems; creativity, by the demonstration that they can solve them in new and surprising ways; and emotion, by defining it as a component of our adaptive "interrupt" system.

Perhaps the only aspect of our unique humanity that Simon does not attack directly is the religious—the possession of a soul. Even there, however, he clearly saw religious faith as an example of bounded rationality, a way of reducing the complexity of the world by establishing a set of givens, and he believed that adherence to the ethical dictates of a religion was yet another example of the human tendency to "identify with subgoals."[56]

Why this seeming compulsion to deflate our claims to uniqueness? One can only speculate in answer, but it is clear that Simon always reacted strongly against anything that smacked of pride. This may appear odd, given that Simon himself was proud and self-confident, even to the point of arrogance. Nevertheless, to him pride was a great sin, whether it was explicit in an individual's assessment of his own self-worth or implicit in a theory that assumed powers of reasoning in man that were in truth only available to God. Even more, to Simon pride was parochial, and the parochial was perilous.

This reaction against the sin of pride also had quite definite political origins and resonance, for to Simon the great sin of central planners was their arrogance in believing that they knew best for all, while the great failing of radical individualists lay in their arrogant belief that the individual could reason as God would. The truth was somewhere in the middle; hence the need for both expert leadership and broad participation in the planning of our lives. From our limited, but not insignificant, powers of reason came not only the need for society but also the need for a specific kind of society—one characterized by "Deweyan" rather than "Miesian" planning.

Experience, Knowledge, and Action

There was something else lurking in the differences between Simon and Dreyfus. Not only did they originate in a fundamental difference over the nature of humans; they also grew from a fundamental difference over the nature of knowledge and its uses. For Simon, the instrumentalist, the "ultimate goal" of knowledge was to "determine relations of known facts of past and present with facts of the future, so that a single possibility is consequential on the present."[57]

For Dreyfus, knowledge was more than data and programs, more than the capacity for prediction. To him, experience altered our being, not just our programs. We change—and we remain the same—in ways that are not captured by changes in our ability to perform a task efficiently. All of this is nebulous and definitely not operational, but that is exactly the point: to Dreyfus, not everything can be made operational, and reduction to mechanism is not the true test of explanation. Does the "explanation" of love, say, or anger as the label we give to the physical sensations associated with certain "sudden, intense" stimuli really explain these emotions? Does it give meaning to them? An inability to process information rapidly enough to perform in a certain task environment may be appropriate to the explanation of stress, perhaps, or anxiety, but it seems inadequate to explain either the logic or the madness of love.

This difference of views about the instrumentality of knowledge resulted in an ironic reversal: an instrumental view of knowledge was oriented toward intervention in the world, but in Simon's information processing model of mind there were substantial gaps between experience and knowledge and between knowledge and action. Curiously, for one who had begun his career so concerned with the problem of the relationship between knowledge and action, the aspects of the human IPS to which Simon pays the least attention are those elements that experience and affect the world directly—the "sensors" and "effectors" that translate experience into information and information into action. As Simon and Newell wrote in "Elements," "if one considers the organism to consist of effectors, receptors, and a control system for joining these, then this theory is mostly a theory of the control system."[58]

The later products of CIP continued to focus on the control system. The Elementary Perceiver and Memorizer, for example, is a theory of the input system in that it is designed to model the processes by which information is stored and indexed in the LTM. It assumes, however, that one is already dealing with the storing and processing of symbols; it is not a theory of how we convert experiences into symbols to be stored and processed.

Simon viewed the "mind-body problem" as having been "solved" by the computer. To him, mind was to body as program was to computer and that was that. But the gaps between experience and knowledge and knowledge and action were not so easily closed. Simon knew this, but he believed he saw the way to bridge those gaps as well. That bridge led him back to familiar terrain: the problems of expertise, planning, and design.

CHAPTER THIRTEEN

Scientist of the Artificial

In his journey through the Garden of Forking Paths—from Milwaukee to Chicago to Berkeley to Pittsburgh—Herbert Simon constructed a sequence of identities for himself: administrative scientist, economist, systems scientist, psychologist, and computer scientist. He wrote influential works in a range of fields, from *Administrative Behavior* to "Rational Choice and the Structure of the Environment" to *Human Problem Solving*. He created an array of partial "models of man" and took a myriad of steps toward synthesis, intellectual and institutional. He reconstructed his theory and practice around the computer.

Simon's career exhibited continuities as well as changes; he possessed a life-long drive toward synthesis and a deep desire to find the pattern hidden beneath the surface of experience. He displayed an enduring personal style, both intellectual and social, characterized by a brash drive, pride, delight in unmasking pride in others, and an evangelical devotion to the spread of a secular gospel. He had remarkable skills for organization, manipulation, and artful compromise, as well as a taste for insurrection. Last, but not least, he retained a continuing concern with linking knowledge and action, research and reform.

In his later career, Simon sought to close this last connection, to complete the circuit of his thought through the construction of a "science of design." This science of design, he hoped, would make the conversion of knowledge into action a science itself. Nowhere is the grand sweep of this vision more apparent than in *The Sciences of the Artificial*.[1] Even more than *Human Problem Solving*, *The Sciences of the Artificial* was the broadest statement of Simon's mature worldview. It was his grandest synthesis, couched in layman's terms, and in it he strove to connect his theory of human problem-solving to the problem that had launched his career: that of translating knowledge into action so that we might make the right choices about our lives and our world.

The Sciences of the Artificial

The Sciences of the Artificial began as a set of lectures Simon delivered at MIT in 1968 as the fifth Karl Compton lecturer.[2] His 1962 essay, "The Architecture of Complexity," was added as a fourth chapter. The book is now in its third edition. It has grown and changed with each revision, but the basic argument has not been altered.[3] In this the book is like Simon himself: although he continually produced new, revised editions of himself, they were revised formulations of the same essential work.

The thesis that Simon develops in *The Sciences of the Artificial* is one that he says has been "central" to "most of his work," whether in organization theory, management science, or psychology. He argues that "certain phenomena are 'artificial' in a very specific sense: they are as they are only because of a system's being molded, by goals or purposes, to the environment in which it lives. If natural phenomena have an air of 'necessity' about them in their subservience to natural law, artificial phenomena have an air of 'contingency' in their malleability by environment." Such phenomena are fascinating and important, Simon observes, but they pose grave problems for the would-be scientist of the artificial. It was not at all clear, for example, "how empirical propositions can be made at all about systems that, given different circumstances, might be quite other than they are."[4]

Reflection on this problem over the years, he states, eventually had led him to see in the problem of artificiality "an explanation of the difficulty that has been experienced in filling engineering and other professions with empirical and theoretical substance distinct from the substance of their supporting sciences. Engineering, medicine, business, architecture, and painting are concerned not with the necessary but with the contingent—not with how things are but how they might be—in short, with design. The possibility of creating a science or sciences of design is exactly as great as the possibility of creating any science of the artificial. The two possibilities stand or fall together."[5]

How, then, to construct such a science of design? As always, Simon begins by laying the conceptual foundations. First, he observes that an "artifact" is a "meeting point—an 'interface' in today's terms—between an 'inner' environment, the substance and organization of the artifact itself, and an 'outer' environment, the surroundings in which it operates." "This way of viewing artifacts applies equally well to many things that are not man-made—to all things, in fact, that can be regarded as 'adapted' to some situation; and, in particular, it applies to the living systems that have evolved through the forces of organic evolution." In Simon's view, rationality plays in the sciences of the artificial the same role that natural selection does in evolutionary biology: it selects those designs that are better suited to their environment, with "better" meaning more efficient in the adaptation of means to ends.[6]

What benefit has one gained by adopting such a perspective? "The first advantage of dividing outer from inner environment in studying an adaptive or artificial system is that we can often predict behavior from knowledge of the system's goals and its outer environment, with only minimal assumptions about the inner environment. An instant corollary is that we often find quite different inner environments accomplishing identical or similar goals in identical or similar outer environments—airplanes and birds, dolphins and tunafish, weight-driven clocks and spring-driven clocks, electrical relays and transistors."[7]

In addition, one can also "characterize the main properties of the system and its behavior without elaborating the detail of either the outer or inner environments. We might look toward a science of the artificial that would depend on the relative simplicity of the interface as its primary source of abstraction and generality."[8]

Such an approach would involve a description of the artifact in terms of its organization and function, which raises the possibility that one can study the artifact and its interactions with its environment through simulation. As shown in chapter 12, the validity of simulation in the analysis of systems long has been controversial. In *The Sciences of the Artificial,* Simon reduces this controversy to a single, crucial question: "How can a simulation ever tell us anything that we do not already know?"[9]

In order to answer that question, Simon turns to a discussion of the most powerful and prevalent of simulation devices, the computer. He observes that there is a direct parallel between the assertions that "a simulation is no better than the assumptions built into it" and that "a computer can do only what it is programmed to do."[10] He admits that both of these statements are true, but he argues that neither proves that simulations cannot teach us something new. In fact, he claims, his experience with simulations is one of constant surprise.

There are two ways in which simulations—or the actions of a programmed computer—can and do teach us new things. First, "even when we have correct premises, it may be very difficult to discover what they imply. All correct reasoning is a grand system of tautologies, but only God can make direct use of that fact." Second, simulation also can tell us a great deal about systems even when we do not have full knowledge of its workings. This is possible because abstraction is the essence of simulation. We are "seldom interested in explaining or predicting phenomena in all their particularity." Rather, the more we "abstract from the detail of a set of phenomena, the easier it becomes to simulate the phenomena . . . we do not have to know, or guess at, all the internal structure of the system."[11]

The reason the computer has become such an important tool for constructing simulations is that "No artifact devised by man is so convenient for this kind of functional description as a digital computer. It is truly Protean, for almost the only ones of its properties that are detectable in its behavior (when it is operating properly!) are the organizational properties." A computer, to Simon, is definable as "an organization of elementary functional components in which, to a high approximation, only the function performed by those components is relevant to the behavior of the whole system."[12] Translated into only slightly different language, the computer is the perfect tool for simulation because it is the perfect bureaucracy.

This peculiarity of computers, in turn, makes possible the creation of "an empirical science of computers—as distinct from the solid-state physics or physiology of their componentry." "Since there are now many such devices in the world, and since the properties that describe them also appear to be shared by the human central nervous system, nothing prevents us from developing a natural history of them. We can study them as we would rabbits or chipmunks and discover how they behave under different patterns of environmental stimulation." For example, time-sharing computer systems have proven to be strikingly complex and unpredictable—so much so that the best way to analyze them is to "build them and see how they behave."[13]

Why is the development of such an empirical science of computer behavior valuable? "If it is the organization of components, and not their physical properties, that largely determines behavior, and if computers are organized somewhat in the image of man, then the computer becomes an obvious device for exploring the consequences of alternative organizational assumptions for human behavior." An empirical science of this universal machine therefore would be an empirical science of the human machine, and psychology could "move forward without awaiting the solutions by neurology of the problems of component design."[14]

Having laid this methodological foundation, Simon then moves to a discussion of what his empirical studies of computer behavior have taught him about human thoughts and actions. First, he notes that "as we succeed in broadening and deepening our knowledge—theoretical and empirical—about computers, we shall discover that in large part their behavior is governed by simple general laws, that what appeared as complexity in the computer program was, to a considerable extent, complexity of the environment to which the program was seeking to adapt its behavior."[15]

Simon then applies this insight to the behavior of humans, arguing that "A man, viewed as a behaving system, is quite simple. The apparent complexity of his behavior over time is largely a reflection of the complexity of the environment in which he finds himself."[16] Simon restates this point several times in the course of the book, with more repetitions in each new edition. He does hedge this claim a bit, noting that it refers to that part of man that is thinking, not feeling or perceiving. Still, it is clear from his other work (such as his article on "Motivational and Emotional Controls," discussed in chapter 12) that he believed the statement applied to these areas as well.

This hypothesis, while startling and perhaps even offensive to those who delight in the marvelous complexity of the human body and mind, is in keeping with Simon's view of the nature and purpose of science. "The central task of a natural science is to make the wonderful commonplace: to show that complexity, correctly viewed is only a mask for simplicity; to find pattern hidden in apparent chaos." This does not destroy wonder or reverence for creation; rather, "when we have explained the wonderful, unmasked the hidden pattern, a new wonder arises at how complexity was woven out of simplicity."[17] A clearer statement of the synthetic, positivist aesthetic underlying Simon's work cannot be found.

Simon then describes the simple principles that underlie even the most complex human behavior: the human information processing system is "basically serial in operation," and "it can process only a few symbols at a time." In addition, the symbols being processed are held in "special, limited memory structures whose content can be changed rapidly," and "the most striking limits on subjects' capacities to employ efficient strategies arise from the very small capacity of the short-term memory structure (seven chunks) and from the relatively long time (five seconds) required to transfer a chunk of information from short-term to long-term memory."[18] From these elemental qualities even the most intricate individuals, the most elaborate organizations, are born.

The Science and the Practice of Design

At this point, Simon may appear to have strayed far from the concern with the "science of design" with which he began *The Sciences of the Artificial.* In the third lecture, however, he closes the circle. Adaptation, he observes, is inextricably linked to purpose and thus to design—to the creation of things that fulfill goals, however well. "Everyone designs who devises courses of action aimed at changing existing situations into preferred ones." Everyone—and everything—who adapts also designs, if they adapt consciously. This statement holds for all systems that adapt and thus is as true for society as it is for the individual: "Engineers are not the only professional designers ... The intellectual activity that produces material artifacts is no different fundamentally from the one that prescribes remedies for a sick patient or the one that devises a new sales plan for a company or a social welfare policy for a state."[19] Professional schools, then, should strive to teach their students the general principles underlying such adaptive endeavors, and these principles constitute the science of design.

For most of the twentieth century, Simon argues, this injunction to teach professionals to design effectively has been understood to mean that they must be taught the fundamentals of natural science. This was only natural, since there was no true science of the artificial until after World War II. Though understandable, this focus on natural science as the core of professional training, particularly in engineering and in medicine, had led to the abandonment of interest in the crucial problem of how to convert the "is-es" of natural science into the "oughts" of practical action. "The professional schools will reassume their professional responsibilities just to the degree that they can discover a science of design—a body of intellectually tough, analytic, partly formalizable, partly empirical, teachable doctrine about the design process."[20]

Such a science, Simon argues, now exists where none had existed before. The ideal curriculum for this new science would have five basic components: First, it would teach the "evaluation of designs," encompassing both "theories of evaluation," such as utility theory and statistical decision theory, and the study of "computational methods," such as algorithms for finding optimal alternatives, when available, and algorithms for choosing satisfactory options when optimization is not possible. Second, it would teach the "formal logic of design" through the study of both declarative and imperative logics. Third, it would instruct professionals about "the search for alternatives," teaching both heuristic search methods (such as factorization and means-end analysis) and methods of allocating resources for search efficiently. Fourth, it would teach the "theory of structure and design organization" through the analysis of that most fundamental of organizational structures, hierarchy. Fifth, it would cover the "representation of design problems," a topic about which Simon says little but which seems to be related to Gestaltist ideas about the importance of mental "sets" for problem-solving.[21]

This curriculum for a science of design, though clearly created with an engineering or business school like Carnegie Tech's Graduate School of Industrial Administration in mind, was not limited to such.[22] Rather, Simon saw the science of design in much more expansive terms. "Many of us have been unhappy about the fragmentation of our society into two cultures. Some of us think there are not just two cultures but a large number of cultures. If we regret that fragmentation, then we must look for a common core of knowledge that can be shared by the members of all cultures ... A common understanding of our relation to the inner and outer environments that define the space in which we live and choose can provide at least part of that significant core."[23]

Simon continues on this theme, writing that "those of us who have lived close to the development of the modern computer ... have been drawn from a wide variety of professional fields ... We have noticed the growing communication among intellectual disciplines that takes place around the computer."

But surely the computer "as a piece of hardware" has "nothing to do" with the matter. Rather, "The ability to communicate across fields—the common ground—comes from the fact that all who use computers in complex ways are using computers to design, or to participate in the process of design." In addition, "we as designers ... have had to be explicit, as never before, about what is involved in creating a design and what takes place while the creation is going on."[24] In the end,

> The real subjects of the new intellectual free trade among the many cultures are our own thought processes ... We are importing and exporting from one intellectual discipline to another ideas about how a serially organized information-processing system like a man—or a computer, or a complex of men and computers in organized cooperation—solves problems and achieves goals in outer environments of great complexity. The proper study of mankind has been said to be man ... If I have made my case, then we can conclude that, in large part, the proper study of mankind is the science of design, not only as the professional component of a technical education but as a core discipline for every liberally educated man.[25]

This belief that the science of design should be a "core discipline" for "every liberally educated man" was, perhaps surprisingly, a democratic as well as an egocentric view. Like his fellow cognitivist George Miller, who wished to "give psychology away," Simon wanted to give design away, to make it a part of every person's armory of adaptive abilities.[26] As he writes in *The Sciences of the Artificial,* we need to "value the search as well as its outcome—to regard the design process as itself a valued activity for those who participate in it."[27]

Because participating in the design process is a valuable activity in itself, he argues, we need to work to develop ways to include others in the process. Take city planning, for instance, his original contribution to expert management: "We have usually thought of city planning as a means whereby the planner's creative activity could build a system that would satisfy the needs of a populace. Perhaps we should think of city planning as a valuable creative activity in which many members of a community can have the opportunity of participating—if we have wits to organize the process that way."[28]

Simon's belief in the value of participation in the design process—particularly in the processes of social design—was no novelty born of the 1960s. Rather, it had been an important element of his political value system since the beginning. In 1951, for example, he wrote that "democratic participation in administration does not find its primary justification as a means toward more efficient administration, but rather as an end that encompasses broad participation in the determination of the organizational goal."[29] Similar passages also can be found scattered throughout *Administrative Behavior* and *Public Administration*.

Simon did not perceive these democratic values to be at odds with his scientific analyses of organizations, but his critics did. Many argued that he had allowed management's objectives to determine his own, that he had lost sight of democracy in his study of the technology of administration.[30] To them, if experts only concerned themselves with the technical, then they were only tools.

To Simon, on the other hand, the expert's job was to represent the larger public interest. The expert could do so in two ways: first, by helping to create a broad design for social institutions that would enable wide participation; and second, by providing the public and its leaders information about the consequences of their design decisions. Without an expert versed in the science of design as one of the participants in the design process, the social organism was destined only to react to immediate crises and to do so in an uncoordinated, purposeless manner. The absence of expertise from the design process was as dangerous as the absence of opportunities for wide participation in choosing social goals.

Simon had tried to transform administrative science so that it could serve this vital enabling role through his work at the International City Manager's Association and the Bureau of Public Administration (see chapters 4 and 5). He continued to do so throughout his career. In the late 1940s, for example, he served in the Economic Cooperation Administration (the Marshall Plan organization)—a brief, but proud moment in his career, in which events conspired to wed vast resources and high ideals before the altar of geopolitical strategy, producing perhaps the most triumphant expression of liberal, managerial ideals.[31]

Simon's later forays into the world of practical politics were less rewarding, however. In 1965, for example, Pennsylvania governor William Scranton appointed a commission to study the costs and benefits of cutting the state subsidy for milk producers. Simon was asked to chair this commission, and he enthusiastically accepted the task of leading this examination of one of the state's most important industries. His commission was part blue-ribbon panel and part representative body, and the interests represented were those of the larger milk producers. Simon's experience on the milk commission severely tested his faith in the power of expertise to lead the public interest to victory over self-interest.

Economic analysis of the milk industry showed clearly that the milk subsidies primarily went into the coffers of the largest producers, to the detriment of both the consumer and the small producer. Because a majority of the commission's members were representatives of the industry, however, the majority report endorsed the continuance of the subsidies. The minority report, authored by Simon, opposed the subsidies, though Simon was politically sensitive enough to do so in less strident language than he used in private.[32] In truth, he was enraged by the naked pursuit of self-interest at the expense of the public interest and by the inability of rational expertise—of a better, more effective design—to win the day.

As he wrote to the president of the Tax-Payers Petition Committee, the recommendations of the "rump" report

> fall far short of the complete removal of price controls which I then thought, and continue to think, was the best solution ... But they seemed to me to form the basis for a reasonable compromise that I could accept, and to remove some of the more vicious aspects of the present law—which provides for control of the industry, by the industry, in the interests of the industry. Alas, I discovered in the committee that very few businessmen really believe in the price mechanism and free competitive markets when applied to their own business. I found myself in the peculiar position of an old-line New Deal Democrat defending free enterprise against businessmen. I continue to find myself in that position on other issues with great frequency ... Frankly, I was outnumbered, outgunned, and outmaneuvered by the combination of milk dealers, dairy farmers who had advantageous marketing positions, misguided dairy farmers who didn't but who thought the milk law was helping them, the Teamsters' Union, and the State Department of Agriculture.[33]

Such defeats of public interest by special interest were disappointing, but they did not cause Simon to lose faith. Expertise still was on the side of the wider public, and he knew that fighting the powers that be always was a long, uphill struggle. The late 1960s and early 1970s, however, confronted Simon with a new, far more disturbing challenge.

During this period, Simon's work as chair of the President's Science Advisory Committee's Committee on Environmental Quality and his prominent role as university leader at Carnegie Mellon brought him into close contact with—and made him a visible target for—radical political activists. Such radicals (which at one time included his own daughter and son-in-law, then at Berkeley) took aim at the most basic of Simon's faiths, challenging the ideals of objectivity, progress, and expert leadership. To them, science was a mask for interest and the expert a tool of "the system."

Such attacks puzzled, alarmed, and angered Simon, although they did not sour his relationship with his children.[34] He, like many aging New Deal liberals, was used to thinking of himself as fighting the good fight against narrow-minded reactionaries. As a result, he was incensed at being lumped in with the very powers that he saw himself as combating, such as the auto industry, which was quite displeased with a report urging emission controls that Simon's subcommittee produced for the President's Science Advisory Committee.

Simon responded to this "crisis of our times" in many venues, including a series of essays in the CMU student newspaper, the *Tartan*. Printed under the heading, "Simon Says," this series of columns followed an acid letter from Simon to the editor: "The *Tartan*," he wrote, "through the non-voluntary contributions of all CMU students . . . subsidizes the propaganda of a little band of self-appointed 'radicals' who use it to preach a muddle of tea-party anarchism." It was "ludicrous that the 'all-powerful' Establishment should foot the bill for the call to revolution." In fact, it was "more than ludicrous," it was "immoral."[35] As a remedy, Simon proposed the "complete cessation of subsidies to student publications," grounding his call in the "rights guaranteed by the First Amendment."

The editors of the *Tartan* responded in kind, calling Simon an "irresponsible lunatic," and various faculty members chimed in on both sides of the issue. Simon then asked if he could "provide the *Tartan*, for five weeks, with some alternative views," asking if the editor was "only able to dish it out, not take it?"[36]

In the columns that followed, Simon put forward his views about university reform, student life, and the place of science and technology in society. He agreed that universities badly needed reforming, but he held that "on our campus and many others, university reform has bogged down in trivialities and a sordid power struggle."[37] This power struggle was based on the mistaken presumption that students and faculty were opponents in a zero-sum game.

Genuine reform, he argued, did not begin with questions of power; rather, it began with trying to clarify goals. The goal of the university was to foster learning, and in the pursuit of this goal students and faculty were allies. Their interests were not identical, so the pursuit of this common goal would require compromise. "Compromising has a bad sound to those who want to turn every question of choice into a moral issue." However, "Those who think they will settle for nothing less than 'best' worlds—for utopias—surrender the chance of working for better worlds—better worlds that can actually be achieved."[38]

Continuing, Simon noted that "on the CMU campus you hear arguments about liberal education versus professional education. Some of us don't buy that 'versus.' We don't believe that an education that doesn't equip students with professional skills is truly liberal. Nor do we believe that an education is truly professional that prepares people only to be guns for hire—technicians applying their skills without concern for the problems they are applying them to." That meant that the scientist or engineer "needs a reasoned belief in the social importance of scientific and technical knowledge for today's and tomorrow's world." Such a belief "wouldn't have been challenged ten years ago. For many who held it, it was more a faith than a belief . . . But today, scientific and technical knowledge isn't treated as an unquestioned good. The Bogeyman of our childhood has been replaced by a demon called 'The Military-Industrial-Scientific-Technological Complex.' "[39]

This critique of science and technology had proceeded to the point where "A man named Toffler" could proclaim, "with great vigor and few facts," that society was "suffering from 'Future Shock'—from experiencing too rapid change produced by technology." Toffler, Simon noted, "conveniently forgets that our grandparents, the generation that moved from farm to city, survived—and quite nicely, thank you—a far more wrenching psychological and social change than any we are experiencing now . . . Nevertheless, we shouldn't overreact for the defense. If science and technology are not the devil, they probably are not God either."[40]

Simon then asks "What are the facts?" regarding science and society. "The first fact is that, without science and technology, the human condition always was, and always would be, desperate." Thanks to science, instead of living in a Malthusian world, "we now have the possibility of a world in which poverty— I mean real hunger and discomfort, not inability to keep up with the

Joneses—will be unknown. To make that possibility a reality will require more science, not less."[41]

Even more, science and technology produced moral as well as material uplift:

> Science and technology, and the affluence they've brought, have had another effect on us that's less often mentioned. We are often told that we are a materialistic society. No one who has visited a really poor country—India, say—will make that mistake ... We are the first peoples of the world, apart from favored aristocracies, who can afford the luxury of rejecting materialism.
>
> Science and technology permit us to reach higher goals—material and moral—hence to set them still higher. Science and technology enable us to see consequences of our acts that were previously hidden from us ... In doing so, they make our task harder. But they also give us reason to believe in the reality and continuing prospect of human progress.[42]

The Architecture of Complexity

Simon concludes *The Sciences of the Artificial* with his 1962 essay, "The Architecture of Complexity." While the topic of this essay at first might seem far removed from the questions of simulation, intervention, and design, "The Architecture of Complexity" was linked to the rest of the book by more than mere editorial convenience. As Simon writes, "The reader will discover, in the course of discussion, that artificiality is interesting principally when it concerns complex systems that live in complex environments. The topics of artificiality and complexity are inextricably interwoven."[43] If the rest of the work asks, what is it about humans that enables them to think and know, this essay asks what it is about the world that allows it to be known?

Simon's answer to this question connects the structure of complex systems of behavior to evolution and to the universal phenomenon of the generation of the complex from the simple. In the end, we discover that if evolution is nature's basic mechanism, hierarchy is its heuristic.

As noted earlier, Simon's world is populated with simple creatures that process information about their world in order to adapt to it. They are able to do so because their small set of elementary information processes enables them to modify their behavior in ways that are contingent upon the outcomes of their actions. In addition, adaptive systems are able to assemble units of behavior into "stable sub-assemblies," a process that is the reverse of the analytical process of factoring problems into subproblems.

This process of assembly, Simon argues, is fundamental to the generation of the complex from the simple in any and all circumstances. If one grants that such stable subassemblies can exist, then the evolution of complex life and complex behavior is so likely as to be almost inevitable. To prove this point, Simon offers the example of a complicated watch, harking back to William Paley's famous argument from design in support of the existence of a divine designer. Contrary to Paley, however, Simon uses the watch example to argue that complexity is extremely likely to evolve spontaneously and so requires no divine intervention.[44]

Granted, if it were necessary for all the thousands of components of a complicated watch to come together perfectly *at once* if they are not to fall apart, then the evolution of such a complex system by chance would be almost inconceivable. But, Simon notes, they do not have to come together all at once. All that is necessary for complexity to evolve is that *two* pieces be able to come together and stay together. If such "stable sub-assemblies" are possible, and empirical evidence demonstrates that they are, then complex organizations are not merely possible but likely products of unthinking nature.

Such a process of generating complexity out of successively larger subassemblies produces hierarchically organized systems. This process, and the hierarchical organization of complex systems, is universal. It applies to the evolution of complex organisms, the evolution of complex behaviors, and the evolution of complex patterns of thought: Simon's examples of the process range from the elementary particle-atom-molecule-chemical compound hierarchy of atomic structure to the nucleus-cell-tissue-organ-system hierarchy of organic life. Hierarchy is nature's heuristic—its shortcut to complexity.

Crucially, hierarchy enables not only the construction of complexity but also its analysis. For Simon homo adaptivus exists in a complex environment about which it has only partial knowledge. Fortunately, it only interacts with certain parts of the larger environment: while everything in the world is interconnected at some level, the majority of those interconnections are irrelevant. The influence of most other systems on any individual system is generally so small as to be negligible. In Simon's phrase, the world is "nearly decomposable." One consequence of this fact is that "Scientific knowledge is organized in levels, not because reduction in principle is impossible, but because nature is organized in levels, and the pattern at each level is most clearly discerned by abstracting from the detail of the levels far below."[45]

The near decomposability of the world is why we are able to draw boundaries around systems. If the elements of one system interacted with elements outside the system as continuously as they did with those inside the system, then they all would be parts of the same system. To use the classic example from chaos theory, a butterfly flapping its wings in South America *could* cause an earthquake in China, but it is rather unlikely that it will do so. The better place to look for the cause of the earthquake, Simon would say, is in China, and a more plausible agency would be something with energy levels comparable to those released by the quake.[46]

An important consequence of the hierarchical structure of the world is that while complete knowledge of the world is impossible, sufficient knowledge for successful adaptation is possible. Such knowledge is not easy to obtain, of course, for even the limited set of relevant subsystems may be fantastically complex, especially if those subsystems are themselves adaptive and therefore able to change their behavior in response to the observer's actions. The good news, according to Simon, is that perfect knowledge is unnecessary. Even partial knowledge is valuable.

Simon goes on to connect the problem of knowledge, the prevalence of hierarchy, the adaptive, problem-solving process, and the topic of design. To do so, he first notes that "The distinction between the world as sensed and the world as acted upon defines the basic condition for the survival of adaptive organisms. The organism must develop correlations between goals in the sensed world and actions in the world of process." Fortunately, "many complex systems have a nearly decomposable, hierarchic structure," which enables us "to understand, to describe, and even to 'see' such systems and their parts."[47] Thus, the organism can understand the world because the hierarchical structure of complex systems enables parsimonious description of both current and goal states. At the same time, its programs of behavior—heuristic programs that mimic nature's own selective search processes—enable the organism to adapt existing means to the achievement of its ends.

Simon concludes by stating that "The correlation between state description and process description is basic to the functioning of any adaptive organism, to its capacity for acting purposefully upon its environment." The essential task of any adaptive organism is thus the same as that of the science of design: that is, "given a blueprint, to find the corresponding recipe."[48]

Adaptation and Design

Simon's new model of the human as homo adaptivus connected organism and environment, theory and practice, mind and mechanism. It was a striking synthesis of the sciences of choice and control, and, as his essay on the "Architecture of Complexity" shows, it even pointed the way toward a "non-trivial" general theory of all adaptive systems, not just humans and their problem-solving behaviors. Homo adaptivus brought together biology and behavior, evolution and the program, making a powerful case for the ultimate reduction of life to mechanism.

As with all of Simon's syntheses, however, *The Sciences of the Artificial* was another "stable sub-assembly," not a final product. His intellectual evolution continued, and his selective search began to focus on the solution of new problems. These solutions could be harmonized with his previous conclusions, but only by reinterpreting them.

The starting point for these new searches was an idea central to Simon's model of homo adaptivus: the idea that the complexities of human behavior come from outside the individual. Homo adaptivus is a simple creature, always equipped with the same basic set of elementary capabilities, but its environment is complex and varied. As a result, human behavior becomes complex and varied—individualized—through people's encounters with the world.

This idea had important implications for Simon's later thoughts on science, expertise, and design. While the elementary processes of the human information processing system are the same in everyone, our experiences, the ways that we learn to organize those experiences, and the heuristics we induce from patterned experiences all are contingent upon our individual histories. This idea that knowledge is both universal and individuated lay at the foundations of Simon's later research into expert systems and the processes of scientific discovery.

CHAPTER FOURTEEN

The Expert Problem Solver

Simon's theory of heuristic problem-solving led him to develop ever more individualized and contextualized accounts of cognition. These new accounts were as mechanist and reductionist and as oriented toward the development of a general theory as his earlier work, but they increasingly focused on domain-specific cognition, usually referred to as *expertise*.

Simon's exploration of how people acquire and employ heuristics led him to posit that there are two kinds of heuristics: "weak" universal ones and "strong" domain-specific ones, the latter depending heavily on the development of domain-specific memory structures. In addition, Simon increasingly emphasized that both heuristics and memory structures evolved through interaction with the environment. The mind was an induction machine. This process of interaction and induction produced evolutionary growth patterns, giving both heuristics and memories a common treelike organization. Thus, for Simon, Darwin's evolutionary tree of life became not only a decision tree but also the tree of knowledge.

At the same time, Simon's interest in defending simulation as a valid research technique led him to study the processes of scientific discovery, to which he applied his new ideas about the development of expertise. Once again, he brought his theories of cognition and his philosophy of science into harmony, using each to support the other. Scientific discovery, to Simon, became an interactive process of induction from experience, and simulation became not merely valid for but essential to human science.

All of these developments reflected Simon's growing appreciation for experience and domain-specific expertise, an appreciation that stood in marked contrast to his youthful enthusiasm for general theory and interdisciplinary research. When one compares Simon's articles from the 1950s with those of the 1990s, for example, one cannot help being struck by the differences in his patterns of citation: his articles from the 1950s cited works from several fields as sources of theories and mathematical methods, adding to them only limited references to empirical studies, the vast majority of which had been conducted by Simon and the Carnegie-RAND group. In the 1990s, however, Simon's articles very consciously constructed a cumulative tradition in cognitive psychology, citing works in the field almost exclusively. In addition, his later articles took pains to integrate the empirical results from a number of research programs in cognitive psychology. While it would be too much to say that Simon had become *only* a cognitive psychologist, it is accurate to say that for the last thirty-plus years of his career, he *primarily* was one.

Simon's increased appreciation for the value of experience and domain-specific expertise likely was connected to his advancing age: youth cannot trumpet experience, but it can push for novelty, while age tends to help one appreciate cumulative endeavor and the value of experience. At the same time, Simon's own adaptive processes were at work. He was a disputatious character, and his instinctive response to every criticism was to turn it into a challenge, thus incorporating it into his own goal structures. For example, his early theories of human problem-solving had been criticized as lacking sensitivity to context, as noted in chapter 12. His later theories attempted to deal with this problem, and he deliberately strove to create a theory of "situated" thinking.

As he repeatedly insisted, these new theories could be incorporated into his broad model of homo adaptivus, but doing so required a series of redefinitions, some small, some large, of what his earlier work had meant. In particular, as his research became more bounded by disciplinary agendas and supported by patrons of "basic science" (NSF, DARPA, NIMH), the distance grew between it and his original concern with reconciling choice and control.

How individuals solved problems was now the problem he sought to solve, not the more immediate problems of human society. Thus, in a seeming paradox, as his work became more empirically based and laboratory oriented, it also became more abstract. To Simon this was no paradox, of course, but simply another consequence of bounded rationality. Fundamental knowledge necessarily was specialized, bounded knowledge. To cut deeply, one needed a knife with a sharp point.

Acquiring Knowledge: Chess, EPAM, and Expertise

The place to begin the study of Simon's ideas about expertise and situated knowledge is the chessboard. He was an inveterate puzzle-solver who enjoyed all manner of intellectual games, from the crosswords he did every morning to the ultimate brain-teaser, chess. Simon once wrote that puzzle-solving was one of his greatest pleasures and the primary motivation behind his scientific work.[1]

The more intricate the puzzle, the better. One of the most intricate and most fascinating of puzzles was the game of chess, which had an almost magnetic attraction for many pioneers in early computer science and AI, from Claude Shannon to Allen Newell to John von Neumann. Simon did not exaggerate when he called chess "the drosophila" for AI—its model organism and central research project.[2]

Chess has a number of qualities that endeared it to researchers interested in the simulation of mind by machine. First and foremost, it is a decidedly intellectual game with high cultural status. Hence, the ability to play chess could be taken to prove, *a fortiori*, the ability to do any number of other complex mental tasks. As Simon wrote, "If one could devise a successful chess machine, one would seem to have penetrated to the core of human intellectual endeavor."[3] Second, although chess requires the ability to think ahead, it also requires the ability to respond to contingent events, making it suitable for the study of adaptive behavior. Third, a chess match involves both rational choice and social behavior: a great deal of chess involves anticipating the moves of a specific opponent, which chess masters are able to do by studying their opponents' moves, demeanor, and chess histories. These social aspects of chess often are forgotten, yet it was precisely the absence of social cues that made it so difficult for Gary Kasparov to defeat his inhuman opponent, Deep Blue.[4]

Fourth, although it is possible to think ahead and thus to evaluate the likely outcomes of some number of moves, chess is simply too complex to be amenable to an algorithmic solution. Not even the fastest computer imaginable could evaluate every possibility contingent on every move. (By one estimate, it would take the fastest computer in the world longer than the age of the universe to calculate all the possible outcomes of a game of chess.) The impossibility of calculating the perfect move in every situation meant to Simon that chess was representative of a wide range of human cognitive tasks: the chess player's ability to make rational choices was bounded, just like the administrator's.

Fifth, and finally, despite the impossibility of calculating the perfect move in every situation, some players are much better than others. This is where the fact that chess is a competitive game with clear final outcomes (win, lose, or draw) matters, for it introduces the element of relative skill into the system. Though all skilled players clearly are intelligent, by most measures at least, not all intelligent people are good chess players. Simon, for instance, was as smart as they come, and he did learn to play good chess, but he was no grandmaster and he knew it. One cannot help but think that part of the motivation for studying chess for Simon—and for many others—was a gut level "how can I be losing to this guy, when I *know* I'm smarter than he is?" Clearly, something more than sheer brainpower was involved in playing good chess, something that developed with experience. Somehow, one *learned* to play good chess.

Simon's fascination with computer chess dated to 1952, the fateful year he met Allen Newell at RAND's Systems Research Laboratory. Newell, like many mathematicians, was interested in chess, and he and Simon soon began to discuss the possibility of creating a computer program that could play the complex game. As a result of this interest, Simon's landmark "Behavioral Theory of Rational Choice" originally contained an appendix describing a computer program that could play chess, and much of Simon and Newell's early work on problem-solving programs focused on the problems of chess.[5] In 1955 they switched their primary efforts to proving geometry theorems and then to proving the theorems of symbolic logic, producing Logic Theorist. They made this switch because a logic machine required very little knowledge about the state of the world, while a chess machine would have to be able to respond to subtle differences in the arrangement of pieces on the board, differences that were difficult to represent in symbolic form.[6]

Simon and Newell did not abandon computer chess when they developed Logic Theorist. Rather, they immediately applied the principles behind Logic Theorist to chess. As they noted at the time, "In a fundamental sense, proving theorems and playing chess involve the same problem: reasoning with heuristics that select fruitful paths of exploration in a space of possibilities that grows exponentially."[7] In truth, it would be more accurate to say that they reapplied the principles of heuristic problem-solving to chess after an excursion into theorem-proving. The manifest impossibility of playing algorithmic chess and the obvious ability of humans to play chess anyway were among the basic facts that had led Simon and Newell to their theory of heuristic problem-solving in the first place.

How do people play chess? Simon and Newell's first formal attempt to answer this question was their 1958 paper, "Chess-Playing Programs and the Problem of Complexity." Their answer focuses on the idea of selective search—the use of heuristics to guide an exploration of possibilities. They begin by observing that because it is a finite game, "chess can be completely described as a branching tree, the nodes corresponding to positions and the branches corresponding to the alternative moves from each position." Yet, the tree contains far too many branches for it to be explored in any depth. Making the machine faster helps a little, but not nearly enough, making the use of heuristics necessary. As they write, "selectivity is a very powerful device and speed a very weak device for improving the performance of complex programs."[8]

Thus, because their rationality was bounded, all chess players engaged in a selective search for good moves. But what about chess masters? If all players necessarily employed selective search techniques, then what made some players better than others? Was this where processing speed came in, with some people being able to carry the same selective searches a step farther than others? Did better players use different, better heuristics, somehow understanding the game better? Or did better players simply know more chess—were they able to recognize and interpret patterns on the board that novices simply did not see?

Simon and his student Edward Feigenbaum took up these questions, in much simplified form, in 1958–59.[9] They began with the observation that skilled readers of a language, like skilled chess players, were able to see patterns in the symbols of that language. Infants and young children, however, could not, just as novice chess players were not able to "see" certain arrangements of pieces as meaningful patterns. They had to learn to see that the letters were parts of words and that the words had specific meanings.[10]

Because verbal learning was a subject with a long experimental history— and thus much readily accessible data—Simon and Feigenbaum sought to find in the processes of verbal learning the answer to the basic question "How do people learn to see patterns in things?" Together they constructed a program that instantiated their theory of learning, dubbing it EPAM, for Elementary Perceiver and Memorizer. EPAM was based on the same core assumptions about homo adaptivus that characterized Simon and Newell's work with Logic Theorist, but whereas LT was a theory of the control system of the human information processor, EPAM was a theory of the perceptual and mnemonic systems.[11]

When Simon first began work on EPAM, he saw it as an adjunct to the central problem-solving "control system," represented by LT and its descendant, the General Problem Solver. Simon understood perception and memory in terms of their contributions to heuristic problem-solving, and EPAM's core elements were defined by their relationship to the central task of heuristic problem-solving. But EPAM, like any good adaptive organism, did not remain in stasis. It continued to evolve, and by its last version (EPAM IV), Simon had developed new ideas about expert memory, ideas that led to significant modifications of EPAM's architecture—and significant changes in his understanding of problem-solving.

Certain things about EPAM did remain constant. From the first version to the last, the core of the EPAM system was a small short-term memory and a "long-term semantic memory accessed from a discrimination net." These three structures—the STM, LTM, and discrimination net—were "operated on by programs of information processes, also stored in long-term memory." Both the "LTM and the discrimination net expand and are modified by learning processes like recognition and recall."[12] Because EPAM modeled learning processes, it had to be able to change and grow.

The most innovative element of EPAM was the concept of a discrimination net. The discrimination net was a treelike sorting mechanism. When a stimulus was perceived, it was sorted through a sequence of nodes at which tests were applied to select the subsequent node to which the stimulus would be passed. "Ultimately, the stimulus reaches a *leaf node,* which serves as an interface between the discrimination net and semantic memory. A leaf node contains no tests, but instead stores a partial image of the stimulus (a *chunk*) together with links (associations) to structures in semantic memory that contain additional information about it."[13] If this network structure sounds rather like the Internet, it should: the designers of the Internet's architecture, many of whom were influenced either directly or indirectly by Simon, shared a similar vision of human memory, which they hoped both to emulate and augment via computer networks.[14]

The discrimination net grows over time as more information is input, organized, and connected. The net grows as a tree grows, creating more leaves and more branches to hold them. Certain types of growth are quicker and easier than others, however: adding a leaf to an existing branch is very rapid, but creating a new branch or building an associative link between two leaves on different branches takes much more time—seconds rather than milliseconds. The ability to perceive new patterns in words (or on chessboards) involved growing new leaves at the margins of a verbal (chess) discrimination net, and people learned to see (and to generate) such patterns inductively, generalizing from experience.

Over the years, Simon returned to EPAM time and again, working with new collaborators each time, modifying and extending the program to account for new phenomena—and to respond to new criticisms. By 1995, he had moved on to EPAM IV, which was a new "version extended to account for expert memory . . . The main modifications of EPAM that are relevant . . . are a schema in long-term memory (called a *retrieval structure*) created by the expert's learning and the addition of an associative search process in long-term memory."[15]

In this new version, "Descriptions (full or partial) of objects . . . are stored as images at leaf nodes of the discrimination net or elsewhere in semantic memory. In particular, they may be stored in a *retrieval structure*, which is a specialized and learned tree of nodes and links."[16] This retrieval structure was in some ways simply an extension of the discrimination net, for it was built by the same processes, but it was a specialized structure possessed by experts and only by experts. Thus, Simon was able to claim that "With its rich memory structures and contents, through which previous experience interacts with current stimuli to determine behavior, EPAM IV is highly responsive to context, and its behavior is strongly 'situated.' Using semantic memory, it models an individual mind that is in close and constant communication with its physical and social environment."[17]

The import of these changes was that experts not only have better heuristics, learned from experience, but that they also are able to recall many more things directly rather than associatively. Their memory and their perceptual skills work differently in areas where they have expertise than in areas where they are merely intelligent novices. To illustrate this point, Simon notes that while chess masters can recognize and recall thousands of positions of chess pieces from game situations, they are only marginally better than novices at recalling random arrangements of pieces that would have no game significance.[18]

Experts, then, have more than a better theoretical understanding of the game; they have masses of data at their fingertips. Ready access to these data means that experts can see things that novices cannot. In addition, possessing this critical mass of information enables experts to induce general rules and to develop powerful heuristics for solving the problems characteristic of their fields.

This new understanding of expertise could be harmonized with Simon's earlier emphasis on the importance of conceptual schemes in organizing and giving meaning to experience, but the emphasis had shifted. It was the same tree of knowledge, but now the tree grew inductively, with the leaves generating the branches, rather than the branches producing the leaves.

Expertise, to the older Simon, was something acquired over time. Never one to leave a conclusion in vague form, Simon even went so far as to specify how much one had to know to become an expert in any field: fifty thousand "chunks" of information. (Simon borrowed the concept of the chunk—a unit of knowledge treated as a discrete entity by the mind's machinery—from George Miller, whose work was discussed in chapter 11.) It took roughly ten years to become an expert at any complex task because that was how long it took to build a semantic net of fifty thousand chunks of information, the critical mass necessary for truly expert performance.

This finding, Simon argued, applied to all areas of challenging endeavor and to all people, no matter what their gifts. Children did not become expert (fluent) speakers of their native language, for example, until they learned about fifty thousand words. Even a prodigy like Mozart did not become an expert until he had been a musician for ten years: according to Simon, his earlier music is interesting only because Mozart wrote it. Only the music he wrote from his teens and later was "world-class." Similarly, Simon argued that even a chess genius such as Bobby Fischer did not become an expert until he had played chess for ten years.

One should note that in these examples Simon used a rather slippery definition of expertise, talking about basic fluency in one example and extraordinary skill in the others. In part, this leveling of the field of expertise was a necessary correlate of his search for universal laws: genius had to be simply "more of the same"—more leaves on a bigger tree. If it were not, then inducing general principles from individual behavior—which is exactly what Simon's simulations aimed to do—was a problematic endeavor at best. Thus, Simon's defense of simulation, his embrace of a more Baconian empiricism, and his growing identification with his new discipline, cognitive psychology, all were connected to his increasing appreciation for the value of experience in both science and life.

Feedback, Simulation, and Surprise

The chain of reasoning linking heuristics, adaptation, and domain-specific knowledge was not the only thing that led Simon from syntax back to semantics. His responses to criticism were equally important in shaping the course of his career. His pattern of intellectual development was very much like a feedback loop, though the goals that drove him to modify or to extend his theories often were implicit rather than explicit. If a criticism came from within cognitive psychology or AI, then he dealt with it directly and explicitly; if it came from outside, then he translated it into his own language and addressed it implicitly.

The most relevant challenges to his early theories of human problem-solving were first, that computer simulation is tautological; second, that human knowledge is situated; and third, that a computer is a fully determinate system. Simulation is tautological, the argument ran, because it can tell us only whether a theory is internally consistent or not. A simulation cannot escape the assumptions that are programmed into it and so merely reproduces those assumptions instead of testing them. Similarly, human thought is always situated and contextualized, and Simon's simulations did not, and perhaps could not, deal with the interaction between an organism and its specific environment. Finally, because a computer is a fully determinate system, it is incapable of telling us anything we do not know already. Every action the computer takes could have been predicted with perfect precision. Hackles raised and teeth bared (in a sharp-edged smile), Simon took these criticisms as challenges and set out to refute them.

Simon's response to the first challenge was to defend computer simulation as "empirical inquiry" (see chapter 13). In his view, simulation forces a researcher to be precise and imposes a logical rigor that mere words do not. As a result, simulation not only helps us discover whether a theory is internally consistent but also helps us see—and thus to test—the implications of our theories and the assumptions upon which they are based. Every theory incorporates assumptions about the world, Simon argued, but only certain kinds of theories and certain ways of expressing them enable those assumptions to be tested. Even extremely sophisticated thinkers often are not able to specify their assumptions or to understand their implications without external assistance, Simon frequently observed, sometimes using his own flawed extrapolations as evidence.[19]

The successive versions of EPAM became Simon's response to the second of these criticisms. By creating a machine that simulated expert memory and expert ability to see patterns in data, Simon believed he had created a model of human problem-solving that was sensitive to context.

The challenge to simulation that Simon took the greatest pains to refute was the third: the idea that because a computer is a determinate system containing only the information provided by the experimenter, it is incapable of telling us something we do not know already. To Simon, all correct reasoning might be "a grand system of tautologies, but only God can make direct use of that fact." The same applies to computer simulations of complex systems. In almost all such cases, Simon argued, the only way for an experimenter to discover the outcome of a simulation was to run it, just as the only way to understand the behavior of complex machines (e.g., large, time-sharing computer systems) or complex organisms (people) was to observe them in action. Continuing the evolutionary analogy that became so central to his work from the late 1950s, Simon argued that the study of adaptive systems was best described as the "natural history" of such systems: "We can study them as we would rabbits or chipmunks and discover how they behave under different patterns of environmental stimulation."[20]

He also highlighted the capacity of simulations to surprise their designers. Beginning in the 1960s, but even more frequently in the 1970s and 1980s, Simon's articles emphasize that the outcomes of his simulations surprised him and his colleagues. He particularly stressed surprising results in simulations of domain-specific, situated knowledge—that is, of expertise—and consistently presented such surprising results as the spurs that had led him to undertake new research projects, the most important of which were attempts to simulate the processes of scientific discovery.

These attempts to simulate the creative processes of a certain kind of expert—the scientist—were parts of a larger project to develop a general theory of discovery as a special form of problem-solving. To Simon, discovery became a problem-solving process driven by surprising data. Thus, his late work on scientific discovery represented yet another synthesis—another stable subassembly—of the multiple aspects of his model of homo adaptivus, the finite problem solver.

BACONian Science: Simulating Scientific Discovery

Fittingly, surprise itself came to play an unexpected role in Simon's evolving conception of science. While the capacity to surprise the designers of a simulation had entered Simon's thought in the late 1950s as a defense of simulation's value as a research method, by the 1980s it had become a central element in his theories of scientific discovery.

Simon became interested in the process of scientific discovery in the 1970s, seeing the topic as a way of validating his methods—simulation—as well as his specific theories of problem-solving. Scientific arguments, this time about science itself, had to be put to the test of experience. A successful simulation of scientific discovery would validate simulation as science at the same time it supported Simon's theories of problem-solving.

Scientific discovery had many virtues as a research topic, for even more than chess, science had to be admitted to be a high intellectual endeavor involving all aspects of human thought processes. In addition, studying scientific discovery offered Simon the opportunity to bring together his theories of heuristic problem-solving and his theories of expert memory. Even more, discovery implied creativity and the solution of ill-structured problems, qualities that chess only possessed to a limited extent. Chess-playing can be creative, but it involves well-structured problems and clear rules. Discovery implied working at the edges of the known, where data is scarce and rules have yet to be found.

Simon's early work on the processes of scientific discovery, like his other work on cognition, focused on the importance of heuristics in the search for problem solutions. In a series of studies in the 1970s and 1980s, Simon set out to discover the heuristics common to scientific discovery, hypothesizing that these were the same universal heuristics common to all human problem-solving—factoring, means-ends analysis, hill-climbing, and so forth. By the 1980s, however, his research on expert problem-solving had made him aware of the power of domain-specific heuristics (e.g., in medical research, the heuristic that germs are the causes of most diseases, which implies that when confronted with a new disease, one should try to find an associated germ). Similarly, his research on expert memory had led him to the conclusion that experts recognized patterns in their environment that amateurs could not see, no matter what their heuristics. Amateurs could become experts, of course, by acquiring the domain-specific knowledge and heuristics, and this process of acquisition was structured by the more universal heuristics that all adaptive systems necessarily employed.

The best summation of Simon's research on scientific discovery is a 1999 article titled "Studies of Scientific Discovery: Complementary Approaches and Convergent Findings," which Simon coauthored with his colleague (and former student), the psychologist David Klahr.[21] Simon and Klahr compared and contrasted a variety of approaches to scientific discovery, focusing on four in particular: historical studies of individual scientists and their discoveries, psychological studies of problem-solving, sociological studies of innovation, and their own computer simulations of specific discoveries.

Much to the historian's relief, historical studies come off quite well from the comparison, having strengths that complement the other approaches. In fact, the complementarity of the four approaches is one of the major arguments of the piece: "The central thesis of this article is that although research on scientific discovery has taken many different paths, these paths show remarkable convergence on key aspects of the discovery processes, allowing one to aspire to a general theory of scientific discovery."[22]

That general theory treats "discovery as a particular species of human problem-solving." To this end, Simon and Klahr argue that "a model of discovery, like any theory, must be factored into two components: (a) its basic mechanisms, retained without alteration from one application to another, and (b) specific knowledge of the content and research methods of each task domain to which it is applied." The various components of expertise "constitute the strong or domain-specific methods." In keeping with Simon's continuing interest in the universal, the processes they choose to focus on are the "weak methods: domain-general, universal, problem-solving processes." Thus, "Although the strong methods used in scientific problem-solving distinguish the content of scientific thinking from everyday thought, we claim that the weak methods invoked by scientists as they ply their trade are the same ones that underlie all human cognition."[23]

Simon and Klahr identify five major weak methods of problem-solving: (1) generate and test (trial and error); (2) hill climbing (measure progress toward a goal and take the step that moves you closer most quickly); (3) means-ends analysis (compare goal state and current state; apply some operator to reduce the difference); (4) planning (form an abstract, simplified version of the problem space; solve the problem there and then translate it back); and (5) analogy (map a new target domain onto a previously encountered base domain).

Because discovery implies that a scientist is working outside the areas where his or her domain-specific knowledge and heuristics are useful, discovery involves the use of weak methods. As a result, in these domains experimentation is "steered by very general hypotheses," not the more targeted, specific hypotheses of work in familiar domains. The result is that "experimental outcomes generally guided theorizing, rather than theory guiding experimental design."[24] This conclusion stands in marked contrast to Simon's 1946 defense of his work in public administration, in which he wrote:

> I should like to assure you most emphatically that the theory generally led to the conclusions, and not the reverse, so far as the development of my own thinking was concerned. If you will accept this explanation, perhaps it will remove your perplexity as to how I "ever thought this stuff up in the atmosphere of 1313." It is erroneous to identify "the deductive method" if that is, indeed, what I use, with Thomism. The trouble with the Thomists is not that they deduce, but that they try to deduce from *a priori* truths rather than from empirical premises ... if there is no more to administrative theory than accumulated empirical observation, then there is nothing at all that we academicians can do for the practitioners than to record practices and to pass them on ... I am optimistic enough to believe that theory—based on correct empirical premises—has something to contribute to practice, as it does in the other sciences.[25]

In keeping with Simon's new view of the role of empirical induction versus theoretical deduction in science, Simon and Klahr make a strong case for understanding experimentation as "exploration," not as theory-testing. In contrast to the Popperian position that science progresses through the testing and elimination of flawed theories, they argue "that much of the important empirical work in science is undertaken—to use Reichenbach's phrase—in the context of discovery rather than the context of verification."[26] In support of this claim, they cite the success of BACON, a program constructed by Simon and his student Patrick Langley to simulate Kepler's discovery of his third law, along with several other major scientific discoveries. BACON was an "inductive machine" that searched for patterns in data and then generated hypotheses based on that data, hypotheses that required new data in order to be tested.[27] Thus, it experimented not only to test hypotheses but also to generate them, and the latter was the more important, more difficult, task.

In another article, Simon and his primary collaborator on computer chess, Fernand Gobet, applied this conclusion in defense of EPAM I-IV and their chess-playing programs, arguing that "Chase and Simon [describing a chess program from the early 1970s] took the usual and justified position that failure to account for all known phenomena is not grounds for rejecting a theory's core but rather an invitation to explain the deviation." Elaborating on the point, Simon and Gobet instruct their critics on the nature of science and discovery: "Experiments explore; they do not just test hypotheses; they often initiate theory rather than follow it." As a result, "the whole sequence of experiments on expert recall of chess positions is most fruitfully viewed as a series of problem-solving explorations that became increasingly interesting as new phenomena were revealed and ties were disclosed to other important phenomena." Further, "the line of research we have reviewed was not a series of isolated experiments—each pronounced a failure by significance tests—but a cumulative problem-solving search, in which early errors and insufficiencies led to new experiments, the discovery of new phenomena, and improved theory."[28]

Simulations, Simon and Klahr argue, also showed that "surprise at unexpected experimental outcomes could provide powerful heuristics for choosing the next steps in search." "A surprise can only occur when expectations that have been formed are violated," indicating that at least some conceptual scheme must be in place for data to have meaning. This conceptual scheme, however, can be quite general—a broad rule induced from general experience—and need not be anything as formal as a theory.

At the end of their article, Simon and Klahr state their final conclusion about science, discovery, and problem-solving: "We come now to our final generalization: the hypothesis that the theory of scientific discovery is a special case of the general theory of problem-solving, the special features being supplied by the strong methods of each discipline and the knowledge and procedures that support them, while the ubiquitous weak methods supply the commonalities ... Here we see the emergence of a general theory of data-driven science based on heuristic search."[29]

This general theory leads them to an interesting conclusion: "If we press to the boundaries of creativity, the main difference we see from more mundane examples of problem-solving is that the problems become less well structured, recognition becomes less powerful in evoking prelearned solutions or powerful domain-specific search heuristics, and more, not less, reliance has to be placed on weak methods. The more creative the problem-solving, the more primitive the tools. Perhaps this is why childlike characteristics, such as the propensity to wonder, are so often attributed to creative scientists and artists."[30]

Applying Simon to Simon

While "childlike" is not a word that leaps to mind when describing Simon, his theories of scientific discovery, and his theories of problem-solving more generally, were intended to apply to himself, as well as to historical figures and unnamed experimental subjects. In some ways they apply better to him than to anyone else: Simon was his own model organism.

There are three elements of Simon's model of homo adaptivus that are particularly helpful in understanding the evolution of Simon's thought: (1) the problem of "perfect adaptation," (2) the idea of heuristic search, and (3) the notion that complex structures, including complex ideas, are constructed out of "stable sub-assemblies" that may be combined, recombined, modified, and applied to new areas.

First, perfect adaptation: in *The Sciences of the Artificial*, Simon observes that if an organism is perfectly adapted to its environment, then when one studies its structures and functions what one really learns about is its environment. No qualities of the organism "show through" unless adaptation is imperfect: when one looks at the shape of Jell-O, all one sees are the effects of the mold.

At first glance, Simon seems in many ways to have been just such an organism, one perfectly adapted to the environment of CMU and the postwar behavioral sciences. He moved from success to success. Almost everything he wrote was published; every grant application funded, sometimes at levels beyond his expectations; and every committee or council or society he wanted to join invited him to do so. Following Simon's reasoning, then, one could argue that the study of Simon's life and work tells us much more about his environment than it does about him. If he wished to pursue interdisciplinary research in the systems sciences, then the environment must have rewarded such work; if he wanted to mathematize social science, to adopt behavioral-functional methods of analysis, and to construct computer simulations of human cognition, then the environment must have rewarded him for doing so. And, indeed, it did.

The seeming implication of such an analysis is that Simon was fundamentally an opportunist, one flexible enough to take on whatever shape the environment demanded of him. The thrust of Simon's argument about perfect adaptation, however, leads in a different direction. Such perfect adaptation, he argues, is almost impossible. No organism is perfectly adapted to its environment, so something always "shows through." This conclusion is fortunate, for it is hard to picture the dynamic, aggressive Simon as a kind of intellectual Jell-O.

Simon's homo adaptivus actively and consciously adapts itself to its environment, yes, but it does so in ways limited by its perceptions of that environment and by its own internal abilities. One vital way that any creature adapts is by shaping its environment to suit itself, a talent humans have in spades: homo adaptivus does not simply build automobiles to extend and amplify its powers of locomotion, it builds a network of roads and highways, remaking the world to suit both automobiles and itself. So too with Simon: he consciously shaped his own "environment of decision" at Carnegie Mellon, attempting to make the GSIA and the Department of Psychology into environments supportive of his kind of social science. In these cases, and in the cases of environments he chose not to try to join, such as the discipline of economics, Simon's personal qualities showed through, revealing him to be a creature of remarkable, but still limited, capacities.

Second, heuristic search: Simon, like all people, chose to follow certain paths and not others. Which paths he chose partly were determined by his basic heuristics: first, that understanding decision-making was fundamental to understanding both organizational structure and individual behavior; second, that human behavior was both rational and social; and third, that the human is an adaptive system, a finite problem-solver in an infinitely complex world. To this third basic heuristic could be added a corollary, the belief that simulation was the proper mode of analysis of complex, adaptive systems.

That Simon and all scientists search selectively for solutions to problems and that this search process is guided both by basic heuristics like the ones above and by "strong" heuristics drawn from specialized domains of knowledge seem plausible starting points for analyzing a scientist's intellectual journey. Seeing intellectual development as a kind of selective search leads one to focus on choice points, on why certain alternatives were considered and others not, and on why one specific path was taken rather than the others under consideration. It also reminds us that choices must be understood from the point of view of the chooser: an alternative path may exist in principle, but if an individual does not see it, then it does not exist for him or her. Alternatives have to be recognized or invented before they can be chosen.

In a similar vein, the limits of human time, energy, and understanding that require scientists to search selectively also require them to choose between different paths toward their goals, with no path necessarily being optimal. The choice of one path over another thus may reflect merely slight, momentary preferences rather than sharp-edged commitments. In addition, the importance of selective search reminds us that where one is going depends largely on where one has been, which is comforting to the historian. It also reminds us that there are multiple ways to move toward an intellectual goal and that sometimes the shortest path to one's destination may involve taking a detour early on.[31]

Third, stable subassemblies: Simon's observations about the importance of stable subassemblies in the evolution of complex structures offer valuable insights into his intellectual development and into the history of ideas more generally. Thoughts, like organizations or technologies, do not come from out of the blue. Rather, they are built out of existing components, which are continually being assembled, rearranged, modified, and applied to new contexts. In the process, ideas and practices often are brought together to form relatively stable groupings, which can be moved and manipulated as units, both by their creators and by others who encounter them. Such units, as Simon's career shows, often can be adapted to multiple uses: witness his multiple applications of near decomposability, tree structures, and heuristic search, not to mention the Yule distribution, the program, and the digital computer, that most protean subassembly of ideas and practices.

Simon's basic method of intellectual advance was to create stable sub-assemblies of ideas and practices and then to try to put them together in new and interesting ways. He did not attempt to assemble a general theory of human cognition all at once, though a unified theory of cognition was his larger goal. Rather, he created a model of one aspect of cognition (human chess-playing, say), extended it to another aspect (verbal learning), and then adjusted it to account for other related phenomena (expert memory), "chunking" together the components of an ever-larger, though never complete, theory.

Because his own research focused on problem-solving, scientific discovery, and cognition, Simon was more deliberate in choosing to build his theory out of such subassemblies than most scientists are. Many of Simon's own generation of behavioral scientists sought to develop a general theory for the behavioral sciences (or for their respective disciplines), usually seeing the concept of the system as the key to doing so. Still, it seems a useful heuristic for the historian to look for such subassemblies of ideas and practices and to explore how different individuals modified them, extended them, and adapted them to the solution of new problems, perhaps redefining their problems so as to be able to apply a subassembly they previously had found useful.

Such a heuristic must be used with some caution, for applying evolutionary parallels to the world of human thoughts and actions is perilous. Discovering the proper unit of selection is often impossible, for example, and cultural evolution is distinctly non-Darwinian in that acquired characteristics are heritable. Still, the general picture of humans as adaptable creatures shaped by their environments but capable of shaping their environments in turn seems a useful corrective to both the environmental determinism of the "strong programme" in the sociology of knowledge and to the "great man" tradition in the history of ideas. Simon, the adaptive scientist, chose to become part of certain fields and certain institutions and not others, selecting ones that he believed he could shape to suit his ideas and his goals. He succeeded in shaping those fields and institutions, but that they were fit to be shaped did not mean that they did not affect Simon in turn. Rather, Simon changed markedly as he adapted to a world he could influence but not control.

There are, of course, limits to the applications of Simon to Simon. Applying Simon to Simon does not give us Simon's picture of Simon, precisely, for he was not perfectly consistent in either his actions or his self-analysis. The endless series of reflections in this analytical hall of mirrors does leave one a bit dizzy, however.

The limits of Simon's analysis of science can be seen most vividly in his autobiography. Published in 1991, *Models of My Life* reflected his later understanding of scientific discovery and expert knowledge—and of the significance of his life's work. The title thus was no piece of whimsy (though it did reflect Simon's dry wit), for it was written to be a case study in support of his theories of problem-solving. It was a verbal model of his own path through the maze of discovery. In this, *Models of My Life* is similar to B. F. Skinner's three-part autobiography, which also was intended to support his views of human behavior. Simon's autobiography is probably truer to his theory, however, for his theory accepted verbal reports (such as autobiographies) as data, provided one interpreted them correctly.

Simon's theories of problem-solving and scientific discovery shaped *Models of My Life* in several ways. First, in it Simon treats his own creative acts as "more of the same"—that is, as another example of heuristic problem-solving. Simon knew that he was "smarter than his classmates" at every level, but he never presented any of his achievements as being the product of genius or of any other unique personal quality. Thus, in *Models of My Life* Simon comes across simply as being skilled at using universal "weak methods" in areas without "strong" heuristics. This skill, combined with exposure to a fortunate set of stimuli—the University of Chicago, the BPA, the Cowles Commission, RAND, the GSIA staff and students, especially Allen Newell, the digital computer—led him to apply those weak methods first to decision-making and then to problem-solving, and to do so with great success.

Second, *Models of My Life* follows Simon's late argument that problem-solving becomes rapid and productive as a rich empirical tradition develops. Knowledge in a specific domain must be accumulated in order for one to generalize and then to theorize. In *Models of My Life,* therefore, Simon's own science appears to be as "data-driven" as he believed Kepler's to be. If Simon had written his autobiography in 1950 or 1960, however, he probably would have emphasized novel concepts and precise, operational definitions as the root sources of discovery, arguing that without them perception was disorganized and research undirected. But *Models of My Life* was written in 1989–90, so it emphasized the data, particularly the access to new data enabled by the computer.

Third, *Models of My Life* is intended to illustrate the importance of a good heuristic. The domain-specific heuristics Simon and Newell developed for their new science—problem-solving is the crucial cognitive process, the human is a finite problem-solver, and simulation is the best method for exploring complex adaptive systems—are shown to be enormously productive, just like the "cell doctrine" in biology, even though they do not explain all relevant phenomena. In fact, Simon's autobiography seems intended to make the case that these heuristics were valuable precisely because they did not explain all relevant phenomena: a heuristic would not be a short-cut to knowledge if it did.

Fourth, and finally, *Models of My Life* exemplifies Simon's belief that problem-solving is individual but not personal. That is, problem-solving depends on an individual's experiences—the data he or she has acquired and the heuristics he or she has learned—but it does not depend on a person's emotional makeup or personality. Thus, in his autobiography Simon comes across as a problem-solving machine, an unusually clear-thinking, fortunate one to be sure, but a machine nonetheless. There are hints here and there at a desired—but unconsummated—affair, and Simon does address one salient feature of his personality in the chapter "On Being Argumentative." The former, however, are but vague references unconnected to the main narrative of his life and work, and his confrontational style is portrayed as a kind of "motivational control" for his cognition, directing his attention and intellectual energies to areas of conflict.

The above is not meant as a criticism of Simon for "leaving the juicy bits out" of his autobiography. Rather, it is meant to emphasize that, to Simon, the "juicy bits" were irrelevant to the central story. For the most part, I have followed Simon in this approach, leaving unexplored his relationship with his wife, Dorothea, and his three children. In part, this course was dictated by the nonexistence or inaccessibility of the necessary sources: when I searched the Simon papers at CMU, I found *one* piece of correspondence between Simon and any member of his family among nearly 200,000 pages of documents.[32] For the most part, however, I chose this approach because I, like Simon, believe that his scientific achievements were significant, not his emotional life.

His emotional life did affect his scientific career, of course. His relationships with his family and close friends primarily were enabling factors rather than directive ones; his family was stable and prosperous and his wife handled the child-rearing, all of which meant that Simon was not unusually burdened by domestic traumas and that he could work eighty to one hundred hour weeks when he wished. Further inferences about the connections between his home life and his work life must remain fairly speculative: unlike B. F. Skinner, he never designed baby-tending devices for his family's use; unlike Paul Samuelson, he never was denied a significant opportunity because of his ethnic background; unlike Alan Turing, he was not tormented by his sexuality.

It is likely that the relative absence of emotional troubles in Simon's life reinforced his confidence in the essentially rational basis of human behavior and in the stability of social systems, and it is possible that his parents' belief in equality reinforced his commitment to finding universals in human behavior. These are plausible connections, though I cannot document them. Still, I think that probably the most important thing about Simon's life outside his work was that it enabled him to be concerned primarily—indeed, almost exclusively—with matters intellectual. Such a situation is extremely rare, making Simon rather unique in his seeming normalcy.

Other, more personal qualities did play a role in Simon's life and work, however, revealing some of the limitations of Simon's view of science. These qualities appear in *Models of My Life* but are relegated to very minor roles. For example, Simon's argumentative nature not only was a part of his attention-direction mechanism but also a part of the reason for his success. Simon fought for his goals, vigorously and relentlessly, intimidating some opponents and overwhelming others. If he had been merely internally argumentative, using conflict to drive himself to devote long hours to his research but not to confront others in person or in print, then his ideas would not have had the influence that they did.

Simon's contentious nature attracted loyal friends and drove away enemies, helping him to reshape Carnegie Mellon, the National Academy of Sciences, and the behavioral sciences more generally. In addition, his particular style of argument—fierce but usually not *ad hominem*—often allowed him to remain friends with those with whom he argued. His usual opening was "Listen, *friend*," after which he proceeded to demolish that friend's argument.[33] By all accounts, Simon meant both words of that phrase quite seriously. He expected one to listen, carefully, but he also expected one to realize that he was attacking an argument, not a person.

The intensity of Simon's commitment to his work also raises questions of motivation that his "model" of his life simply does not answer. Simon clearly loved his work more than anything—and perhaps more than anyone. He worked long hours and thought his work was "great fun." He never was content to let a question rest or a challenge go unanswered. If he needed to know something in order to solve a problem, he learned it, even to the extent of mastering entire new disciplines. If he needed something new, a programming language, say, or a technique for dealing with multiple variables in a dynamic system, he invented it. If he made an assumption, he formalized it, elaborated it, and tested its limits. In short, his work was not so much "data-driven" as simply driven.

Simon's was a restless mind, always seeking to extend its reach, always looking to grow. As his colleague David Klahr recalls, one night he and Simon were driving back from a conference. Klahr observed that the cars were few and far between at that hour of night, which "got Herb started." Simon then began to attempt to estimate the average density of cars on the interstate for the country, starting with some simplifying assumptions about the shape of the nation: "assume the U.S. is a rectangle, 2500 miles long by 1200 miles wide . . . Estimate that the interstates are distributed on a grid, with the east-west routes approximately every 200 miles apart and the north-south routes approximately 300 miles apart . . ." The problem kept Simon pleasantly occupied for hours, leading Klahr to remark that "the man never stopped thinking. Ideas were his passion."[34]

Klahr's is a common testimonial made by colleagues of famous intellectuals and so should not be taken simply at face value. Such praise is intended not merely to honor a person but also to instruct others by celebrating certain values and certain qualities. Such a statement also fits certain conventions of academic praise and so may be generated almost automatically rather than as the product of measured reflection. Yet having a passion for ideas should not be dismissed as a motivating force for intellectual endeavor, especially by historians. Most of us chose our field because we had a passion for ideas that was stronger than our passions for money or power or fame, else we would not have become academic scholars, to whom money, power, and fame are but nodding acquaintances. It is unfair to refuse to attribute to others the same "noble" goals that we attribute to ourselves, just as it is unwise to refuse to analyze one's own motives the way one analyzes those of others.

But why the drive, why the passion? Simon at heart was a prophet and evangelist, a man who had glimpsed a deep truth and felt compelled to pull away the veils that hid its full depth and beauty. To Simon, as to any true believer, there was no topic that did not relate to his faith. (Hence his inability to make normal small talk without transforming the conversation into an experiment in cognition.) He had Good News, and he wanted to shout it from the mountaintop. That news was simple, but profound: the world can be known, despite its complexity and despite our limits, and in that knowledge lies our hope for a better world.

A Model Scientist

In this book I have told the story of the life and work of an individual scientist, Herbert Alexander Simon. This project did not begin as the story of an individual, however. Rather, it began as an attempt to understand the connections among the many changes that took place in the social sciences after World War II. My attempt to understand broad patterns eventually led me, like Simon, to study an individual and the relationship between his thoughts and his experiences.

This choice of focus has had consequences, of course. The events of Simon's life provided a narrative structure different from what I would have chosen for a study of an intellectual movement, such as the behavioral revolution. Furthermore, a biography necessarily must explore the personal qualities and experiences that made its subject extraordinary, unusual, individual. I have emphasized the evangelical character of Simon's belief that human reason was both bounded and powerful, his remarkable self-confidence and intellectual ambition, his contentious nature, and his unusual eagerness to apply his theories to himself and his own institutional environment. Taken together, these qualities gave Simon's work a strongly recursive character, with each step in his intellectual journey being influenced not only by his vision of his ultimate destination but also by his conscious analysis of feedback from every previous step. Like his autobiography, Simon's science had to be able to explain his life.

Similarly, I have emphasized Simon's experiences at the University of Chicago, the Bureau of Public Administration, Illinois Institute of Technology, the Cowles Commission, Carnegie Tech, and RAND as the events that shaped him, distinguishing him from his fellow social scientists. Ironically, while Simon believed that his interactions with these complex, changing environments were what made him unique, those very interactions were also what enabled me to connect him to the broader intellectual and institutional transformation of the social sciences. Simon would not have been surprised by this seeming paradox, for he believed that the study of an individual organism's relationship to its environment could reveal a great deal about that environment. Like Simon, I too believe that the study of an individual can reveal a great deal about the world in which that person moved, if one asks the right questions.

Simon's Evolution

The larger patterns that connect to Simon's life and work include several of the central themes in the history of twentieth-century human science: the rise of functionalism, behavioralism, and mathematical analysis; the institutional transformation of the social sciences concomitant with the rise of new patrons, especially the Rockefeller and Ford Foundations, the military research agencies, and, later, the NSF and NIH; and the advent of new research practices, including laboratory studies of human behavior, as well as computer modeling and simulation. Because Simon was so eclectic in his interests and influences, the study of his life and work illuminates these developments as they played out in a range of fields, from political science to psychology, sociology, and economics.

Simon's work also was part of an ongoing attempt by social scientists to respond to the challenges of change, interdependence, and subjectivity posed by modern society, challenges that produced two distinct approaches to human science in response. One of these approaches emphasized the ability of the individual human actor to make free rational choices; the other emphasized the plastic nature of humans and the limits to their reason. The first of these approaches characterized the *sciences of choice*, such as neoclassical economics, game theory, and decision theory, while the latter belief characterized the *sciences of control*, such as sociology and social psychology.

Simon sought to bring these two approaches to human science together, eventually developing a new model of the human organism, its environment, and the science that would study the relationship between the two. His new model of the human organism was homo adaptivus, which depicted humans as adaptive, problem-solving organisms of finite powers that moved in an infinitely complex world. His model of the environment was the hierarchically organized complex system, typically depicted as a tree structure. Although Simon saw the environment as infinitely complex, its treelike structure meant that it was fit to be known and thus fit to be shaped, even by the bounded reason of homo adaptivus. Simon's model of human science followed naturally from these models of nature and humanity: human science was the study of homo adaptivus in relation to its environment. Its basic orientation was behavioral and functional; its characteristic method was simulation; and its essential formalism was the program.

Simon's central goal was to bring together the sciences of choice and control and so to create a new model of human behavior and a new method for human science. Simon's basic assumptions about science and the world set the parameters for this new model and method. First, he believed that there was an order to nature, even human nature. All behavior was caused, and the causes were lawful, ordered. Second, he assumed that this order was universal, meaning that the complex and the local always were manifestations of the simple and the global. The reduction of complex phenomena to the simple mechanisms by which they were generated thus was the basic task of science. Third, he held that this order was accessible to humans through reason, not revelation. A rigorous science of human behavior was possible, if the proper concepts could be found and the proper methods employed.

Fourth, and finally, Simon believed that a theoretically robust and practically useful science of human behavior had to embrace both the power of the environment to shape human thoughts and actions and the power of humans to shape their environment. A true science of human behavior had to be able to describe actions as rational, within certain bounds, and to be able to specify the nature of those limits. In addition, it needed to account for the responsiveness of human actions to a changing environment, meaning that it had to deal with contingent actions. It had to account for responses to those actions as well, meaning that it needed to deal with sequences of contingent actions unfolding in time.

Simon's first attempt to develop a human science that met these criteria centered on a model of the human actor that he called homo administrativus, or administrative man. Homo administrativus was an organization man. He was a member of a number of formal organizations, each of which specified a set of premises on which he based his decisions. He identified with his organizational home and accepted the authority of his superiors, within the bounds set by his allegiances to other organizations, including the superorganization called society. Homo administrativus was rational, but his rationality was severely bounded by his organizational affiliations. At the same time, those affiliations and the limits they placed on his reason were what made rational action possible.

Simon developed this model of human behavior during his years as a graduate student at the University of Chicago, his work at the Bureau of Public Administration at the University of California at Berkeley, and as a member of the Department of Political Science at the Illinois Institute of Technology. It continued to play a major role in his thought in his early years at Carnegie Tech, but by the time he reached Carnegie Tech, Simon already had begun to look for ways to incorporate the powerful mathematical tools of the sciences of choice into his work. With such tools, he might be able to specify how homo administrativus behaved within the bounds of its reason.

At Carnegie Tech, Simon began to search for ways to broaden his model of human behavior and thus bring a new unity to social science. This synthesis would take place on many levels, he hoped, bringing together not just choice and control but also theory and application. Intellectually, he wanted to bring together the sciences of choice and control through the study of decision-making in social systems. Institutionally, he sought to create a set of interdisciplinary research centers and a network of patrons that would support social science that was mathematical, behavioral-functional, and problem-centered.

Simon's efforts to develop and integrate the multiple "islands of theory" in social science led him to a new synthesis. He developed a new model of man to replace both the perfectly rational, perfectly free homo economicus of the sciences of choice and the perfectly malleable, perfectly docile homo administrativus of the sciences of control. This new model of man was homo adaptivus, the human as a limited but still capable problem-solver, with problem-solving being understood as a process of conscious adaptation to an environment.

Simon's work on administrative decision-making revealed to him the importance of the principle of bounded rationality and the centrality of the problem of choice, and it taught him to see both the individual and the organization as decision-making machines. From cybernetics and servomechanism theory he added the ideas that organisms, organizations, and adaptive machines were functionally equivalent, not merely similar; that feedback was an essential component of all adaptive systems, organic and mechanical; and that adaptive systems could evolve enormously complex behaviors by nesting rather than chaining simple behavior mechanisms. Here the term *evolve* is particularly important, for, in Simon's view, a process akin to natural selection was the mechanism that produced both adaptive behaviors and their organization into a hierarchical system of behavior.

Finally, from Gestalt theory Simon learned to think of learning and problem-solving as processes of cognitive adaptation. Homo adaptivus adapted to its environment by learning to construct simplified mental models of that environment, models that served as the reference points not only for decisions as to how to achieve its goals but also as the basis for defining the goals themselves.

These ideas all came together for Simon in 1952 at RAND's Systems Research Laboratory. At the SRL, Simon first encountered the digital computer, the concept of the program, the technique of simulation, and the mind of his intellectual soul mate, Allen Newell. Together, Simon and Newell developed the model of homo adaptivus, focusing on its information-processing capabilities, and created the programs that would serve as the exemplars for cognitive psychology from the late 1950s to the present.

In the course of developing his model of homo adaptivus, Simon reoriented both his intellectual and his technical practice around the computer. On the technical level, digital computers became a focal point of his work as he attempted to program them to simulate human behaviors. On the intellectual level, his basic questions—and his answers—all were translated into the language of information processing. On the institutional level, the adoption of the computer as his primary research instrument also involved the creation of a new set of professional relations, including a new professional identity for Simon and a reconfigured network of patrons for his work.

Simon's own evolution did not cease with homo adaptivus, however. Rather, Simon's theories of heuristic problem-solving, as modified in response to certain discoveries and certain critiques, led him to develop ever more individualized and contextualized accounts of cognition. These new accounts were as mechanist and reductionist and as oriented toward the development of a general theory as his earlier work, but they increasingly focused on domain-specific cognition, usually described as expertise. If the shift from his early work in public administration to his mature work in psychology can be described as a move from homo administrativus to homo adaptivus, with the former being a special case of the latter, the changes in Simon's later work can be described as the elaboration of homo adaptivus into homo adaptivus expertus, the expert problem solver.

This evolution reflected Simon's growing appreciation for experience and domain-specific expertise, which stood in marked contrast to his youthful enthusiasm for the development of theory and for interdisciplinary research. As his work became more bounded by disciplinary agendas and supported by patrons of "basic science" (NSF, DARPA, NIMH), the distance grew between it and his original concerns with reconciling not only choice and control but also research and reform. The Simon who studied homo administrativus was "an intensely political animal" who created not only a new theoretical structure for administrative science but also a new rationale for an active—but accountable—government. The Simon who studied homo adaptivus no longer read the newspaper, except to solve the crossword puzzle.

Beyond Simon

What does this story of Simon's life and work tell us about his world? Was he typical, extraordinary, aberrant? Did he march at the vanguard of a revolutionary movement; was he carried along by a rising tide; or was he simply one of many ships passing in the night?

Simon was unique in the degree to which his science took as its subject the very organizational revolution that shaped his outlook and in the degree of consistency he demanded from his philosophical, theoretical, institutional, and methodological agendas. He was unusual in the breadth of his interests, remarkable in the power of his mind, titanic in his ambitions, and mildly eccentric in his daily habits. The sum was a man who, to adapt a phrase from Orwell, was more unique than most.

At the same time, Simon's ideas won wide acceptance and had such a marked influence because he articulated a coherent vision of humanity, nature, and knowledge that organized and gave meaning to ideas and experiences shared by many intellectuals in the middle third of the twentieth century. This set of common ideas was rooted in the bureaucratic worldview: the prefiguration of the world as an adaptive, hierarchic system.[1] Understanding this shared worldview helps link Simon, the unique individual, to the larger world. In so doing, it helps connect some of the most significant intellectual trends in human science to one of the most important developments in modern history, the rise of large-scale bureaucratic organizations.

Early in the twentieth century, the Dutch philosopher and historian E. K. Dijksterhuis wrote a famous book in which he argued that one of the crucial aspects of the intellectual transformation usually called the scientific revolution was the adoption of a mechanical worldview. In this mechanical worldview, the workings of complicated machines, especially automatic machines such as clocks and water-powered automata, were the essential reference points for descriptions of both nature and society.[2]

While there has been much debate regarding the exact nature of the components of this mechanical worldview and the ways it influenced seventeenth-century understandings of nature, the core of Dijksterhuis's argument has survived. From Lewis Mumford, Carolyn Merchant, and other critics of the mechanical worldview to Stephen Shapin, Simon Shaffer, and other critics of the idea of the scientific revolution, one finds wide agreement that a new philosophy emerged in the seventeenth century and that this philosophy was predicated on the assumption that the world was a complicated machine.[3]

Scholars have made similar arguments linking other significant changes in worldview to large-scale shifts in technological structures, with new technologies becoming sources of new models and metaphors, as well as new experiences. For example, Jay Bolter has written on the crucial importance of the spinning wheel in the shaping of ancient Greek and Roman thought, Otto Mayr has argued that automatic machinery played a critical role in the development of new ideas about liberty and authority in the early modern period, and Norton Wise and Crosbie Smith have explored the ramifications of the momentous shift from the balance to the engine as the fundamental metaphorical referent underlying both political economy and physics in the nineteenth century.[4]

Similarly, Anson Rabinbach has argued that the idea of the "human motor" lay at the heart of a host of new ideas, attitudes, and institutions in the late nineteenth century, and Laura Otis has noted the vital role that the telegraph and telephone have played as models for all sorts of systems, biological, physical, and social, especially the human nervous system.[5] The computer and, more recently, the Internet likewise have become central supports for many new ideas about mind, body, nature, and society, as Paul Edwards, N. Katherine Hayles, Manuel Castells, and a host of other scholars have shown.[6]

To this listing of metaphorically resonant technologies one must add organizational technologies, such as the factory and the assembly line. Like their more material counterparts, these organizational technologies have been shown to be intimately connected to broader ideas about mind, body, nature, and society. Simon Schaffer, for example, has argued that Charles Babbage and his intellectual allies saw the human mind as a factory and mathematical logic as rational production system.[7] Similarly, Emily Martin has found that women's bodies in the mid-twentieth century typically were understood as factory-like production systems, particularly by physicians interested in managing their reproductive processes.[8] Simon and Newell's use of "production systems" as the format for describing both human and computer operations thus was in keeping with a long tradition. Simon attested to this connection, arguing on more than one occasion that Adam Smith was the discoverer of the essential principle at the heart of computing: the division of labor.[9]

The factory and the assembly line, however, are but two of the three central organizational technologies of the modern era. The third is the bureaucracy, which has not yet been recognized as being a fertile source of metaphors in science. In some ways this lack of recognition is surprising, for historians and social scientists long have held that the central development of the second Industrial Revolution was the creation of large-scale bureaucratic organizations, specifically the great industrial corporations and the governmental bureaucracies that were created to regulate them. From Max Weber to Talcott Parsons to Alfred Chandler to Robert Wiebe and the pioneers of the "organizational synthesis" of American history, the bureaucracy has been seen to be one of the defining social forms of the twentieth century—for good or for ill.[10]

In other ways, however, the relative lack of attention to the conceptual significance of bureaucracy is not surprising. Bureaucracies are intangible things, after all, despite the myriad monuments of glass and steel that have been built to house them. As a result, one might expect that bureaucracy would be a less vivid metaphorical referent than the potter's wheel, the clock, or the computer.

In addition, in all of the studies mentioned above, the technologies in question acquired the ability to serve as heuristic models for thinking in a variety of fields because they changed understandings of the human body, which was and is a basic metaphorical referent for human thought, as George Lakoff and Mark Johnson have shown.[11] Humans may not be the measure of all things, as E. M. Forster wrote, but they are the ones who measure and thus the ones who count. Until one could build a connection between the bureaucracy and the body, then, bureaucracy likely would be only a weak support for metaphors and analogies.

The first step in linking the bureaucracy and the body was the idea that the organism is an organization, and vice-versa, an idea that flourished in nineteenth- and early-twentieth-century biology, especially physiology. Herbert Simon and the systems scientists of his generation took this idea two steps further, first by formalizing it in the language of cybernetics and systems dynamics, and then by instantiating it in the digital computer, a device that they found fascinating because it was defined almost entirely by its organization.

Thus, one way of understanding Simon's unique contributions is to see him as a theorist who made the links between organism and organization formal and explicit and who elaborated their implications by reference to a very material, very powerful technology, the digital computer. To Simon, the programmed computer was an electronic bureaucracy, which made it a model organization, a model organism, and a model mind. Through the digital computer, ideas about mind and body could be connected to ideas about bureaucracy, changing the way people in many fields thought about themselves and their world by giving them a new way to organize their experience.[12] In short, one reason for the influence of Simon's ideas in so many fields was that he built a connection between the bureaucracy and the body, using the digital computer as the bridge.

The wide influence of Simon's ideas, coupled with the broad, intense fascination with cybernetics and the systems sciences at midcentury suggests that Simon was not alone in adopting a bureaucratic worldview. It is not too much to suggest that the bureaucracy played a similar role in the scientific revolution of the mid-twentieth century to that played by the machine in the scientific revolution of the seventeenth century, with the caveat that, in most cases, a material bridge (the digital computer) was necessary to carry the bureaucratic worldview to new fields.

The essential components of this bureaucratic worldview, if Simon is any guide, were (1) a prefiguration of all subjects of study as adaptive, hierarchic systems, with a correlate emphasis on the behavioral-functional analysis of organizational properties, especially those properties (such as communications and control systems) that enable internal coordination of the system so as to maintain equilibrium; (2) an acceptance of some form of analytic realism, since a bureaucracy is a mental construct used to describe a system of intangible (but very functional) relationships;[13] (3) an acceptance of at least a weak holism, since new properties often emerge at successively higher levels in the organizational hierarchy; (4) an idealization of formal, instrumental reason, exemplified by the development of procedural logic and by the exaltation of objectivity, as in the development of ostensibly value-neutral algorithms, protocols, and programs; and (5) an interest in the means by which systems store and process information about themselves and their environments, often expressed through the formal analysis of information and the symbols that represent it (e.g., information theory, communications theory, Chomskyian linguistics, analytic philosophy).

The increased interest since the mid-1970s in networks (rather than systems), chaos and complexity (as opposed to organization and hierarchy), flexibility and the spontaneous production of order from disorder (as opposed to stability produced through continuous management), and contextual, situated knowledge (as opposed to formal, instrumental knowledge) thus can be seen as a reaction against the narrower forms of a bureaucratic worldview, a reaction concurrent with broader public disenchantment with bureaucratic rigidities. The adherents of the bureaucratic worldview marched under the sign of the Tree; today the Net seems a more congenial symbol.

To document this broad claim and to explore its ramifications in detail would be the task of another book, perhaps several, so my claim at present is no more than a hypothesis. Yet it is one worth exploring.

Legacy

Science. Objectivity. Expertise. Progress. For Simon these were words to conjure with, but to humanist scholars in the twenty-first century they summon not allegiance but knowing smiles. We who have tasted the fruit of the postmodern tree of power/knowledge believe ourselves wiser about the ways of experts than was Simon's generation, but with this wisdom has come a certain sadness, a certain loss of faith not only in science and experts but in reason and, indeed, in democracy. What, then, is Simon's legacy to the postmodern world? Can we learn from him as well as about him?

As befits a man who moved from field to field, Simon's legacies are multiple. During the 1950s and 1960s, the growing community of behavioral scientists shared Simon's bureaucratic worldview and aspired to create a behavioral-functional, mathematical, problem-oriented social science. Political science experienced a behavioral, economics an econometric, sociology a functionalist, and psychology a cognitive revolution, all linked by their shared vision of a new science of adaptive systems.

This community diversified as it grew, however, and academia bought synthesis in much the same way Simon did, building an ever-expanding structure of specialized compartments. In each of the behavioral sciences, and in the "management science" that was built on them, this compartmentalization led their leaders in the 1970s and 1980s to proclaim that their fields were "adrift," "fragmented," in "crisis." Some blamed Simon and his behavioralist allies for the "descent" of political theory, the "crisis in public administration," the "coming crisis of Western sociology," or the "misdirection" of psychology.[14] (The "crisis in economic theory" usually has been laid at other feet.)[15] Others, however, found virtue in fragmentation, holding that a community encompassing both Kantians and rational choice theorists, cognitive scientists and psychoanalysts, was stronger for the diversity.

With this diversification has come a divergence from the path Simon charted. This divergence is most notable in economics, where homo economicus now appears to reign triumphant, clothed in the calculus of subjective expected utility theory. Here Simon's legacy is like that of a grain of sand in an oyster: an irritant, yet to produce the pearl of a new paradigm.[16] Perhaps his calls for an empirically based microeconomics will be heeded by more than a minority in the new millennium, and the awarding of the 2002 Nobel Prize in Economics to Daniel Kahneman, a psychologist very much in the mold of Simon, gives Simonians reason for hope. Still, the enormous success of the economists over the past thirty years has given them great professional self-confidence, which makes major conceptual and methodological reforms appear unlikely, at least in the near future.[17]

More fundamentally, the new hopes for synthesis in recent years have been the concepts of the network rather than the system, of complexity rather than hierarchy, flexibility rather than stability, and contextual rather than formal knowledge. Simon himself anticipated this shift to what one might call a "networked worldview" in his landmark essay "The Architecture of Complexity," but, as the title indicates, it was the architecture of complex systems that still held his interest.[18] While Simon saw organization and complexity as two sides of the same coin, others took the discovery of complexity to be a repudiation of organization or, more precisely, of intervention. Since the 1970s, many social scientists, particularly but not exclusively neoconservative ones, have become entranced by the "law of unintended consequences" that supposedly dooms all efforts to intervene in our complex, interdependent society, attacking "Deweyan" planning along with the "Miesian."[19]

At the same time, the excesses of both bureaucratized warfare and bureaucratized welfare have inspired many to look for new models of humanity and society. After a brief flirtation with more organic models in the late 1960s and 1970s, public discourse in America seems to have returned to technological models, especially following the sudden advent of the Internet in the 1990s. In many ways, however, this new networked worldview is another stage in the evolution of the bureaucratic worldview, not a new species of thought descended from another line: the Internet, after all, has a decentralized hierarchic structure, not a fully distributed one. Similarly, the network is a generalization of the concept of the system, not a repudiation of it; complexity often turns out to have a hierarchic structure, as evidenced by the prevalence of "self-similar" structures; flexibility frequently is valued for its contributions to maintaining stability; and contextual knowledge can be every bit as instrumental as formal, procedural knowledge. Thus, this new species of ideas has evolved through the adaptation of older forms to a new environment.

Lessons Learned

I began my study of Herbert Simon, years ago, as an instinctive supporter of his views on economics and critic of his psychology and his political science. I did not have much of an opinion regarding his ideas in computer science, thinking the subject too arcane for me to understand. I still find his economic views appealing, but I have come to be much more sympathetic to his work in psychology and political science. (I still find his computer science rather arcane, though it seems more fruitful to me than it once did.)

In part this change is due to my discovery that Simon held many of the same political ideals that I do, supporting equal rights for all, a federal government that actively supports those rights, stewardship of the earth's resources, and rational tolerance of different peoples, cultures, faiths, and political views. Hence, his science did not lead him to embrace the political ideals that I once had thought were tightly associated with strong positivism and reductionism. Simon was no authoritarian believer in centralized power, nor was he a proponent of technological fixes to social problems (at least, not of simple fixes), nor did the idea that the human is a biological machine lead him to justify treating other people as means rather than ends. He was no humanist, but he was a very humane mechanist.

How well an idea meshes with one's political values is not the best basis for judging scientific contributions, of course, but scholars should be honest about their biases: in this case, I came to think better of Simon as a consequence of discovering that my biases had led me to make some false assumptions.

Becoming a parent also persuaded me of the utility of Simon's psychological ideas, much to my surprise. Watching my infant son acquire new skills and assemble them into larger complexes of behaviors has been particularly illuminating. While I would not call his attempts to walk "ultrastable" as yet, he certainly is a highly adaptive little organism, acutely attuned to his environment.

Perhaps the most useful thing I have learned from Simon is that fantastic complexity and diversity in behavior can be the product of a small set of simple, common mechanisms interacting with a complex, varied environment. This idea seems a useful heuristic for historians interested in understanding the relationship between the local and the global, the individual and the universal. Translated into a historian's language, this principle might read "people in different times and places do share important similarities, making it possible for us to find meaningful patterns in history, but among the most important of these similarities is that people everywhere must adapt themselves to their environments, no two of which are identical, meaning that context and contingency are powerful agents in history." A corollary to this view is that individuals do have agency but that the extent of this agency can vary;

different environments allow different degrees of freedom and so are fit to be shaped in different ways and to different degrees.

While Simon's theories of human problem-solving do seem to shed light on significant areas of human experience, there are areas where homo adaptivus strikes me as less useful. The most obvious of these areas is emotion: Simon's theories of human problem-solving simply do not help me (or many others, to judge by current research in the field) understand any emotion other than the frustration caused by cognitive overload. While it seems perfectly plausible that emotions are adaptive and thus rational in an evolutionary sense, such knowledge is rather difficult to apply to relationships in the present. Simon was well aware that emotion was a vital part of human experience, and he thought that his theories were compatible with his emotional experiences, which may well be true, but problem-solving is a long way from loving or hating. There is a reason why means-ends analysis is thought unromantic and cost-benefit analysis is not the typical basis for religious faith.

In the end, however, Simon's vision of humans as creatures of bounded but still meaningful rationality seems a sounder guide to designing our political environment than do its competitors. His basic understanding of humans as both limited and capable, plastic and purposed, certainly fits my understanding of my own mind. Our reason is limited but still powerful. Our knowledge of the world always will be incomplete, bounded, and biased, but it can become more complete, more expansive, and less parochial, and to these goals we should aspire. Only by such striving can we enlarge the bounds of reason and design a better world.

Patrons of the Revolution

Postwar social science was shaped by two distinct, successive patronage systems. The first of these thrived from 1945 to the mid-1960s, while the second began to take shape in 1958, grew throughout the 1960s, and became clearly dominant in the early 1970s. The two systems overlapped between 1958 and 1968, a period of enormously rapid growth in funding for social science research.

As discussed in chapter 7, the first of these systems was shaped by the program officers (and their advisers in academia) at several foundations, the Social Science Research Council, and a range of military research agencies, all of whom consciously sought to promote research that was behavioral-functional, mathematical, problem-centered, and interdisciplinary. The primary foundations involved were the Carnegie Corporation, the Rockefeller Foundation, and the Ford Foundation, with the latter being by far the largest of the three. The primary military research agencies sponsoring social science research were the Office of Naval Research, the Air Force Office of Scientific Research, RAND, and various research units in the U.S. Army, including the Operations Research Office and several units that performed psychological research. Military agencies also often sponsored social science research as a part of large technical projects, such as SAGE, making it difficult to separate out sums for social science research. The advisory boards for these agencies and the foundations interlocked.

The National Science Foundation provided some small funds for social science research from 1954 onward, as did the National Institute of Mental Health, but neither of these federal, civilian agencies played a significant role in the support of social science research until 1958. NSF grants were highly prestigious, so its influence was somewhat greater than its dollar presence would indicate, but the total influence of the NSF before 1958 was small nonetheless.

The second of these patronage systems began to take form in 1958 in response to the Soviet launch of the Sputnik satellite. As its central institutions grew in the 1960s, they came to have increasing influence on funding policy generally. This second system was shaped primarily by scientists and program officers at civilian federal agencies, primarily the NSF and NIMH but also the Defense Advanced Research Projects Agency (DARPA) and the other institutes of the NIH. I include DARPA under the same rubric as NSF and NIMH because DARPA was established to support research that was more "basic" and more discipline-oriented than the mission-oriented military research agencies. In addition, the relevant programs at DARPA were under civilian control. The program officers at these agencies consciously sought to promote research that would advance the several social sciences as disciplines.

Typically, these agencies embraced work that developed specialized methods, techniques, and technologies appropriate to specific disciplines and that tended toward the "basic" end of the spectrum between basic and applied science. In addition, while rapid growth in funding for social science research characterized the first system and the overlap period of the 1960s, the second system was characterized by stagnant (and sometimes declining) funding, when one accounts for inflation.

Neither the military patrons of social science research nor the foundations disappeared, with the exception of the Ford Foundation. Their effects became much more localized within social science, however. The ONR, for example, continued to provide support to certain strands of cognitive psychology in the late 1960s and 1970s, but its influence on psychology as a whole dwindled relative to that of the NSF and NIMH. Military research in the social sciences became much more separated from mainstream research in the wake of Project Camelot, the Mansfield amendment, and a general concern, shared by both the left and the right (though for different reasons), that the military was funding research not closely related to its mission. The Ford Foundation abandoned its program in the behavioral sciences at the end of 1957, shifting its resources to the support of "area studies," further reducing the foundations' impact on social science (except in those fields that received funds for area studies, such as human geography and the study of developing nations).

This second system overlapped with the first from roughly 1958 to the mid-1960s, but by the late 1960s the balance had shifted decisively in favor of federal, civilian support for relatively "pure" science, as the established leaders of the social science disciplines defined it. Changes in the leadership of the relevant programs at the civilian, federal agencies accentuated this shift in the relative weight of the different patrons. While many of the key program officers for the behavioral and social sciences in the NSF, NIH, and DARPA between 1958 and 1964 maintained a philosophy similar to that of the leaders of the first system, a new generation of program officers took over between 1964 and 1972. The primary exceptions to this rule were the various agencies associated with the War on Poverty, especially the Office of Economic Opportunity, which funded a new wave of interdisciplinary research projects and institutes related to poverty from 1964 to the early 1970s. Thus, in areas related to poverty studies, the period of overlap lasted until the early 1970s, at which point the second, discipline-based system began to assert itself.

This shift in patronage systems and its connections to the ideas, practices, and institutions of social scientists can be seen clearly in the career of the over 250 interdisciplinary social science research institutes created in the first two decades after the war. A great many of them experienced a similar life cycle: establishment and growth between 1945 and the mid-late 1960s followed by fragmentation and redefinition, usually between the mid-1960s and early 1970s.

The Reagan-era cuts in funding for social science research (especially in 1983) prompted a new round of much more vigorous (and fairly effective) lobbying for federal support for social science research. The changing fortunes of the social sciences in the 1980s, combined with the multitude of laments regarding the "crisis" in every discipline (with the crises usually being associated with hyperspecialization—and with increased conflicts over how to divide a pie that never seemed to grow), spurred a revival of interest in interdisciplinary research. Because the interdisciplinary ideal has been most strongly associated with mission-oriented basic research, I suspect that interdisciplinary rhetoric was more common than interdisciplinary research in the late 1980s and 1990s, though my evidence for this is only anecdotal.

The change in patronage systems and the concomitant shifts in the fortunes of interdisciplinary research institutes are also apparent in the natural sciences, with two important qualifications: first, the continued importance of the military in supporting research in many areas of physical science enabled the continuation of problem-oriented (and therefore more interdisciplinary) research; second, the growing importance of large-scale technologies in the conduct of research in the physical and biological sciences encouraged interdisciplinary work oriented around the use of those technologies.

Some Thoughts on Patronage

While there are many studies in the history of science that discuss patronage, many do so in a loose and vague manner. In particular, many studies are quite unclear about the nature of the influence that a patronage system is supposed to have exerted upon an individual or institution, and many studies do not differentiate between the effects that patronage had upon an individual versus the effects of patronage upon different groups.

There are several distinct ways a patronage system can affect a science. At one level, patrons can provide differential support for or against certain philosophical stances (e.g., the unity of science, reductionism), methodological approaches (e.g., quantitative analysis, simulation, experimentation), research topics (e.g., human-machine interaction, cognitive processes), institutional or organizational forms (e.g., interdisciplinary research institutes, team research), research practices (e.g., conducting surveys, running simulations, programming computers, interviewing subjects), or research products (e.g., specific technologies or techniques, such as command-and-control technologies), as well as differential support for (or against) certain groups defined by social criteria such as race, gender, geographic location, or class. All these effects are the consequences of deliberate decisions made by patrons or their agents, though some criteria are likely to be implicit rather than explicit factors in decisions.

The social and political values of patrons are expressed through the above categories, though it certainly is possible for a patron to demand adherence to certain social values or ideologies as specific products, rather than the more general kinds of products suggested above: for instance, the Manhattan Institute is not going to support research that implies that the federal government should spend more on social welfare programs, no matter how congenial they find every other aspect of the project. Most academic researchers consider such a specific value- or ideology-based demand for a research product to be illegitimate, but that does not mean that there are not a number of institutions that make such demands and support such work, nor does it mean that there are not a number of professionally trained social scientists who will conduct such work.

At another level, a patronage system can have certain effects that are largely independent of the specific choices of patrons regarding individual projects.

The most important of these are the effects due to the general scale of support and to the structure of the patronage system. The two most significant scale effects have been specialization, which generally has increased as the size of the market for social science research has grown, and expanded capabilities, especially access to expensive equipment, which has enabled certain kinds of work (e.g., computer simulation) that would have been impossible or very rare at lower levels of funding. Another scale effect has been the ability to conduct large-scale research, such as large social surveys or social experiments.

An increase in the scale of support for a field also has tended to increase its prestige. Increased prestige brings increased support in many cases, creating a benign regress for that field. In the case of the postwar social and behavioral sciences, winning support from powerful patrons typically came first, with an increase in prestige being a consequence rather than a cause of increased funding. Patrons had to respect a field enough to see *potential* before they began to fund research in it, but they did not require a long record of concrete achievement. In the first postwar period, powerful patrons saw great potential in certain individuals and certain approaches, not in any discipline as a whole. Military and foundation-based patronage thus elevated certain individuals and approaches within the social science disciplines. In so doing, they raised the status of the disciplines as well.

Another relevant structural effect has had to do with the relative centralization of decision-making regarding support and the organization of the review process. A distinguishing feature of the second postwar patronage system was the relative concentration of resources: there was roughly the same amount of money to be had as in the first system, but the great bulk of it now came from two sources rather than ten. In addition, the institutions at the second system's center (the NSF and NIH) usually organized their grants programs and review panels along disciplinary lines, even though a number of individuals at both the NSF and the NIH may have favored interdisciplinary work. Because of the structure of the review system, however, such interdisciplinary values tended to be muted, a fact recognized in recent efforts to create "cross-cutting" funding programs at both the NSF and NIH.

A set of charts related to funding for the behavioral and social sciences (BASS) follows. Figure 1 shows the accelerating growth of federal BASS support from 1952 to 1972 and the subsequent leveling-off of support, once inflation is taken into account. Figure 2 shows the rapid increase in membership of the main disciplinary associations in the social sciences in the 1940s–1960s. Figure 3 provides estimates of the relative contributions of federal, university, and foundation sources for BASS funding between 1939 and 1980. Figure 4 breaks down the relative contributions of federal, university, and foundation sources to different kinds of recipient organizations, focusing on one year, 1967, which was at or near the peak of the boom in BASS funding.

Figure 1 (opposite) Federal Funding for Behavioral and Social Science Research (BASS), 1952–2000

Source National Science Foundation, "Federal Funds for Research and Development, Detailed Historical Tables: Fiscal Years 1951–2001" (Washington, DC: NSF, 2001); National Science Foundation, "Federal Funds for Research and Development, Fiscal Years 1970–2001, Federal Obligations for Research by Agency and Detailed Field of Science and Engineering" (Washington, DC: NSF, 2001); Congressional Research Service, "Research Policies for the Social and Behavioral Sciences, Science Policy Study Background Report No. 6" (Washington, DC: Task Force on Science Policy, Committee on Science and Technology, U.S. House of Representatives, 1986). These figures almost certainly underestimate federal funding for BASS research, as much BASS research was funded as an adjunct to large military projects or through the NIH, whose support for social science research appears to be systematically underrepresented in NSF statistics. (For example, in every one of the past five years, the NIH has claimed to have spent more on BASS research than the NSF believes the entire federal government has spent.) An educated guess is that the underestimate is greatest for the 1960s and 1990s and least for the 1970s (when military support for BASS research was curtailed), meaning that the decline in funds available for BASS research in the 1970s is even sharper than the figure indicates.

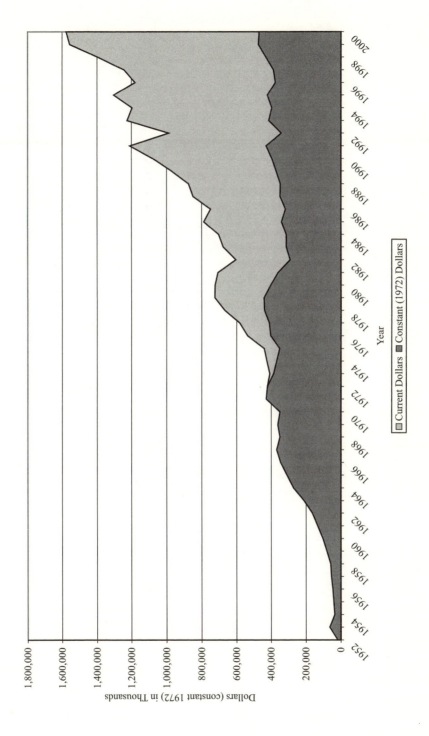

Dollars (constant 1972) in Thousands

Year

■ Current Dollars ■ Constant (1972) Dollars

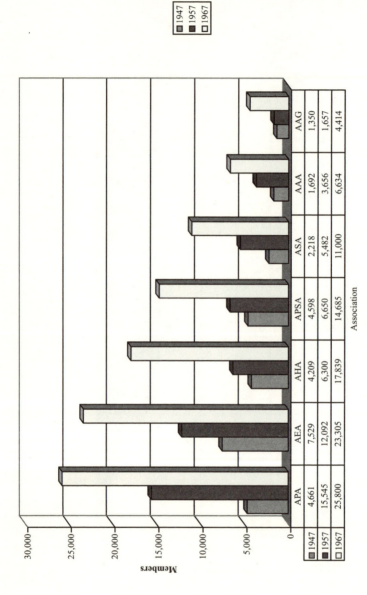

Association	APA	AEA	AHA	APSA	ASA	AAA	AAG
1947	4,661	7,529	4,209	4,598	2,218	1,692	1,350
1957	15,545	12,092	6,300	6,650	5,482	3,656	1,657
1967	25,800	23,305	17,839	14,685	11,000	6,634	4,414

Legend: ■ 1947 ■ 1957 □ 1967

Figure 2 Membership in the Major Social Science Professional Associations, 1947–67

Source National Academy of Sciences Behavioral and Social Science Survey Committee, *The Behavioral and Social Sciences: Outlook and Needs* (Englewood Cliffs, NJ: Prentice-Hall, 1969), p. 23. APA=American Psychological Association; AEA=American Economic Association; AHA=American Historical Association; APSA=American Political Science Association; ASA=American Sociological Association; AAA=American Anthropological Association; AAG=American Association of Geographers.

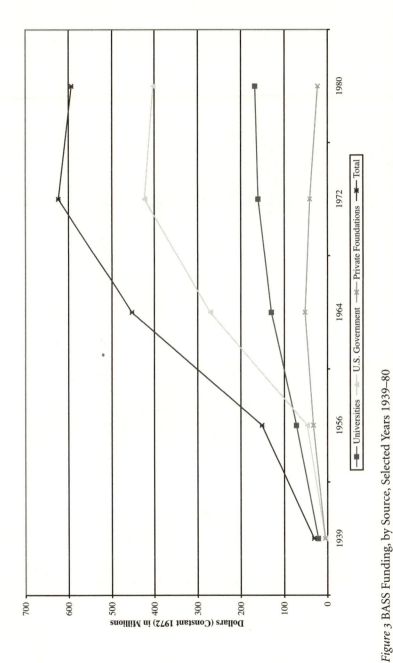

Figure 3 BASS Funding, by Source, Selected Years 1939–80

Source National Science Foundation, "Federal Funds for Research and Development, Detailed Historical Tables: Fiscal Years 1951–2001"; National Science Foundation, "Federal Funds for Research and Development, Fiscal Years 1970–2001"; Congressional Research Service, "Research Policies for the Social and Behavioral Sciences"; Roger Geiger, "American Foundations and Academic Social Science, 1945–1960," *Minerva* 26 (1988): 315–41; and Roberta Miller et al., "Research Support and Intellectual Advance in the Social Sciences," *SSRC Items* 37, nos. 2–3 (1983): 33–49.

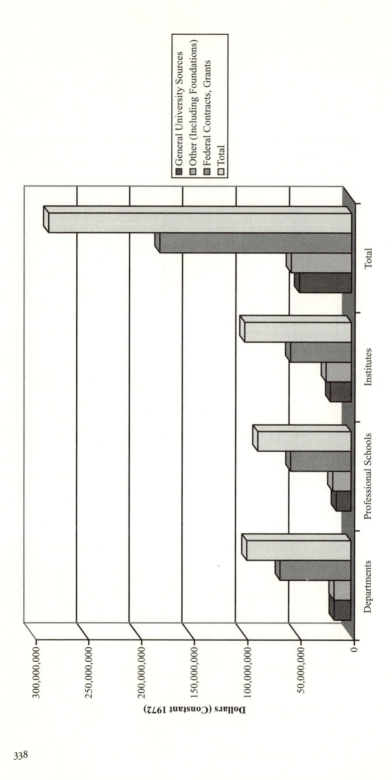

Figure 4 BASS Funding, by Source and Recipient, 1967

Source Behavioral and Social Science Survey Committee, *The Behavioral and Social Sciences*, p. 156. (Data has been converted to constant 1972 dollars for consistency.)

Notes

Abbreviations

CMP   Charles Merriam Papers, University of Chicago Archives

CMU Archives   Carnegie Mellon University Archives, Pittsburgh, PA

GSIA Papers   Graduate School of Industrial Administration Papers, CMU Archives

HSP   Herbert Simon Papers, CMU Archives

RAC   Rockefeller Archives Center, Tarrytown, NY

SSRC Papers   Social Science Research Council Papers

INTRODUCTION: (Un)Bounded Rationality

1. Simon reports that his student Edward Feigenbaum remembers this moment vividly; Herbert Simon, *Models of My Life* (New York: Basic Books, 1991), p. 206.

2. Bertrand Russell, "Letter to Herbert Simon," 11/2/56, HSP, Box 61, ff: Autobiography—Materials for Autobiography—1982.

3. Herbert Simon, *Administrative Behavior* (New York: Macmillan, 1947); James G. March, Herbert Simon, and Harold Guetzkow, *Organizations* (New York: Wiley, 1958).

4. Herbert Simon, *Models of Man: Social and Rational. Mathematical Essays on Rational Behavior in a Social Setting* (New York: Wiley, 1957), p. vii.

5. Herbert Simon, "Some Strategic Considerations in the Construction of Social Science Models," 1951, HSP, Box 4, ff 120, p. 3.

6. The terms *sciences of choice* and *sciences of control* are my own invention, though they mirror the divergent approaches to behavioral science Simon discusses in "Strategic Considerations."

7. Herbert Simon, "Letter to Brother Benedict," 11/22/47, HSP, Box 5, ff 198, p. 1.

8. Herbert Simon, "Letter to George A. Miller (1955b)," 7/25/55, HSP, Box 6, ff 212, p. 1.

9. Ward Edwards, "Letter to Herbert Simon," October 1954, HSP, Box 5, ff 203, p. 1.

10. Herbert Simon, "The Axioms of Newtonian Mechanics," HSP, Box 4, ff 122.

11. This quotation serves as the epigraph for Simon, *Models of Man.*

12. Ibid., pp. 198, 199.

13. Herbert Simon, "Administrative Behavior: A Study of Decision-Making Processes in Administrative Organization," 1945, Box 20, ff: Administrative Behavior—Preliminary ed.—1945, p. 84.

O N E : The Garden of Forking Paths

1. Herbert Simon, *Models of My Life* (New York: Basic Books, 1991), pp. 175–88.

2. Herbert Simon, "Rational Choice and the Structure of the Environment," *Psychological Review* 63 (1956): 129–38.

3. Simon, *Models of My Life*, pp. 175–77. For the stories by Borges, see Donald Yates and James Irby, eds., *Labyrinths: Selected Stories and Other Writings by Jorge Luis Borges* (New York: New Directions, 1964).

4. Simon, *Models of My Life*, pp. 177–78.

5. Jorge Luis Borges, "The Library of Babel," in *Labyrinths: Selected Stories and Other Writings by Jorge Luis Borges*, ed. Donald Yates and James Irby (New York: New Directions, 1964), p. 53.

6. Ibid., p. 54.

7. Simon, *Models of My Life*, pp. 180–81.

8. Ibid., pp. 187–88.

9. Ibid., p. 188.

10. For "monomania," and for "whole career as a gloss," see Herbert Simon, *Models of Man: Social and Rational. Mathematical Essays on Rational Behavior in a Social Setting* (New York: Wiley, 1957), p. vii; for "obvious responses," see Simon, *Models of My Life*, pp. xvii–xviii.

11. Simon, *Models of My Life*, p. 177.

12. Ibid., pp. 3–5, xxv–xxvi.

13. Ibid., pp. 3–6, 14, 22.

14. Ibid., p. 22.

15. Ibid., p. 108.

16. Ibid., p. 109.

17. Ibid., pp. 19–21.

18. Ibid., p. 22. On the professions in both Germany and America, see the Essay on Sources.

19. Ibid., p. 5.

20. Herbert Simon, interview by Hunter Crowther-Heyck, 10/19/97.

21. Simon, *Models of My Life*. Also see Pamela McCorduck, "Transcript of Interview with Herbert Simon, April 1975," 4/9/75, HSP, Box 52, ff: Pamela McCorduck Interviews—1975, pp. 14–15.

22. On gender in human science, see the Essay on Sources.

23. Simon, *Models of My Life*, pp. 7, 19, 23.

24. Herbert Simon, "What It Means to Me to Be Jewish," essay dated 6/9/94, contained in a personal communication to the author, dated 4/24/00. This essay was published as Herbert Simon, "What It Means to Me to Be Jewish," in *Jewish: Does It Make a Difference?*, ed. Elvira Nadin and Mihai Nadin (Middle Village, NY: Jonathan David Publishers, 2000).

25. Ibid.

26. David Tyack, *The One Best System: A History of American Urban Education* (Cambridge, MA: Harvard University Press, 1974), title page.

27. I borrow the distinction between "shop" and "school" cultures from Monte Calvert, *The Mechanical Engineer in America, 1830–1910: Professional Cultures in Conflict* (Baltimore: Johns Hopkins University Press, 1967).

28. George Marsden, *The Soul of the American University: From Protestant Establishment to Established Nonbelief* (New York: Oxford University Press, 1994); George M. Marsden and Bradley J. Longfield, *The Secularization of the Academy* (New York: Oxford University Press, 1992); Edward J. Larson, *Summer for the Gods: The Scopes Trial and America's Continuing Debate over Science and Religion* (Cambridge, MA: Harvard University Press, 1998).

29. Dorothy Ross, *G. Stanley Hall: The Psychologist as Prophet* (Chicago: University of Chicago Press, 1972).

30. David A. Hollinger, *Science, Jews, and Secular Culture: Studies in Mid-Twentieth-Century American Intellectual History* (Princeton, NJ: Princeton University Press, 1996); Marsden, *The Soul of the American University*.

31. Simon, "What It Means to Me to Be Jewish."

32. Herbert Simon, "Letter to Brother Benedict," 11/22/47, HSP, Box 5, ff 198, p. 1.

33. Simon, *Models of My Life*, p. 179.

34. Ibid., pp. 5, 17, xxiii.

35. Simon, "What It Means to Me to Be Jewish."

36. Simon, *Models of My Life*, p. 23.

37. McCorduck, "Interview with Herbert Simon," p. 12. Simon and Newell also had a policy of alternating presentations on their work and of only one of them attending a conference (usually). Simon adhered to a similar policy with James March and Harold Guetzkow: the book *Organizations*, as a result, is often thought of as March's book even though Simon headed the project that produced it, wrote several key chapters, and edited the others.

38. Ibid., p. 19.

39. Simon, *Models of My Life*, pp. 8–9, 18.

40. McCorduck, "Interview with Herbert Simon," p. 20.

41. Herbert A. Simon, Peter Drucker, and Dwight Waldo, "'Development of Theory of Democratic Administration': Replies and Comments," *American Political Science Review* 46, no. 2 (1952): 494–503, at 501.

42. Simon, *Models of My Life*, p. 144.

43. Herbert Simon, *Models of Discovery: And Other Topics in the Methods of Science* (Boston: D. Reidel, 1977), p. xv; Herbert Simon, "Letter to George Madow," 7/25/55, HSP, Box 35, ff: Madow, William G.—Correspondence, 1951–55; Herbert Simon, "Letter to Dwight Waldo," 3/10/53, HSP, Box 6, ff 222, p. 1.

44. Simon, *Models of My Life*, p. 65.

45. McCorduck, "Interview with Herbert Simon," p. 14.

46. Simon, *Models of My Life*, p. 14; Herbert Simon, "The Proverbs of Administration," *Public Administration Review* 6 (1946): 53–67; Allen Newell and Herbert Simon, "Heuristic Problem Solving," *Journal of the Operations Research Society of America* 6 (1958): 1–10. Simon claimed that the predictions in the ORSA paper—

which provoked a storm of controversy—were not meant to do so. Whether that is true or not, he certainly enjoyed the notoriety they brought him. McCorduck, "Interview with Herbert Simon," pp. 21–32.

47. Simon, *Models of My Life*, p. 48.

48. Ibid., p. 268. Some of the more lengthy public exchanges were those with Dwight Waldo on political science, Chris Argyris on organization theory, and with the editors of the CMU student paper on technology, society, and the war in Vietnam. See Dwight Waldo, "Development of Theory of Democratic Administration," *American Political Science Review* 46, no. 1 (1952): 81–103; Herbert Simon, "Letter to Taylor Cole of the APSR," 4/16/52, HSP, Box 10, ff: APSR Correspondence: 1950–74. Also see the exchanges between Argyris and Simon in the *Public Administration Review* 33 (1973); Herbert Simon, "Prometheus Unbound" and "Reflections on the Revolution of Our Times" (both 1970) and his letter to the editors of the CMU *Tartan* of 10/24/70, all found in HSP, Box 61, ff: Autobiography—Source Documents 1969–1988.

49. Simon, *Models of My Life*, pp. 9–10.

50. Herbert Simon, "The Reminiscences of Herbert Simon," 1979, Columbia University Oral History Research Office, New York, p. 42.

51. McCorduck, "Interview with Herbert Simon," p. 15.

52. Ibid.

53. Simon, *Models of My Life*, p. 10.

54. Herbert A. Simon, "Letter to Kathie and David," 5/23/65, HSP, Box 61, ff: Materials for Autobiography—1982.

55. Simon, *Models of My Life*, pp. 3, 28.

T W O : The Chicago School and the Sciences of Control

1. William Cronon, *Nature's Metropolis: Chicago and the Great West* (New York: W. W. Norton, 1991), esp. pp. 55–93.

2. Martin Bulmer, *The Chicago School of Sociology: Institutionalization, Diversity, and the Rise of Sociological Research* (Chicago: University of Chicago Press, 1984), pp. xv–xvi.

3. Ibid., p. 13.

4. Carl Condit, *The Chicago School of Architecture* (Chicago: University of Chicago Press, 1964).

5. Carl Sandburg, *Chicago*, quoted in Bulmer, *The Chicago School of Sociology*, p. xvi.

6. The terms *gemeinschaft* and *gesellschaft* were used by Ferdinand Tönnies to denote the differences between traditional "communities" characterized by personal, face-to-face interactions and modern "societies" characterized by impersonal, formal-bureaucratic interactions. See Ferdinand Tönnies, *Community and Society*, trans. Charles Loomis (East Lansing, MI: Michigan State University Press, 1957).

7. Max Weber, quoted in Bulmer, *The Chicago School of Sociology*, p. xv.

8. On the Chicago Schools of Sociology and Political Science and their place in twentieth-century human science, see the Essay on Sources.

9. Harold F. Gosnell, *Machine Politics Chicago Model*, 2nd ed. (Chicago: University of Chicago Press, 1968).

10. Lincoln Steffens, quoted in Bulmer, *The Chicago School of Sociology*, pp. xv–xvi.

11. The "received view" is that the turn toward scientism was a rejection of social science's reform-driven past. Recent work has made clear, however, that social scientists did not give up their ambitions to reform society when they embraced the research ethos. See the Essay on Sources for a discussion of this issue.

12. On the importance of Johns Hopkins in the history of American higher education, see George Marsden, *The Soul of the American University: From Protestant Establishment to Established Nonbelief* (New York: Oxford University Press, 1994); Dorothy Ross, *The Origins of American Social Science* (Cambridge: Cambridge University Press, 1991); and Lawrence Veysey, *The Emergence of the American University* (Chicago: University of Chicago Press, 1965).

13. On the differences in outlook between locally and nationally oriented elites, see Robert H. Wiebe, *Self-Rule: A Cultural History of American Democracy* (Chicago: University of Chicago Press, 1995). On the creation of the University of Chicago, see Daniel Lee Meyer, "The Chicago Faculty and the University Ideal: 1891–1929" (Ph.D. diss., University of Chicago, 1994). Meyer's introduction and first chapter discuss Harper's ideals and his connections to Chautauqua. Harper was head of the humanities division of the summer college at Chautauqua, which aided his recruiting efforts enormously.

14. Bulmer, *The Chicago School of Sociology*, pp. 19–20.

15. On the Chicago Exposition, see Robert W. Rydell, *All the World's a Fair: Visions of Empire at American International Expositions, 1876–1916* (Chicago: University of Chicago Press, 1984).

16. Meyer, "The Chicago Faculty and the University Ideal," introduction.

17. Barry D. Karl, *Charles E. Merriam and the Study of Politics* (Chicago: University of Chicago Press, 1974), p. x.

18. Meyer, "The Chicago Faculty and the University Ideal," introduction and chap. 1.

19. Ibid. For discussion of the many links between the university and the city, see Bulmer, *The Chicago School of Sociology*, p. 22, and Karl, *Charles E. Merriam and the Study of Politics*.

20. Meyer, "The Chicago Faculty and the University Ideal," chaps. 1 and 2.

21. Dorothy Ross, *G. Stanley Hall: The Psychologist as Prophet* (Chicago: University of Chicago Press, 1972), chaps. 11–14.

22. Meyer, "The Chicago Faculty and the University Ideal," pp. 97–98; Ross, *G. Stanley Hall*, pp. 220–30.

23. Meyer, "The Chicago Faculty and the University Ideal," chap. 1.

24. Robert B. Westbrook, *John Dewey and American Democracy* (Ithaca, NY: Cornell University Press, 1991).

25. The Laboratory School was created by Dewey and others to put his progressive educational ideas in practice at the elementary-school level. It also was a perfect example of the Harper-era commitment to linking the university and the city in a way that joined research and practice. See ibid.

26. Chauncey Samuel Boucher, *The Chicago College Plan* (Chicago: University of

Chicago Press, 1935). Also, see the annual *Announcements* (the course catalog) of the University of Chicago from the 1930s, University of Chicago Archives.

27. Herbert Simon, *Models of My Life* (New York: Basic Books, 1991), p. 36.

28. Karl, *Charles E. Merriam and the Study of Politics.*

29. See ibid., generally, quote from p. 40.

30. Ibid., p. 171. The final product was Charles Edward Merriam, *Civic Education in the United States* (New York: Charles Scribner's Sons, 1934).

31. Gladys Bryson, "The Comparable Interests of the Old Moral Philosophy and the Modern Social Sciences," *Social Forces* 11, no. 1 (1932): 19–27. Ross, in *The Origins of American Social Science,* and Haskell, in *The Emergence of Professional Social Science,* also locate the origins of social science in moral philosophy.

32. See David Roberts, "Mathematics and Pedagogy: Professional Mathematicians and American Educational Reform, 1893–1923" (Ph.D. diss., Johns Hopkins University, 1997).

33. Fritz Ringer, "The German Academic Community," in *The Organization of Knowledge in Modern America, 1860–1920,* ed. Alexandra Oleson and John Voss (Baltimore: Johns Hopkins University Press, 1979), p. 426.

34. Ross, *The Origins of American Social Science.*

35. Robert N. Proctor, *Value-Free Science? Purity and Power in Modern Knowledge* (Cambridge, MA: Harvard University Press, 1991).

36. On Ogburn and value-neutrality, see Mark C. Smith, *Social Science in the Crucible: The American Debate over Objectivity and Purpose, 1918–1941* (Durham, NC: Duke University Press, 1994), pp. 142–49. Also see Dennis Smith, *The Chicago School: A Liberal Critique of Capitalism* (New York: St. Martin's Press, 1988), chap. 9.

37. My notion of instrumentalism comes from many sources and from none, for I intend my argument about the shift from the idealist to the instrumentalist understanding of knowledge both to encompass and to redefine several related arguments about the changing nature of turn of the century social thought. See the Essay on Sources for a discussion of instrumentalism in the historiography of human science.

38. Boucher, *The Chicago College Plan.*

39. On the Hutchins years at the University of Chicago, see William McNeill, *Hutchins' University: A Memoir of the University of Chicago, 1929–50* (Chicago: University of Chicago Press, 1991); Bulmer, *The Chicago School of Sociology;* Karl, *Charles E. Merriam and the Study of Politics;* and Edward Shils, ed., *Remembering the University of Chicago: Teachers, Scientists, and Scholars* (Chicago: University of Chicago Press, 1991). For Hutchins' educational philosophy, see his famous report: Robert Maynard Hutchins, *The Higher Learning in America* (New Haven, CT: Yale University Press, 1936).

40. McNeill, *Hutchins' University,* preface.

41. Mortimer Adler, "Lecture on Religion, from Systematic Social Science," April 1935, Charles Merriam Papers, University of Chicago Archives (CMP), Box 120, ff: "Adler on Religion," p. 4.

42. Edward A. Purcell Jr., *The Crisis of Democratic Theory: Scientific Naturalism and the Problem of Value* (Lexington: University Press of Kentucky, 1973).

43. Simon, *Models of My Life,* p. 38.

44. Ibid., pp. 36–42.

45. Ibid., p. 40.

46. Ibid., p. 50.

47. Ibid., p. 42.

48. Ibid., p. 50.

49. As Simon notes in his autobiography, he and his wife and friends were "intensely political animals" at this time. Simon recalls staying up all night reading transcripts of the Soviet show trials of the late 1930s, trying to figure out what was going on. He later wrote that he did not understand until he read Koestler's *Darkness at Noon;* then their meaning became clear to him (ibid., p. 119, 121–22).

50. Ibid., pp. 42–50.

51. H. Stuart Hughes, *Consciousness and Society: The Reorientation of European Social Thought, 1890–1930* (New York: Vintage, 1958).

52. Albert Somit and Joseph Tanenhaus, *The Development of Political Science: From Burgess to Behavioralism* (Boston: Allyn and Bacon, 1967), pp. 102, 107.

53. Chicago's prominence continued through the 1940s and 1950s, though the department changed its orientation almost 180 degrees: Hans Morgenthau and Leo Strauss now carried the department's banner into battle *against* the very behavioral approach their predecessors in the department had pioneered. See Ted V. McAllister, *Revolt against Modernity: Leo Strauss, Eric Voegelin, and the Search for a Postliberal Order* (Lawrence: University Press of Kansas, 1996); Hans Joachim Morgenthau, *Scientific Man vs. Power Politics* (Chicago: University of Chicago Press, 1946).

54. Bulmer, *The Chicago School of Sociology;* Mary Jo Deegan, *Jane Addams and the Men of the Chicago School, 1892–1918* (New Brunswick, NJ: Transaction Books, 1988); Smith, *The Chicago School: A Liberal Critique of Capitalism.*

55. For an influential statement of the Chicago School's views, see Charles Merriam, *New Aspects of Politics* (Chicago: University of Chicago Press, 1925).

56. Simon, *Models of My Life,* p. 60.

57. Gosnell's work displayed a sophisticated understanding of statistics, but he was unusual in that regard among the political scientists.

58. Karl, *Charles E. Merriam and the Study of Politics,* p. 80.

59. See the Essay on Sources for a discussion of liberal managerialism and the organizational revolution.

60. Daniel T. Rodgers, *Atlantic Crossings: Social Politics in a Progressive Age* (Cambridge, MA: Belknap Press of Harvard University Press, 1998); Daniel T. Rodgers, "In Search of Progressivism," *Reviews in American History* 10 (1982): 113–32; Martin J. Schiesl, *The Politics of Efficiency: Municipal Administration and Reform in America, 1800–1920* (Berkeley: University of California Press, 1977); Guy Alchon, *The Invisible Hand of Planning: Capitalism, Social Science, and the State in the 1920s* (Princeton, NJ: Princeton University Press, 1985).

61. Leonard Dupee White, *The City Manager* (Chicago: University of Chicago Press, 1927). Also see City Club of Chicago, "A City Manager for Chicago," CMP, Box 41, ff 6.

62. The majority of city managers were engineers, another third were business-men, the majority of the rest were professionals of one sort or another (chiefly lawyers or accountants). For data on the backgrounds of city managers, see White, *The City Manager*.

63. The history of the ICMA and the other "1313" organizations are related in C. Herman Pritchett, "1313: An Experiment in Propinquity," CMP, Box 49, ff 3.

64. On the ICMA and PACH, see ibid. Also see Harold A. Stone, Don K. Price, and Kathryn H. Stone, *City Manager Government in the United States: A Review after Twenty-Five Years* (Chicago: Committee on Public Administration of the Social Science Research Council by Public Administration Service, 1940); Harold A. Stone, Don K. Price, and Kathryn H. Stone, *City Manager Government in Nine Cities* (Chicago: Committee on Public Administration of the Social Science Research Council by Public Administration Service, 1940).

65. Harold Gosnell, *Machine Politics: Chicago Model* (Chicago: University of Chicago Press, 1937).

66. Ibid.

67. "Chicago's Report to the People: 1933–1946," Mayor's Office, City of Chicago, 1947. This report is obviously a political document aimed at bolstering Boss Kelly's reputation, but it does reveal what Kelly thought was worth boasting about. The report repeatedly emphasizes the low per capita cost of public services in Chicago and the various administrative reorganizations (as in the police force) that have made the city more efficient.

68. Simon, *Models of My Life*, p. 119.

69. Karl, *Charles E. Merriam and the Study of Politics*, pp. 248–59.

70. John Dewey, cited in ibid., p. 259.

71. Paul Samuelson and Alvin Hansen were just two of the many young economists who cut their teeth on applied social science with the NRPB. Philip W. Warken, *A History of the National Resources Planning Board, 1933–1943* (New York: Garland, 1979).

72. Ibid. Also see Karl, *Charles E. Merriam and the Study of Politics*, p. 264.

73. The "third" way refers to a way between capitalism and socialism. "Planning" is the term most consistently associated with attempts to find this third way. Some, such as Josef Schumpeter and Friedrich von Hayek, believed that such talk of planning was simply the sheep's clothing under which the wolf of socialism lay hidden. Joseph A. Schumpeter, *Capitalism, Socialism, and Democracy*, 3rd ed. (New York: Harper & Row, 1962); Friedrich A. Hayek, *The Road to Serfdom* (Chicago: University of Chicago Press, 1950). Barry Karl notes that Merriam regarded Hayek as one of his greatest enemies: cf. Karl, *Charles E. Merriam and the Study of Politics*, p. 290.

74. The historical approach most interested in the rise of large-scale organizations and the managed society is aptly termed the "organizational synthesis," which is dis-cussed in the Essay on Sources.

75. This argument is inspired by Ross, *The Origins of American Social Science*. Ross is absolutely correct that a change in historical consciousness was a vital part of the turn to scientism in American social science, but she may overemphasize the contri-bution of the idea of "American exceptionalism" (the idea that the United States occu-

pies a unique position in history) to this changed consciousness. See the Essay on Sources for a discussion of Ross, American exceptionalism, and scientism in the historiography of the social sciences.

76. Ibid., p. 8.

77. University of Chicago, "Announcements," 1936, University of Chicago Archives.

78. For gemeinschaft to gesellschaft, see Tönnies, *Community and Society.* For "status to contract," see Sir Henry Maine, *Ancient Law: Its Connection with the Early History of Society, and Its Relation to Modern Ideas* (Tucson: University of Arizona Press, 1986).

79. Frederick Jackson Turner, *The Frontier in American History* (New York: Henry Holt, 1920).

80. Ross, *The Origins of American Social Science,* pp. 312–19. Also see Cynthia E. Russett, *The Concept of Equilibrium in American Social Thought* (New Haven, CT: Yale University Press, 1966). Perhaps the best example of the analysis of process in equilibrium systems in social science was Arthur Bentley, *The Process of Government: A Study of Social Pressure* (Chicago: University of Chicago Press, 1908).

81. Ross, *The Origins of American Social Science,* pp. 388–90.

82. Thorstein Veblen, "Is Economics an Evolutionary Science?" in *The Place of Science in Modern Civilization and Other Essays* (New Brunswick, NJ: Transaction Publishers, 1990).

83. Ross, *The Origins of American Social Science,* p. 388.

84. Charles Merriam, and Harold Gosnell, *Non-Voting: Causes and Methods of Control* (Chicago: University of Chicago Press, 1924); Harold Lasswell, *Politics: Who Gets What, When, How* (New York: P. Smith, 1950); Harold Lasswell and Abraham Kaplan, *Power and Society: A Framework for Political Inquiry* (New Haven, CT: Yale University Press, 1950).

85. Neoclassical economics, for example, becomes almost totally ahistorical in reaction to the German Historical School. See Jurgen Herbst, *The German Historical School in American Scholarship: A Study in the Transfer of Culture* (Ithaca, NY: Cornell University Press, 1965); Josef Schumpeter, *History of Economic Analysis* (London: Routledge, 1954); Ross, *The Origins of American Social Science;* A. K. Dasgupta, *Epochs of Economic Theory* (New York: Basil Blackwell, 1985); Phyllis Deane, *The Evolution of Economic Ideas* (Cambridge: Cambridge University Press, 1978); and Philip Mirowski, *More Heat Than Light: Economics as Social Physics, Physics as Nature's Economics* (New York: Cambridge University Press, 1989).

86. Leonard Dupee White, *The Federalists: A Study in Administrative History* (New York: Macmillan, 1948).

87. Ross, *The Origins of American Social Science,* pp. 58, 66–77, 282–88.

88. Mary Furner, *Advocacy and Objectivity: A Crisis in the Professionalization of American Social Science, 1865–1905* (Lexington: University Press of Kentucky, 1975), p. 166.

89. Ross, *The Origins of American Social Science,* p. 321. Similarly, Herbert Feigl notes that he and his fellow members of the Vienna Circle referred to the days before the war as "prehistoric." Herbert Feigl, "The Wiener Kreis in America," in *The Intellectual Migration: Europe and America, 1930–1960,* ed. Donald Fleming and

Bernard Bailyn (Cambridge, MA: Belknap Press of Harvard University Press, 1969), p. 631.

90. The classic example of this phenomenon is the reluctance of Congress to pass a bill it knows the president will veto.

91. See, for example, Herbert Simon, "Prediction and Hindsight as Confirmatory Evidence," *Philosophy of Science* 22 (1955): 227–30; Herbert A. Simon, "Bandwagon and Underdog Effects of Election Predictions," *Public Opinion Quarterly* 18 (fall 1954).

92. *Statics* refers to an analysis of a system in which time is not a significant variable: for example, calculating the distance from the center point of a beam at which two weights should be placed in order to balance each other. *Comparative statics* refers to the comparison of two states of a system at different times. By comparing such different states, one can gain a better idea about the properties of the system without having to know the actual sequence of events that transformed the system from one state into the other. A *dynamic* theory describes the production of one state of a system out of an earlier state, as evolution does in the case of ecological succession, for example.

93. Simon later pursued the relationship of comparative statics and dynamics in economics in other work, finding Paul Samuelson's formulation of the problem fascinating. Simon, *Models of My Life,* p. 102; Paul Samuelson, "The Stability of Equilibrium: Comparative Statics and Dynamics," *Econometrica* 9 (1941): 97–120. Simon also discusses the problem of statics and dynamics with regard to game theory in Herbert Simon, "Letter to Professor Oskar Morgenstern," 4/23/45, HSP, Box 6, ff 213.

94. This argument is based on Thomas Haskell, *The Emergence of Professional Social Science: The ASSA and the Nineteenth Century Crisis of Authority* (Urbana: University of Illinois Press, 1977). Haskell's emphasis on the importance of interdependence is right on target, but he tends to ascribe all things to interdependence (the discovery of which follows the railroad quite directly). See the Essay on Sources for a discussion of Haskell's work. As regards the importance of the division of labor in early twentieth century social thought, one should note that it loomed large in the thought of all of the social thinkers (Marshall, Durkheim, Weber, Pareto) analyzed in Talcott Parsons, *The Structure of Social Action* (Glencoe, IL: Free Press, 1937). *The Structure of Social Action* would play an important role in shaping Simon's early work on organizations, as will be discussed in chapter 5.

95. Charles Austin Beard, *An Economic Interpretation of the Constitution of the United States* (New York: Macmillan, 1913). Another prominent example was E. R. A. Seligman, *The Economic Interpretation of History* (New York: Columbia University Press, 1902).

96. On psychology in the early twentieth century, see the Essay on Sources.

97. Note, for example, Talcott Parsons's interest in creating a *voluntaristic* theory of social action in *The Structure of Social Action.* Similarly, the first chapter of Chester Barnard's influential *Functions of the Executive* (Cambridge, MA: Harvard University Press, 1938) begins with a discussion of free will and its relationship to decision-making. Chester A. Barnard, *The Functions of the Executive.* Herbert Simon cites Barnard and Parsons as two of the three major sources of ideas

for his thesis (and first book) *Administrative Behavior* (New York: Macmillan, 1947), preface.

98. Social Science Research Council, "Minutes of Meeting, Board of Directors," SSRC Papers, Rockefeller Archives Center (RAC), Tarrytown, NY, Box 357, ff 2099, AC 1, Series 9, p. 40.

99. Haskell, *The Emergence of Professional Social Science,* pp. 236–37.

100. William Archibald Dunning, *Essays on the Civil War and Reconstruction and Related Topics* (New York: Macmillan, 1898); William Graham Sumner, *Folkways: A Study of the Sociological Importance of Usages, Manners, Customs, Mores, and Morals* (Boston: Ginn, 1907).

101. Durkheim invented the term *anomie* in order to describe the confusion that resulted from this dissolution of traditional sources of value and meaning. Emile Durkheim, *Suicide: A Study in Sociology* (Glencoe, IL: Free Press, 1951). For insight into how American social scientists understood Durkheim, see Parsons, *The Structure of Social Action.*

102. Karl, *Charles E. Merriam and the Study of Politics,* pp. 123–35.

103. Committee on Scientific Method in the Social Sciences Social Science Research Council, *Methods in Social Science: A Case Book* (Chicago: University of Chicago Press, 1931).

104. See SSRC Papers, RAC, Box 704, ff 10266, Series 4, Subseries 1, for correspondence between R. T. Crane, director of the SSRC, and various members on this issue (especially with E. B. Wilson).

105. Bulmer, *The Chicago School of Sociology,* pp. 129–48.

106. This situation was not unique to Chicago: economics always was a difficult discipline to bring into such institutional syntheses, as Talcott Parsons found at Harvard and Herbert Simon later found at Carnegie Tech.

107. The "1313" agencies were the International Association of Chiefs of Police, the American Public Works Association, the Municipal Finance Officers' Association, the Civil Service Assembly, the International City Managers' Association, the Governmental Research Association, the National Association of State Auditors, Comptrollers, and Treasurers, the American Municipal Association, the American Legislators' Association, the American Public Welfare Association, the Public Administration Clearing House, the Council of State Governments, the National Association of Housing Officials, the Public Administration Service, the National Association of Assessing Officers, the American Society of Planning Officials, and the Federation of Tax Officers. See Louis Brownlow, "National Governmental Organizations," 1938, CMP, Box 159, ff 1, p. 1.

108. Pritchett, "1313: An Experiment in Propinquity," introduction.

109. This argument is the center of Hughes, *Consciousness and Society.* Also see Dorothy Ross, *Modernist Impulses in the Human Sciences, 1870–1930* (Baltimore: Johns Hopkins University Press, 1994), introduction; Purcell, *The Crisis of Democratic Theory,* chap. 1.

110. Hughes, *Consciousness and Society.* Some exemplary works include Sigmund Freud, *Civilization and Its Discontents,* trans. James Strachey (New York: W. W. Norton, 1961); Vilfredo Pareto, *Mind and Society* (New York: Harcourt, Brace, 1935). Also note that the central argument of Talcott Parsons in *The Structure of Social Action*

is that the discovery of the nonlogical sources of social action led to the demise of utilitarian social thought: cf. Parsons, *The Structure of Social Action.*

111. Robert E. Park, Ernest W. Burgess, and Roderick D. McKenzie, ed., *The City* (Chicago: University of Chicago Press, 1967); William I. Thomas and Florian Znaniecki, *The Polish Peasant in Europe and America: A Monograph of an Immigrant Group* (Boston: R. G. Badger, 1918–20); William Ogburn, *Social Change with Respect to Culture and Original Nature* (London: G. Allen and Unwin, 1923).

112. For Merriam and Lasswell's interest in psychology, see Karl, *Charles E. Merriam and the Study of Politics*, pp. 106–7, 171. Also see Harold Lasswell, *Psychopathology and Politics* (Chicago: University of Chicago Press, 1930); Merriam and Gosnell, *Non-Voting: Causes and Methods of Control.*

113. Walter Lippmann, *Public Opinion* (New York: Harcourt, Brace, 1922); Walter Lippmann, *Drift and Mastery: An Attempt to Diagnose the Current Unrest* (New York: Kennerly, 1914).

114. Purcell, *The Crisis of Democratic Theory*, esp. chaps. 2–3, 10.

115. Ross, *The Origins of American Social Science*, pp. 247–56.

116. On the history of objectivity, see Theodore Porter, *Trust in Numbers: The Pursuit of Objectivity in Science and Public Life* (Princeton, NJ: Princeton University Press, 1995); Proctor, *Value-Free Science;* Allan Megill, "Introduction: Four Senses of Objectivity," *Annals of Scholarship* 8, nos. 3–4 (1991): 301–20; Peter Novick, *That Noble Dream: The "Objectivity Question" and the American Historical Profession* (Cambridge: Cambridge University Press, 1988); and Steven Shapin, *A Social History of Truth: Civility and Science in Seventeenth-Century England* (Chicago: University of Chicago Press, 1994).

117. This shift in the bases of authority is captured well by Richard Hofstadter, *The Age of Reform* (New York: Vintage Books, 1955); Thomas L. Haskell, *The Authority of Experts: Studies in History and Theory* (Bloomington: Indiana University Press, 1984); Robert Wiebe, *The Search for Order, 1877–1920* (New York: Hill and Wang, 1967); Wiebe, *Self-Rule: A Cultural History of American Democracy;* Burton Bledstein, *The Culture of Professionalism: The Middle Class and the Development of Higher Education in America* (New York: W. W. Norton, 1976).

118. On Hall, see Ross, *G. Stanley Hall.* On Titchener and experimentalism in psychology, see Edwin G. Boring, *A History of Experimental Psychology,* 1st ed. (New York: D. Appleton-Century, 1929); Mitchell G. Ash, *Gestalt Psychology in German Culture, 1890–1967: Holism and the Quest for Objectivity* (New York: Cambridge University Press, 1995). For Fisher's views, see Irving Fisher, *Mathematical Investigations in the Theory of Value and Price* (1892; reprint, New York: A. M. Kelly, 1961). Mitchell's most famous work was *Business Cycles* (Berkeley: University of California Press, 1913). Mitchell was instrumental in the creation of the quantitatively oriented National Bureau of Economic Research.

119. Karl Pearson, *The Grammar of Science* (New York: Charles Scribner's Sons, 1892), p. 12.

120. Perhaps the most perfect example of such efforts at procedural reform was R. A. Fisher's science of statistical inference. For a description of how statistical inference became *the* symbol of scientific experimental design, see Harry Marks, *The Progress of Experiment: Science and Therapeutic Reform in the United States, 1900–1990*

(New York: Cambridge University Press, 1997). On the statistical turn in modern science more generally, see Gerd Gigerenzer et al., *The Empire of Chance: How the Science of Probability Changed Science and Everyday Life* (Cambridge: Cambridge University Press, 1989), and Ian Hacking, *The Taming of Chance* (Cambridge: Cambridge University Press, 1990).

T H R E E : Mathematics, Logic, and the Sciences of Choice

1. University of Chicago, "Announcements," 1936, University of Chicago Archives. Henry Simons wrote a memo in July 1937 arguing for upgrading this requirement (adding another two courses), but this did not become a formal requirement until the 1940s. Henry Simons, "Memorandum on Mathematics Requirements," July 1937, Papers of the Department of Economics, University of Chicago Archives, Box 41, ff 12. On the SSRC Institutes, see the Committee on Mathematical Training for Social Scientists Social Science Research Council, "Minutes of the Subcommittee of the Committee on Mathematical Training for Social Scientists," 2/18–19/53, SSRC Papers, Rockefeller Archives Center (RAC), Tarrytown, NY, Box 180, ff 1064, Accession 1, Series 1, Subseries 19.

2. Paul Samuelson, "Economics in a Golden Age," in *Paul Samuelson and Modern Economic Theory,* ed. E. Cary Brown and Robert M. Solow (New York: McGraw-Hill, 1983).

3. Philip Mirowski, *More Heat Than Light: Economics as Social Physics, Physics as Nature's Economics* (New York: Cambridge University Press, 1989); Philip Mirowski, "The When, the How, and the Why of Mathematical Expression in the History of Economic Analysis," *Journal of Economic Perspectives* 5 (winter 1991): 145–57.

4. Mary S. Morgan and Margaret Morrison, eds., *Models as Mediators: Perspectives on Natural and Social Science* (Cambridge: Cambridge University Press, 2000); Herbert Simon, *Models of My Life* (New York: Basic Books, 1991).

5. Dorothy Ross, *The Origins of American Social Science* (Cambridge: Cambridge University Press, 1991).

6. Leonard White notes that where only 25 economists had been employed in all the federal government in 1896 (all statisticians), 848 were in 1928–31, and a great number of these could be found in the bureau. There are 305 listed in agricultural economics. Leonard Dupee White, *Trends in Public Administration* (New York: McGraw-Hill, 1933), p. 272.

7. Carl F. Christ, "The Cowles Commission's Contributions to Econometrics at Chicago, 1939–1955," *Journal of Economic Literature* 32 (March 1994): 30–59.

8. Robert L. Church, "Economists as Experts: The Rise of an Academic Profession in the United States, 1870–1920," in *The University in Society,* ed. Lawrence Stone (Princeton, NJ: Princeton University Press, 1974); Ross, *The Origins of American Social Science.*

9. Curiously, neoclassical economics received a boost from such interest, even though it was much more closely allied with mathematical modeling than with statistical analysis.

10. On the marginalist revolution, see A. K. Dasgupta, *Epochs of Economic Theory*

(New York: Basil Blackwell, 1985); Phyllis Deane, *The Evolution of Economic Ideas* (Cambridge: Cambridge University Press, 1978); Josef Schumpeter, *History of Economic Analysis* (London: Routledge, 1954); Mirowski, *More Heat Than Light.*

11. John Bates Clark, *The Distribution of Wealth: A Theory of Wages, Interest, and Profits* (New York: Macmillan, 1899).

12. On the "battle of the schools," see Charles Camic, "Introduction: Talcott Parsons before the Structure of Social Action," in *Talcott Parsons: The Early Essays,* ed. Charles Camic (Chicago: University of Chicago Press, 1991); Philip Mirowski, "The Philosophical Bases of Institutionalist Economics," *Journal of Economic Issues* 21 (September 1987): 1001–38; Donald K. Pickens, "Clarence E. Ayres and the Legacy of German Idealism," *American Journal of Economics and Sociology* 46 (July 1987): 287–98.

13. Probably the most important single text in the mathematization of economics was Paul Samuelson, *Foundations of Economic Analysis* (Cambridge, MA: Harvard University Press, 1947). The most important institutional loci of mathematization were the Econometric Society (founded 1932) and the Cowles Commission for Research in Economics (also founded 1932), both of which were affiliated with the University of Chicago until the 1950s.

14. John Maynard Keynes, *The General Theory of Employment, Interest, and Money* (London: Macmillan, 1936).

15. Karl Pearson was a well known philosopher and eugenicist who made enormous contributions to the development of statistics. See Gerd Gigerenzer et al., *The Empire of Chance: How the Science of Probability Changed Science and Everyday Life* (Cambridge: Cambridge University Press, 1989), and Ian Hacking, *The Taming of Chance* (Cambridge: Cambridge University Press, 1990). As Dorothy Ross has shown, Pearson's *The Grammar of Science* (New York: Charles Scribner's Sons, 1892) was a central text for social scientists in the late-nineteenth century, for it argued that the unity of science lay in its method, not its subject matter; Ross, *The Origins of American Social Science,* pp. 156–57.

16. Henry Schultz, *The Theory and Measurement of Demand* (Chicago: University of Chicago Press, 1938), 10–12. Note also that Schultz derives both a "Special" and a "General Theory of Related Demands" in this text.

17. Ibid., 12. Percy Bridgman, *The Logic of Modern Physics* (New York: Macmillan, 1927); Maila L. Walter, *Science and Cultural Crisis: An Intellectual Biography of Percy Williams Bridgman (1882–1961)* (Stanford, CA: Stanford University Press, 1990).

18. For some examples of Bridgman's influence upon psychology, see Clark Hull, *The Principles of Behavior: An Introduction to Behavior Theory* (New York: D. Appleton-Century, 1943), 30–31; S. S. Stevens, "The Operational Basis of Psychology," *American Journal of Psychology* 43 (1935): 323–30; S. S. Stevens, "The Operational Definition of Psychological Concepts," *Psychological Review* 42 (1935): 517–42.

19. Talcott Parsons, *The Structure of Social Action* (Glencoe, IL: Free Press, 1937), pp. 37–41.

20. Schultz, *The Theory and Measurement of Demand,* pp. 12, 666; emphasis in original. The term *synthetic economics* refers to Henry Moore's book of the same name:

Henry L. Moore, *Synthetic Economics* (New York: Macmillan, 1929). *Synthetic* indicated agreement with the main tenets of logical empiricism, also known as logical positivism.

21. Simon, *Models of My Life*, pp. 51–2.

22. Herbert Simon, "Letter to Howard Cirker of Dover Publications," 8/31/53, HSP, Box 5, ff 202. Alfred Lotka, *Elements of Physical Biology* (Baltimore: William and Wilkens, 1925).

23. Sharon Kingsland, "Economics of Evolution," in *Natural Images in Economic Thought: Markets Read in Tooth and Claw,* ed. Philip Mirowski (New York: Cambridge University Press, 1994), quote from p. 232.

24. Ibid., p. 233. Also see Howard Odum, *Environment, Power, and Society* (New York: Wiley-Interscience, 1971).

25. Kingsland, "Economics of Evolution," p. 239.

26. Wiener and his fellow cyberneticists did add two vital concepts: *feedback* as the mechanism for self-regulation of equilibrium systems and *information* as a measurable quantity.

27. Kingsland, "Economics of Evolution," pp. 236–73.

28. Herbert Simon, *Models of Discovery: And Other Topics in the Methods of Science* (Boston: D. Reidel, 1977), p. xv.

29. These two metaphors are discussed at length in Yaron Ezrahi, *The Descent of Icarus: Science and the Transformation of Contemporary Democracy* (Cambridge: Harvard University Press, 1990). A classic example of this cross-fertilization is Karl W. Deutsch, "Mechanism, Organism, and Society: Some Models in Natural and Social Science," *Philosophy of Science* 18, no. 3 (1951): 230–52. For a discussion of how the computer led to the reintroduction of mind and purpose into psychology, see Hunter Crowther-Heyck, "George A. Miller, Language, and the Computer Metaphor of Mind," *History of Psychology* 2, no. 1 (1999): 37–64.

30. JoAnne Brown, for example, argues in *The Definition of a Profession* that applied psychologists attempted to link their work to that of both physicians and physicists by portraying psychology as both social medicine and social engineering. JoAnne Brown, *The Definition of a Profession: The Authority of Metaphor in the History of Intelligence Testing, 1890–1930* (Princeton, NJ: Princeton University Press, 1992).

31. John Louis Parascandola, "Lawrence J. Henderson and the Concept of Organized Systems" (Ph. D. diss., University of Wisconsin, 1968); Barbara Heyl, "The Harvard Pareto Circle," in *Talcott Parsons: Critical Assessments,* ed. Peter Hamilton (New York: 1992).

32. Walter B. Cannon, *The Wisdom of the Body* (New York: W. W. Norton, 1932).

33. Philip J. Pauly, *Controlling Life: Jacques Loeb and the Engineering Ideal in Biology* (New York: Oxford University Press, 1987).

34. Arturo Rosenblueth, Norbert Wiener, and Julian Bigelow, "Behavior, Purpose, and Teleology," *Philosophy of Science* 10 (1943): 18–24; Erwin Schrödinger, *What Is Life? The Physical Aspect of the Living Cell* (New York: Macmillan, 1946); Evelyn Fox Keller, *Refiguring Life: Metaphors of Twentieth Century Biology* (New York: Columbia University Press, 1995).

35. Anson Rabinbach, *The Human Motor: Energy, Fatigue, and the Origins of Modernity* (Berkeley: University of California Press, 1990); Crosbie Smith and Norton Wise, "Work and Waste: Political Economy and Natural Philosophy in Nineteenth-Century Britain (I) and (II)," *History of Science* 27 (1989): 263–301, 391–449; Crosbie Smith and Norton Wise, "Work and Waste: Political Economy and Natural Philosophy in Nineteenth Century Britain (III)," *History of Science* 28 (1990): 221–61.

36. Rabinbach, *The Human Motor*; Mirowski, *More Heat Than Light*.

37. Mirowski, *More Heat Than Light*.

38. Samuelson, *Foundations of Economic Analysis*. The continuing importance of concepts from thermodynamics/energetics in the post-WWII is evidenced by the pervasive power of the concept of *information*, defined by Claude Shannon and other systems scientists as the negative of *entropy*—that is, they defined information as a measure of *order* in a system. Warren Weaver and Claude E. Shannon, *The Mathematical Theory of Communication* (Urbana: University of Illinois Press, 1949).

39. Schultz cites Pareto countless times in his *Theory and Measurement of Demand*. For Pareto's importance to Lotka, see Kingsland, "Economics of Evolution," p. 237. For Pareto's importance to Henderson, see Heyl, "The Harvard Pareto Circle," and Parascandola, "Lawrence J. Henderson and the Concept of Organized Systems."

40. Lawrence J. Henderson, *Pareto's General Sociology: A Physiologist's Interpretation* (Cambridge, MA: Harvard University Press, 1935).

41. It is interesting to note that a similar zero-sum conservation principle appears in the work of the famous Kantian ethicist, John Rawls. Rawls begins his landmark *Theory of Justice* by stating that a maximum state of justice in a society is one in which no individual's position could be improved without resulting in an equal or greater decline in justice for someone else; John Rawls, *A Theory of Justice* (Cambridge, MA: Belknap Press of Harvard University Press, 1971).

42. Alfred Lotka, *Elements of Mathematical Biology* (New York: Dover, 1956). On the creation of the program in mathematical biology, see Nicholas Rashevsky, "The Rise of Mathematical Biology," n.d., Rashevsky Papers, University of Chicago Archives, Box 2, ff: "Rise of Mathematical Biology."

43. Simon, *Models of My Life*, p. 51. It did Simon's later career in science policy no harm that he was known and respected by such prominent physicists as Householder and Weinberg.

44. Nicholas Rashevsky, "Memorandum on Possible Practical Aspects of Research in Mathematical Biology of the Central Nervous System," 1947–48, Rashevsky Papers, Box 1, ff: Correspondence 1947–48.

45. See the Essay on Sources for the accounts of the Vienna Circle that I have found most useful. I also have drawn on the correspondence of Charles Morris with Otto Neurath and Rudolf Carnap in the Papers of the Unity of Science Movement, University of Chicago Archives. Probably the best short exposition of the ideas of the Vienna Circle is A. J. Ayer, *Language, Truth, and Logic* (London: V. Gollancz, 1936). Simon continued to assign Ayer to his graduate students into the 1990s.

46. Ludwig Wittgenstein had an ambiguous relationship to the Circle. His logical analysis of language had a powerful influence on the Circle, especially on Carnap, but

the logical empiricists always had the feeling that he was not entirely on their side. Neurath and Carnap even suspected that he was making fun of them at times, which he was, according to Janik and Toulmin: Allan Janik and Stephen Toulmin, *Wittgenstein's Vienna* (New York: Simon & Schuster, 1973). Also see Herbert Feigl, "The Wiener Kreis in America," in *The Intellectual Migration, Europe and America, 1930–1960*, ed. Donald Fleming and Bernard Bailyn (Cambridge, MA: Belknap Press of Harvard University Press, 1969), p. 638; Carl E. Schorske, *Fin-De-Siécle Vienna: Politics and Culture* (New York: Vintage Books, 1981).

47. Quoted in Peter Galison, "Aufbau/Bauhaus: Logical Positivism and Architectural Modernism," *Critical Inquiry* 16 (summer 1990): 709–52, p. 736.

48. Ibid., p. 742.

49. Hans Hahn, "Logic, Mathematics, and Knowledge of Nature," and Rudolf Carnap, "The Task of the Logic of Science," both in *Unified Science: The Vienna Circle Monograph Series,* ed. Brian McGuinness (Boston: D. Reidel, 1987).

50. Carnap, "The Task of the Logic of Science." Also see Rudolf Carnap, *Der Logische Aufbau Der Welt* (Berlin, 1928).

51. The term *transparent construction* is Peter Galison's from "Aufbau/Bauhaus: Logical Positivism and Architectural Modernism."

52. Hence, they termed their movement the "Unity of Science Movement." On the importance of the concept of unified science to the logical empiricists, see the introduction to McGuinness, *Unified Science.*

53. Carnap, "The Task of the Logic of Science."

54. Ibid.

55. Herbert Simon, "Letter to Professor Rudolf Carnap," 8/2/37, HSP, Box 61, ff: Autobiography—Source Documents—1942–82.

56. For the original title, see ibid. For the original outline, see Herbert Simon, "Outline for the Logical Structure of an Administrative Science," 7/28/37, HSP, Box 61, ff: Autobiography—Source Documents—1942–82.

57. Simon, *Models of My Life,* p. 53; Simon, *Models of Discovery,* p. xv.

58. See, for example, Herbert Simon, "On the Definition of the Causal Relation," *Journal of Philosophy* 49, no. 16 (1952).

59. Simon, "Letter to Professor Rudolf Carnap."

60. Simon, *Models of My Life,* 53.

61. Simon, *Models of Discovery,* p. xv.

FOUR: Research and Reform

1. Herbert Simon, *Models of My Life* (New York: Basic Books, 1991), p. 78.

2. Ibid., p. 79.

3. State Relief Administration State of California, "Unemployment Relief in California: Monthly Bulletin of the SRA" (Sacramento: State of California, 1939). My own calculations based on examination of caseloads for 1939 (as reported in this volume) and total numbers of persons receiving relief lead to an estimate of between three and four persons per "case."

4. Simon, *Models of My Life,* p. 74. Simon says little about his relationship with his wife in his autobiography, though there was clearly a deep bond of affection between

the two. He also notes that her "smile made up for my sometime prickliness" in social situations. None of Simon's personal correspondence with his wife or children was saved in his manuscript collections, except for one letter to his children regarding the "troubles" of the 1960s.

5. Dorothea Simon was very active as a political researcher and activist at this time as well, working with the League of Women Voters in the preparation of the *California Voter's Handbook* and finishing up her master's thesis (supervised by Charles Merriam). Dorothea P. Simon, "Letter to Charles Merriam," 10/7/40, CMP, Box 46, ff: Correspondence—S. Merriam wrote "more power to you" in reply.

6. Theodore Porter, *Trust in Numbers: The Pursuit of Objectivity in Science and Public Life* (Princeton, NJ: Princeton University Press, 1995). For other approaches to the rise of the professions and the role of abstract or scientific knowledge within them, see the Essay on Sources.

7. Clarence Ridley and Herbert Simon, *Measuring Municipal Activities: A Survey of Suggested Criteria and Reporting Forms for Appraising Administration* (Chicago: International City Managers' Association, 1938). Though Ridley is listed as the first author, Simon wrote the majority of the report.

8. The first municipal research agencies were created in the early 1900s, as were the first college programs in public administration and in business management. The Brookings Institution—the first governmental research agency for national-level issues—was founded in 1916. On municipal research agencies and other urban reform efforts, see the Essay on Sources.

9. Clarence Ridley and Herbert Simon, *Measuring Municipal Activities: A Survey of Suggested Criteria for Appraising Administration*, 2d ed. (Chicago: International City Managers' Association, 1943), pp. ix, 1.

10. Ibid., pp. x, 1–2. Simon and Ridley were not alone in their efforts to shift debate to services rendered rather than costs alone: see Lent Upson, "The Other Side of the Budget," *National Municipal Review* (1923).

11. Ridley and Simon, *Measuring Municipal Activities*, 2d ed., p. vii.

12. Ibid., p. viii.

13. In a wonderful example of protesting too much, the word "practical" appears dozens of times in the introduction, chap. 1, and chap. 2 of the report.

14. Andrew Abbott, *The System of Professions: An Essay on the Division of Expert Labor* (Chicago: University of Chicago Press, 1988). Abbott argues that the possession of a body of abstract knowledge is the key to a profession's claim to authority—or jurisdiction—over some social function. See the Essay on Sources for a discussion of Abbott and other works on the professions.

15. Ridley and Simon, *Measuring Municipal Activities*, 2d ed., p. 6.

16. Porter, *Trust in Numbers*.

17. Herbert Simon, *Models of Discovery: And Other Topics in the Methods of Science* (Boston: D. Reidel, 1977), introduction.

18. See the Essay on Sources for some recent works on the rise of the economists.

19. Clarence E. Ridley and Orin F. Nolting, "The Municipal Yearbook" (Chicago: International City Managers' Association, 1936–39). For Ridley's influence on Simon, see Simon, *Models of My Life*, p. 65.

20. Simon, *Models of My Life*, p. 65.

21. Samuel May was a member of the ICMA and had known Merriam and Brownlow well for many years, co-authoring a report with them in 1926 on creating an institute for research in local government. Correspondence between May and Merriam is in CMP, Box 160.

22. Simon, *Models of My Life*, pp. 74–77.

23. The Rockefeller Foundation has been the subject of almost as much research as it has funded. See the Essay on Sources for some key works on the foundation.

24. Beardsley Ruml was still active in the affairs of the foundation, even after leaving the presidency of its sister philanthropy, the Spelman Fund, in 1931. Ruml also was a key member of the National Resources Planning Board, discussed in chapter 2.

25. Raymond Seidelman, *Disenchanted Realists: Political Science and the American Crisis, 1884–1984* (Albany: State University of New York Press, 1985); Martin J. Schiesl, *The Politics of Efficiency: Municipal Administration and Reform in America, 1800–1920* (Berkeley: University of California Press, 1977).

26. Louis Wirth, "Report on the History, Activities, and Policies of the Social Science Research Council," August 1937, SSRC Papers, RAC, Box 704, ff 10276, AC 2, Series 4, Subseries 1. Also, SSRC, "Minutes of Meeting, Board of Directors," 9/12–14/44, SSRC Papers, RAC, Box 357, ff 2098, AC 1, Series 9, pp. 22–25. The first SSRC committee in the area was established in 1928 under the leadership of John Gaus, who performed a survey of the status of research in the field. The Public Administration Clearing House (one of the "1313" organizations) was established in 1930 in response to Gaus's report. The Gaus committee was only temporary; a permanent committee on public administration was established in 1933, under Louis Brownlow's direction. Brownlow was, at the time, also director of the PACH.

27. Bureau of Public Administration, "Governmental Research Organizations in the Western States," (Berkeley, CA: Bureau of Public Administration, 1935). Also see the IGS web page, www.igs.org.

28. Simon, *Models of My Life*, p. 80.

29. In J. Robert Oppenheimer's infamous clearance hearing, mention is made of a known "leftist" at Berkeley named Sam May.

30. Simon, *Models of My Life*, pp. 80, 124–25.

31. Ibid., p. 93.

32. Ronald Shephard, "The Incidence of Taxation Studied in a Simplified Economic System" (Ph.D. diss., University of California, 1941).

33. Herbert Alexander Simon et al., *Determining Work Loads for Professional Staff in a Public Welfare Agency* (Berkeley, CA: Bureau of Public Administration, 1941). On SRA caseloads, see the monthly bulletins of the California State Relief Administration, titled *Unemployment Relief in California*.

34. Simon et al., *Determining Work Loads for Professional Staff in a Public Welfare Agency*, preface by Frank Hoehler, p. v.

35. Jackson Putnam, *Modern California Politics* (San Francisco: Boyd and Fraser, 1980), p. 27.

36. California State Legislature, "Report of the Joint Legislative Fact-Finding Committee on Employment," (California State Legislature, 1940), pp. 20–29, quote on p. 21. This report shows clearly that the battle over the SRA was very important to Speaker Garland.

37. Simon et al., *Determining Work Loads*, p. 30.

38. California State Legislature, "Report of the Joint Legislative Fact-Finding Committee on Employment," p. 21.

39. Simon et al., *Determining Work Loads,* p. 30.

40. Ibid., p. 65.

41. Ibid., p. 27. On the Hawthorne studies, see Richard Gillespie, *Manufacturing Knowledge: A History of the Hawthorne Experiments* (Cambridge: Cambridge University Press, 1991).

42. Simon et al., *Determining Work Loads,* pp. 27, 30.

43. Ibid., p. 31.

44. Herbert Simon, Frederick Sharp, and Ronald Shephard, "Fire Losses and Fire Risks," (Berkeley, CA: Bureau of Public Administration, 1942), p. vii.

45. Herbert Simon, *Fiscal Aspects of Metropolitan Consolidation* (Berkeley, CA: Bureau of Public Administration, 1943).

46. Ibid., foreword by Samuel May, p. vii.

47. Ibid., pp. 54, 56.

48. Ibid., p. 6.

49. Ibid., p. 1; emphasis in original.

50. Herbert A. Simon, "The Incidence of a Tax on Urban Real Property," *Quarterly Journal of Economics* 57, no. 3 (1943): 398–420.

51. Simon, *Fiscal Aspects of Metropolitan Consolidation,* p. 5.

52. Note that this analysis presumes that people capitalize taxes to a far greater extent than services (i.e., that tax rates enter more fully into purchasing decisions than do service levels).

53. The quoted phrase comes from Herbert A. Simon, "The Planning Approach in Public Economy," *Quarterly Journal of Economics* 55, no. 2 (1941): 325–30, 325. Though Simon's advocacy of the planning approach is less bold in *Fiscal Aspects* than in this article, the thrust of both is the same.

54. Ibid., p. 330.

55. Ibid., p. 326.

56. James McCamy, "Letter to Herbert Simon," 6/6/47, HSP, Box 6, ff 213, p. 14.

FIVE: *Homo Administrativus,* or Choice under Control

1. Herbert Simon, "Administrative Behavior: A Study of Decision-Making Processes in Administrative Organization," 1945, HSP, Box 20, ff: Administrative Behavior—Preliminary ed.—1945, pp. 200–201. This "preliminary edition" of *Administrative Behavior* is almost completely unchanged from the original version of the thesis. The published version of 1947 is broadly similar to this preliminary edition, but there are significant differences between the two. For example, in the 1947 version one chapter is eliminated and the others are heavily rewritten and put in a new order. This preliminary edition will be cited as Simon, "Administrative Behavior—Preliminary Edition."

2. Luther Halsey Gulick and Lyndall F. Urwick, *Papers on the Science of Administration,* 2nd ed. (New York: Columbia University Institute of Public Administration, 1937).

3. Simon, "Administrative Behavior—Preliminary Edition," pp. 200–201.

4. Ibid., pp. 202–5.

5. Though Simon does not call them proverbs in the thesis, he does so in a contemporaneous article: Herbert Simon, "The Proverbs of Administration," *Public Administration Review* 6 (1946): 53–67.

6. Simon, "Administrative Behavior—Preliminary Edition," pp. 217–21.

7. Ibid., p. 218.

8. Ibid., p. 219.

9. For Simon, these authors' most significant works were Talcott Parsons, *The Structure of Social Action* (Glencoe, IL: Free Press, 1937); Edward C. Tolman, *Purposive Behavior in Animals and Men* (New York: Century Co., 1932); Chester A. Barnard, *The Functions of the Executive* (Cambridge, MA: Harvard University Press, 1968); John Dewey, *Human Nature and Conduct: An Introduction to Social Psychology* (New York: Carlton House, 1922).

10. Talcott Parsons and Edward Shils, eds., *Toward a General Theory of Action* (Cambridge, MA: Harvard University Press, 1951).

11. Herbert Simon, "Letter to Professor Rudolf Carnap," 8/2/37, HSP, Box 61, ff: Autobiography—Source Documents—1942–1982.

12. Simon, "Administrative Behavior—Preliminary Edition," p. 20.

13. Ibid., pp. 25, 220.

14. On the relationship between behaviorism and logical positivism, see the discussion of logical positivism in chapter 3 and in the Essay on Sources. Also see Hunter Crowther-Heyck, "George A. Miller, Language, and the Computer Metaphor of Mind," *History of Psychology* 2, no. 1 (1999): 37–64; Laurence D. Smith, *Behaviorism and Logical Positivism: A Reassessment of the Alliance* (Stanford, CA: Stanford University Press, 1986).

15. Tolman, *Purposive Behavior in Animals and Men.* This book and Tolman's later article on "Cognitive Maps in Rats and Men," *Psychological Review* 55, no. 4 (1948): 189–208, were extremely influential sources of ideas for the first generation of cognitive psychologists, such as George A. Miller. See Crowther-Heyck, "George A. Miller, Language, and the Computer Metaphor of Mind," p. 54.

16. Tolman, *Purposive Behavior in Animals and Men,* pp. 2, 204, xi, 10.

17. Ibid., pp. 14, 206, 210.

18. Ibid., pp. 26.

19. Simon, "Administrative Behavior—Preliminary Edition," p. 1. Simon was not the only political scientist to take interest in the problem of decision-making: see also Edwin O. Stene, "An Approach to a Science of Administration," *American Political Science Review* 34, no. 6 (1940): 1124–37. Simon does not cite Stene as a source of his interest in decision-making, though he later welcomed Stene's interest in the subject. Edwin O. Stene, "Letter to Herbert Simon," 6/9/44, HSP, Box 6, ff 219; Herbert Simon, "Letter to Dr. Edwin Stene," 6/28/44, HSP, Box 6, ff 219, p. 1.

20. James McCamy, "Letter to Herbert Simon," 6/6/47, HSP, Box 6, ff 213, p. 1.

21. Barnard, *The Functions of the Executive,* pp. 8, 10, 15.

22. Ibid., p. 296.

23. Simon, "Administrative Behavior—Preliminary Edition," pp. 3–4.

24. Ibid.

25. Ibid., pp. 5–7.

26. Herbert Simon, "On the Definition of the Causal Relation," and "Causal Ordering and Identifiability," both reprinted in Herbert Simon, *Models of Discovery: And Other Topics in the Methods of Science* (Boston: D. Reidel, 1977).

27. Tolman, *Purposive Behavior in Animals and Men,* p. 7. Simon also cites J. H. Woodger, *The Axiomatic Method in Biology* (Cambridge: The University Press, 1937), as a vital source for the formal definition of hierarchy. Simon, "Administrative Behavior—Preliminary Edition," p. 102. Hierarchy also was a vital principle in Barnard's analysis of management: see Barnard, *The Functions of the Executive.* On Woodger, see Joe Cain, "Woodger, Positivism, and the Evolutionary Synthesis," *Biological Philosophy* 15 (2000): 535–51.

28. Parsons, *The Structure of Social Action,* pp. 48, 75, 251.

29. Simon, "Administrative Behavior—Preliminary Edition," p. 6.

30. Ibid., p. 197.

31. Ibid., p. 6.

32. On these aspects of Dewey's thought, see Robert B. Westbrook, *John Dewey and American Democracy* (Ithaca, NY: Cornell University Press, 1991); Alan Ryan, *John Dewey and the High Tide of American Liberalism* (New York: W. W. Norton, 1995); James Kloppenberg, *Uncertain Victory: Social Democracy and Progressivism in European and American Thought, 1870–1920* (New York: Oxford University Press, 1986); Andrew Feffer, *Chicago Pragmatists and American Progressivism* (Ithaca, NY: Cornell University Press, 1993); James Livingston, *Pragmatism and the Political Economy of Cultural Revolution, 1850–1940* (Chapel Hill: University of North Carolina Press, 1994). Also see Dewey himself, especially *Human Nature and Conduct* and *Problems of Men* (New York: Philosophical Library, 1946), esp. chaps. 8 and 9, "Authority and Resistance to Social Change" and "Liberty and Social Control," pp. 93–110, 111–25.

33. The political goals and ideals of Wilson and other pioneers of public administration in the United States are best expressed in Woodrow Wilson, "The Study of Administration," *Political Science Quarterly* 2 (1887): 197–220.

34. Simon, "Administrative Behavior—Preliminary Edition," pp. 154, 152.

35. Stephen Waring, *Taylorism Transformed: Scientific Management Theory since 1945* (Chapel Hill: University of North Carolina Press, 1991), p. 62.

36. Sherman Krupp, *Pattern in Organization Analysis: A Critical Examination* (Philadelphia: Chilton, 1961), p. 99.

37. Ibid., p. 25.

38. Parsons, *The Structure of Social Action,* pp. 6, 7.

39. Ibid., pp. 21, 9.

40. Ibid., pp. 10, 28; Lawrence J. Henderson, "An Approximate Definition of a Fact," *University of California Studies in Philosophy,* vol. 4 (1932), pp. 179–99.

41. Kuhn's *The Structure of Scientific Revolutions* (Chicago: University of Chicago Press, 1962) was published as part of a monograph series on the unity of science.

42. Herbert A. Simon, "Letter to William Cooper (January)," 1/17/46, HSP, Box 1, ff 26, p. 1.

43. Herbert Simon, "Letter to Dr. James Fesler," 5/20/44, HSP, Box 5, ff 204, p. 2.

44. Simon, "Letter to William Cooper (January)," p. 1.

45. President's Research Committee on Social Trends, *Recent Social Trends in the United States: Report of the President's Research Committee on Social Trends* (New York: McGraw-Hill, 1933).

46. On this fascination with measurement, especially via machinery, see Allan Megill, "Introduction: Four Senses of Objectivity," *Annals of Scholarship* 8, nos. 3–4 (1991): 301–20; Theodore Porter, "Economics and the History of Measurement," *History of Political Economy* 33, Annual Supplement (2001): 4–22; Lorraine Daston and Peter Galison, "The Image of Objectivity," *Representations* 40 (fall 1992): 81–128.

47. The quote is from Lord Kelvin, and it is mentioned in almost every discussion of the social sciences at Chicago. See, for example, Martin Bulmer, *The Chicago School of Sociology: Institutionalization, Diversity, and the Rise of Sociological Research* (Chicago: University of Chicago Press, 1984). The uses of this quote are discussed in Robert K. Merton, David L. Sills, and Stephen M. Stigler, "The Kelvin Dictum and Social Science: An Excursion into the History of an Idea," *Journal of the History of the Behavioral Sciences* 20 (1984): 319–31.

48. Bernard Barber, *Science and the Social Order* (Glencoe, IL: Free Press, 1952), p. 13.

49. Ibid., p. 18.

50. Mitchell G. Ash, ed., *Forced Migration and Scientific Change: Émigré German-Speaking Scientists and Scholars after 1933* (Washington, DC: German Historical Institute; New York: Cambridge University Press, 1996); Donald Fleming and Bernard Bailyn, eds., *The Intellectual Migration, Europe and America, 1930–1960* (Cambridge, MA: Belknap Press of Harvard University Press, 1969).

51. Alfred North Whitehead, *Science and the Modern World* (New York: Macmillan, 1925).

52. Carl F. Christ, "The Cowles Commission's Contributions to Econometrics at Chicago, 1939–1955," *Journal of Economic Literature* 32 (March 1994): 30–59; emphasis added.

53. Simon, "Administrative Behavior—Preliminary Edition," pp. 35, 28.

54. Ibid., pp. 35–42, 53.

55. Ibid., pp. 42–43.

56. Ibid., p. 45.

57. Ibid., pp. 63–64.

58. Ibid., pp. 45–46. A system in which there is close interdependence among a few variables and only very loose connections between them and other "external" variables is "nearly decomposable." As noted earlier, there is a strong connection between hierarchical structure and near decomposability. See Simon, *Models of Discovery*.

59. Simon, "Administrative Behavior—Preliminary Edition," pp. 82–84, 86.

60. Ibid., pp. 94, 100, 110.

61. Ibid., pp. 98, 94.

62. Ibid., pp. 99–100.

63. Ibid., pp. 162, 167.

64. Ibid., pp. 167–73.

65. Ibid., pp. 71–72.

66. Ibid., pp. 72–74, 80.

67. Ibid., pp. 84, 83.

68. Barnard, *The Functions of the Executive*, p. xxix; Talcott Parsons, "Lecture Notes for the Sociology of Institutions," Parsons Papers, Harvard University Archives, Box "Talcott Parsons, Sociology 6: Lecture Notes 1930s", ff: "General Introduction—lecture notes, Soc 6 1930s."

69. Simon, "Administrative Behavior—Preliminary Edition," pp. 83–84.

70. Ibid., p. 83.

71. The reader will note the masculine article: its use is intentional, for masculinity was an important part of professional identity then as now. For an interesting look at this issue, see Mary Jo Deegan, *Jane Addams and the Men of the Chicago School, 1892–1918* (New Brunswick, NJ: Transaction Books, 1988); JoAnne Brown, *The Definition of a Profession: The Authority of Metaphor in the History of Intelligence Testing, 1890–1930* (Princeton, NJ: Princeton University Press, 1992). One should note that Brown's book demonstrates the power of gender more than the "authority of metaphor"—the women she describes used the same metaphors to describe their work that men did, but only the male-dominated occupations achieved professional status during the period she describes. For more on gender in human science, see the Essay on Sources.

72. Herbert Simon, "Current Research and Its Relation to Organization Theory," n.d. [presumably late 1951], HSP, Box 4, ff 126, p. 1. Although this document postdates the period under discussion, the phrase captures his argument in *Administrative Behavior* very well.

73. Later, Simon would go a step further, defining emotion as an "interrupt mechanism" triggered by "sudden, intense stimuli." Herbert Simon, "Motivational and Emotional Controls of Cognition," *Psychological Review* 74 (1967): 29–39.

74. Herbert Simon, "Letter to Professor James Mccamy of the University of Wisconsin," 12/2/47, HSP, Box 6, ff 213, p. 2.

s i x : Decisions and Revisions

1. Herbert Simon, *Models of My Life* (New York: Basic Books, 1991), p. 84.

2. Herbert Simon, "Memorandum: Mathematical Training of Social Scientists," 9/13/52, HSP, Box 4, ff 121, p.1.

3. Herbert Simon, "Letter to Grace Knoedler," 1/3/43, HSP, Box 61, ff: Autobiography—Source Documents—1942–82, pp. 1–2.

4. Simon, *Models of My Life*, p. 84.

5. Simon, "Letter to Grace Knoedler," p. 1. In his autobiography, Simon states that he believes that a faculty member should be able to teach "almost any undergraduate course," a testament to his commitment to interdisciplinarity—and to his immense self-confidence. Simon, *Models of My Life*, p. 100.

6. Simon, "Letter to Grace Knoedler," pp. 1–2.

7. Herbert A. Simon, "The Meaning of 'Democracy' in American Political Thought," 1946–47, HSP, Box 1, ff 32.

8. Ibid., p. 21.

9. Herbert A. Simon, Donald Smithburg, and Victor Thompson, *Public Administration*, 2nd ed. (New York: Knopf, 1956), p. 3. Except for pagination, the 1956 printing is the same as the 1950 original.

10. Karl Menger, *Reminiscences of the Vienna Circle and the Mathematical Colloquium* (Dordrecht: Kluwer Academic, 1994).

11. Franz Schulze, *Mies Van Der Rohe: A Critical Biography* (Chicago: University of Chicago Press, 1985).

12. Herbert A. Simon, "The Planning Approach in Public Economy," *Quarterly Journal of Economics* 55, no. 2 (1941): 325–30, p. 326.

13. Simon, "Letter to Grace Knoedler," p. 1.

14. Simon, *Models of My Life*, p. 98.

15. Herbert A. Simon, "Lecture on 'Economics of City Planning'" n.d. [c. 1943], HSP, Box 1, ff 31, p. 3.

16. Ibid.

17. The Cowles Commission has not received sufficient attention from historians of economics as yet. For a member's account of the early years, see Carl Christ, "History of the Cowles Commission, 1932–1952," in *Economic Theory and Measurement: A Twenty Year Research Report* (Chicago: Cowles Commission for Research in Economics, University of Chicago, 1952); Carl F. Christ, "The Cowles Commission's Contributions to Econometrics at Chicago, 1939–1955," *Journal of Economic Literature* 32 (March 1994): 30–59.

18. Christ, "History of the Cowles Commission," pp. 5–6.

19. Ibid., pp. 19–26. In 1956, the commission moved to New Haven and became affiliated with Yale, helping make it a leader in economics and political economy in the late 1950s and 1960s.

20. Cowles Commission for Research in Economics, "Economic Theory and Measurement: A Twenty Year Research Report," (Chicago: Cowles Commission for Research in Economics, University of Chicago, 1952), Appendix I: "Biographies of Staff, Fellows, and Guests," pp. 111–50. For Simon's experiences with the commission, see Simon, *Models of My Life*, pp. 101–7.

21. The mathematically skilled economists-by-training at the commission almost all came from the University of Chicago, with a few coming from Columbia University as well. Cowles Commission for Research in Economics, "Economic Theory and Measurement," 111–50.

22. As David Jardini, Paul Edwards, Henry Aaron, and Ida Hoos have shown, it was not uncommon for researchers to travel in the opposite direction in the 1960s, as many designers of defense systems attempted to tackle social systems, usually with limited results. David Jardini, "Out of the Blue Yonder: The Rand Corporation's Diversification into Social Welfare Research, 1946–1968" (Ph.D. diss., Carnegie Mellon University, 1996); Paul Edwards, *The Closed World: Computers and the Politics of Discourse in Cold War America* (Cambridge, MA: MIT Press, 1996); Henry J. Aaron, *Politics and the Professors: The Great Society in Perspective* (Washington, DC: Brookings Institution, 1978); Ida Hoos, *Systems Analysis in Public Policy: A Critique* (1972; rev. ed., Berkeley: University of California Press, 1983).

23. Cowles Commission for Research in Economics, "Economic Theory and Measurement: A Twenty Year Research Report," Preface.

24. Ibid., p. 61.

25. Ibid., p. 65.

26. Ibid., p. 31.

27. There have been several studies recently of OR's origins, all of which empha-

size the context of WWII as vital for its development. Clearly, the war was central to OR's emergence, but it did not create interest in these problems de novo. Rather, the war (and military patronage) selected and amplified several already existing lines of research, organizing them around a specific set of problems. On the origins of OR, see the Essay on Sources.

28. Ralph Lapp, "Nuclear Weapons: Past and Present," in *Alamogordo Plus Twenty-Five Years,* ed. Richard S. Lewis and Jane Wilson (New York: Viking, 1970), p. 248.

29. Herbert A. Simon, "Effects of Increased Productivity upon the Ratio of Urban to Rural Population," *Econometrica* 15, no. 1 (1947): 31–42; Herbert Simon, "Invention and Cost Reduction in Technological Change: Cowles Commission Discussion Paper: Economics No. 247," HSP, Box 2, ff 70.

30. Simon, "Effects of Increased Productivity." On the Yule distribution, see also Herbert A. Simon, "On a Class of Skew Distribution Functions," *Biometrika* 42 (December 1955), reprinted in *Models of Man: Social and Rational. Mathematical Essays on Rational Behavior in a Social Setting* (New York: Wiley, 1957), pp. 145–164. Also see Herbert Simon, "Letter to Stuart Dodd of the Washington Public Opinion Laboratory," 2/8/55, HSP, Box 5, ff 202, p. 1; "Letter to Benoit Mandelbrot," 11/9/54, HSP, Box 6, ff 213; Herbert Simon, "Letter to Robert Solow," 9/21/53, HSP, Box 6, ff 219; "Letter to George A. Miller (1955a)," 4/4/55, HSP, Box 6, ff 210; and "Letter to George A. Miller (1955b)," 7/25/55, HSP, Box 6, ff 212.

31. Samuel Schurr, Jacob Marschak, and Herbert Simon, *Economic Aspects of Atomic Power* (Princeton, NJ: Princeton University Press, 1950).

32. Lewis Strauss, quoted in Spencer Weart, *Nuclear Fear: A History of Images* (Cambridge, MA: Harvard University Press, 1988), p. 166.

33. Herbert Simon, "Letter to Eugene Zuckert," 1/14/61, HSP, Box 11, ff: Nuclear Science and Engineering Corporation—Correspondence—1960–64.

34. Tjalling C. Koopmans, "Letter to Herbert Simon," 3/23/49, HSP, Box 6, ff 226; Herbert Simon, "Progress Report: Research on Decision-Making under Uncertainty under ONR-Cowles Commission Contract," 4/11/53, HSP, Box 6, ff 226.

35. Menger, *Reminiscences of the Vienna Circle.*

36. Simon, *Models of My Life,* pp. 100–101; "Letter to Professor Karl Menger," 6/26/47, HSP, Box 2, ff 55.

37. Simon, "Letter to Professor Karl Menger," p. 1. This seminar series appears to have been the point at which Simon was introduced to the movement to "axiomatize" various branches of science, for example J. H. Woodger, *The Axiomatic Method in Biology* (Cambridge: The University Press, 1937).

38. Herbert Simon, "The Classical Concepts of Mass and Force," fall 1947, HSP, Box 2, ff 55; "Equality of Inertial and Gravitational Mass," n.d. [1947], HSP, Box 1, ff 38. Simon published the ideas set forth in these drafts in "The Axioms of Newtonian Mechanics," *Philosophical Magazine* 30 (1947): 888–905.

39. The "identification problem" has to do with determining which variables should be treated as independent and which as dependent. See Herbert Simon, "On the Definition of the Causal Relation," *Journal of Philosophy* 49, no. 16 (1952): 517–28, and "Causal Ordering and Identifiability," in *Studies in Econometric Method,* ed. William C. Hood and T. C. Koopmans (New York: Wiley, 1953). Both articles were reprinted in Simon, *Models of Man.*

40. For further discussion of the connection between operationalist philosophy, the computer, and the redefinition of mind, see chaps. 9–12 and Hunter Crowther-Heyck, "George A. Miller, Language, and the Computer Metaphor of Mind," *History of Psychology* 2, no. 1 (1999): 37–64.

41. Correspondence related to Simon's thesis is found in HSP, Box 1, ff 26.

42. Some of the more enlightening commentaries are James McCamy, "Letter to Herbert Simon," 6/6/47, HSP, Box 6, ff 213; John A. Vieg, "Letter to Herbert Simon," 8/11/45, HSP, Box 1, ff 26; Herbert Simon, "Letter to Paul Appleby," 2/23/48, HSP, Box 5, ff 197; Chester A. Barnard, "Letter to Herbert Simon (May)," 5/11/45, HSP, Box 1, ff 26. Also see Herbert A. Simon, "Letter to William Cooper (January)," 1/17/46, HSP, Box 1, ff 26.

43. McCamy, "Letter to Herbert Simon." Similar views can be seen in Barnard, "Letter to Herbert Simon (May)"; Simon, "Letter to William Cooper (January)"; and James Fesler, "Review of Administrative Behavior," *Journal of Politics* 10, no. 1 (1948): 187–89.

44. Barnard, "Letter to Herbert Simon (May)"; Chester A. Barnard, "Letter to Herbert Simon (June)," 6/24/45, HSP, Box 1, ff 26.

45. Herbert A. Simon, "Letter to Chester Barnard (July)," 7/16/45, HSP, Box 1, ff 26, p. 1.

46. Barnard, "Letter to Herbert Simon (May)," p. 1.

47. Ibid., p. 1.

48. Barnard, "Letter to Herbert Simon (June)," p. 15.

49. Barnard, "Letter to Herbert Simon (May)," p. 5.

50. Barnard, "Letter to Herbert Simon (June)," p. 2.

51. Ibid., p. 5.

52. Barnard, "Letter to Herbert Simon (May)," p. 3.

53. Barnard, "Letter to Herbert Simon (June)," p. 1.

54. Ibid., p. 18.

55. Barnard, "Letter to Herbert Simon (May)," pp. 5–6.

56. Barnard, "Letter to Herbert Simon (June)," pp. 11–12, 17.

57. Simon does protest that "those pages . . . were written with quite another thought in mind"—that of convincing skeptical students of public administration that pursuing efficiency does not imply neglecting human values; Simon, "Letter to Chester Barnard (July)," p. 1.

58. Simon, "Letter to William Cooper (January)," p. 1.

59. Ibid. Cutting the third chapter, which outlined Simon's theory of choice, was particularly difficult.

60. See, for example, Fesler, "Review of Administrative Behavior"; John D. Millett, "Review of Administrative Behavior," *Political Science Quarterly* 62, no. 4 (1947): 621–22.

61. For a dismissive review, see Lloyd Short, "Review of *Administrative Behavior*," *American Political Science Review* 41, no. 6 (1947): 1215–16. For a critical review that takes Simon very seriously, see Dwight Waldo, "Development of Theory of Democratic Administration," *American Political Science Review* 46, no. 1 (1952): 81–103.

62. David Truman, *The Governmental Process: Political Interests and Public Opinion* (New York: Knopf, 1951); Harold Lasswell and Abraham Kaplan, *Power and Society: A*

Framework for Political Inquiry (New Haven, CT: Yale University Press, 1950). On the behavioral revolution in political science, see the Essay on Sources.

63. Herbert A. Simon, Peter Drucker, and Dwight Waldo, "'Development of a Theory of Democratic Administration': Replies and Comments," *American Political Science Review* 46, no. 2 (1952): 494–503.

64. Ibid., p. 494.

65. Ibid., pp. 495–6.

66. John Gunnell, *The Descent of Political Theory: The Genealogy of an American Vocation* (Chicago: University of Chicago Press, 1993), p. 224.

67. Herbert Simon, Donald Smithburg, and Victor Thompson, "A Manual for Teachers Using *Public Administration*," 3/11/52, HSP, Box 28, ff: "A Manual for Teachers Using *Public Administration*—1952," p. 3.

68. Eugene Jacobson, "Letter to Herbert Simon," 9/21/48, HSP, Box 6, ff 209.

69. Simon, Smithburg, and Thompson, *Public Administration*, p. 3.

70. Donald Smithburg, "Letter to Herbert Simon (February)," 2/10/50, HSP, Box 1, ff 35, p. 1.

71. Donald Smithburg, "Letter to Herbert Simon (March)," 3/4/50, HSP, Box 1, ff 35, p. 1.

72. Donald Smithburg, "Letter to Herbert Simon," 10/18/51, HSP, Box 4, ff 134.

73. Bert F. Green, interview by Hunter Crowther-Heyck, 4/28/99.

74. Henry C. Hart, "Review of Public Administration," *Journal of Politics* 13, no. 2 (1951): 294–97, p. 296.

75. Chester A. Barnard, *The Functions of the Executive* (Cambridge, MA: Harvard University Press, 1968), pp. 294–95.

76. Michael A. Bernstein, "American Economics and the National Security State, 1941–53," *Radical History Review* 63 (1995): 8–26; Michael Bernstein, *A Perilous Progress: Economists and Public Purpose in Twentieth-Century America* (Princeton, NJ: Princeton University Press, 2001); Robert Collins, *The Business Response to Keynes, 1929–1964* (New York: Columbia University Press, 1981); Robert M. Collins, *More: The Politics of Economic Growth in Postwar America* (New York: Oxford University Press, 2000). This tension is particularly evident if one looks at economic analyses of advertising: billions of dollars are spent each year on advertising that seeks to manipulate audiences into making irrational choices. If such advertising works, then individual consumers are not free, rational choosers. If it does not work, then the firms that pay for such campaigns are equally irrational.

77. Quoted in Barnard, *The Functions of the Executive*, final page (not numbered).

SEVEN: Structuring His Environment

1. Simon was appointed to a newly created Faculty Committee on Academic Policy in January 1947 and helped organize a faculty representative organization called the Faculty Council in the mid- to late 1940s as well. He was appointed full professor in March 1947. He was selected as department chair in advance of being made full professor but did not assume the position officially until the fall of 1947.

2. IIT was not in Hyde Park, but the Simons lived there anyway. They remembered it fondly from their graduate school days at the University of Chicago.

3. The Simons' children were named Kathie, Peter, and Barbara. They were born in 1942, 1944, and 1946, respectively.

4. Morton Grodzins, "Letter to Herbert Simon," 11/4/54, HSP, Box 5, ff 205. Ithiel de Sola Pool approached Simon about coming to MIT in 1953. Ithiel de Sola Pool, "Letter to Herbert Simon," 3/6/53, HSP, Box 6, ff 216. George A. Miller, Wassily Leontief, Frederick Mosteller, and Freed Bales (among others) all asked Simon to come to Harvard in 1957. Herbert A. Simon, "Letter to Robert F. Bales," 11/1/57, HSP, Box 16, ff: Correspondence—B—1953–59.

5. Herbert Simon, *Models of My Life* (New York: Basic Books, 1991), p. 260.

6. David Klahr, interview by Hunter Crowther-Heyck, 7/15/2003.

7. The "weaker medicine" phrase is from a letter from V. O. Key to Simon, undated, commending him on *Administrative Behavior;* V. O. Key, "Letter to Herbert Simon," 1948, HSP, Box 6, ff 210.

8. Joseph A. Schumpeter, *Capitalism, Socialism, and Democracy,* 3rd ed. (New York: Harper & Row, 1962).

9. These laboratories all focused on applied research in areas of greatest interest to their local industrial patrons. For example, Alcoa supported the Metals Research Laboratory, Gulf Oil sponsored the Molecular Structure Laboratory, and United States Steel, GE, Koppers, and Westinghouse supported the Coal Research Laboratory. Arthur W. Tarbell, *The Story of Carnegie Tech* (Pittsburgh, PA: Carnegie Institute of Technology, 1937), pp. 107–14. Also see Glen U. Cleeton, *The Story of Carnegie Tech II: The Doherty Administration* (Pittsburgh, PA: Carnegie Institute of Technology, 1965).

10. Robert E. Doherty, "The First Fifty Years," 10/27/50, CMU Archives, President's Papers—Doherty, Box 60, ff: "President—Doherty, Robert E. Speeches, 1949–50."

11. Ibid. Also see Carnegie Institute of Technology, "The Carnegie Plan of Professional Education in Engineering and Science," 12/01/48, CMU President's Papers—Doherty, Box 60, ff: "President—Doherty, Robert E., Carnegie Plan—General Characteristics, Journal Articles, 1948–57."

12. Doherty, "The First Fifty Years."

13. On Doherty's interest in economics and management, see Robert Gleeson and Steven Schlossman, "George Leland Bach and the Rebirth of Graduate Management Education in the United States, 1945–1975," *Selections* (1996): 8–46, and Robert Gleeson, "The Rise of Graduate Management Education in American Universities, 1908–1970" (Ph.D. diss., Carnegie Mellon University, 1997), esp. chaps. 3–4.

14. Gleeson and Schlossman, "George Leland Bach," pp. 9–12.

15. Ibid.

16. Ibid., pp. 11–15. Also see William Cooper, "Proposed Sequence in Quantitative Controls in Business," 4/3/47, HSP, Box 4, ff 200, and Herbert Simon, "Letter to William Cooper (April)," 4/24/46, HSP, Box 4, ff 200.

17. Carter A. Daniel, *MBA: The First Century* (Lewisburg, PA: Bucknell University Press, 1998), pp. 149, 196–97.

18. GSIA Ph.D.s were in such demand at other business schools in the 1960s that it was not uncommon for them to receive job offers from leading schools not only before

they had completed their degrees but even before they had applied for the jobs! Klahr, interview.

19. John Servos, "Changing Partners: The Mellon Institute, Private Industry, and the Federal Patron," *Technology and Culture* 35, no. 2 (1994): 221–57.

20. Daniel, *MBA: The First Century,* pp. 140–41; Gleeson and Schlossman, "George Leland Bach," pp. 14–15.

21. Gleeson and Schlossman, "George Leland Bach," p. 14.

22. Herbert Simon, *Administrative Behavior* (New York: Macmillan, 1947).

23. On the historiography of the systems sciences, see the Essay on Sources.

24. From the patron's perspective, one way to describe such work would be "mission-oriented basic research," a phrase used to describe the kind of work the ONR, ARO, and AFOSR supported in the behavioral sciences during the 1950s and 1960s. Congressional Research Service, "Research Policies for the Social and Behavioral Sciences, Science Policy Study Background Report No. 6" Task Force on Science Policy, Committee on Science and Technology, U.S. House of Representatives, 1986.

25. W. W. Cooper, David Rosenblatt, and Herbert Simon, "Memorandum: Research Program of the School: Project on Intra-Firm Planning," 2/21/50, HSP, Box 61, ff: Materials for Autobiography—1982. Simon's views are also clearly expressed in his "Letter to Bernard Berelson, December 1951," 12/10/51, HSP, Box 4, ff 119. Also see his handwritten comments in the margins of Bernard Berelson, "The Ford Foundation Behavioral Sciences Program: Proposed Plan for the Development of the Behavioral Sciences Program—Confidential and Preliminary Draft," 1951, in HSP, Box 4, ff 136.

26. National Academy of Sciences Behavioral and Social Science Survey Committee, *The Behavioral and Social Sciences: Outlook and Needs* (Englewood Cliffs, NJ: Prentice-Hall, 1969). A second wave of institutes for social research followed in the mid- to late 1960s; like the first wave, these centers received a great deal of their funding from the federal government, but unlike their predecessors, their funds came from civilian agencies associated with the Great Society and War on Poverty programs, not from the military.

27. In the first annual report the school is already identified as the GSIA, and the research focus is apparent. G. L. Bach, "First Annual Report, 1949–50," 7/1/50, GSIA Papers, CMU Archives, Box 4, ff: "GSIA First Annual Report 1949–50, July 1, 1950."

28. George L. Bach, "Letter to Herbert Simon," 4/23/49, HSP, Box 5, ff 198. Also see Bach, "First Annual Report, 1949–50." G.L. Bach, "Second Annual Report, 1950–1951, School of Industrial Administration," 7/1/51, GSIA Papers, Box 4, ff: "Second Annual Report, 1951."

29. Carnegie Institute of Technology, "Bulletin of the School of Industrial Administration," 12/01/49, CMU President's Papers—Doherty, Box 60, ff: "Doherty, Robert E., Speeches 1949–50."

30. Harold Guetzkow, "Letter to Herbert Simon," 7/28/51, HSP, Box 2, ff 74.

31. Bach, "Second Annual Report, 1950–1951, School of Industrial Administration."

32. Selected statistics related to the growth of the behavioral and social sciences after World War II can be found in the Appendix.

33. On the relationship between the patrons of postwar social science and the behavioral revolution, see the Appendix and Hunter Crowther-Heyck, "Patrons of the Revolution: Ideas and Institutions in Postwar Behavioral Science," paper delivered to the University of Oklahoma Colloquium, December 12, 2003.

34. Some of the other key figures in this network of patrons were Mina Rees, who headed the Office of Naval Research's Mathematics Division; Warren Weaver, who headed the Rockefeller Foundation's natural science programs and who sat on the advisory board of the ONR; and H. Rowan Gaither, who headed the planning teams for the Ford Foundation and RAND and was an important advisor to the Air Force Office of Scientific Research.

35. See Robert M. Thrall, Clyde Coombs, and Robert L. Davis, eds., *Decision Processes* (New York: Wiley, 1954); Vernon L. Smith, "Game Theory and Experimental Economics: Beginnings and Early Influences," in *Toward a History of Game Theory: Supplement to the History of Political Economy,* ed. E. Roy Weintraub (Durham, NC: Duke University Press, 1992); M. M. Flood, "Report of a Seminar on Organization Science, Rm-709," 10/29/51, HSP, Box 28, ff: "The RAND Corporation—'Report of a Seminar on Organization Science'—1951." Other attendees included Robert Bush, Clyde Coombs, William Estes, Leon Festinger, Clifford Hildreth, Samuel Karlin, Tjalling Koopmans, Jacob Marschak, Oskar Morgenstern, Roy Radner, and Lloyd Shapley.

36. Herbert Simon, "Letter to Quincy Wright," 4/8/49, HSP, Box 6, ff 222.

37. On "big science," see the discussion of patronage in the Essay on Sources.

38. Harvey Sapolsky is one of the few to note the importance of these interlocking committee memberships, but even he discusses this phenomenon primarily in terms of whether or not an elite "inner circle" dominated postwar science. It is quite clear that there was an inner circle of policy-makers in the postwar social and behavioral sciences; what I am arguing here is that the way that this inner circle was constructed—as a network of interlocking memberships on advisory committees—rewarded skills of brokers and entrepreneurs. Harvey Sapolsky, *Science and the Navy: The History of the Office of Naval Research,* (Princeton, NJ: Princeton University Press, 1990), p. 99.

39. Bernard Berelson, "The Ford Foundation Behavioral Sciences Division Report, June 1953," 6/53, HSP, Box 7, ff 236.

40. Francis X. Sutton, "The Ford Foundation: The Early Years," *Daedalus* 116, no. 1 (1987): 41–91. The foundation received control of around 90 percent of the Ford Motor Company's assets (in the form of stock). That the foundation received stock rather than cash meant that the value of the endowment fluctuated with the fortunes of the company. By the end of the 1950s, the stock had increased in value well beyond expectations.

41. On Gaither's role at RAND, see David Hounshell, "The Cold War, Rand, and the Generation of Knowledge, 1946–1962," *Historical Studies in the Physical and Biological Sciences* 27, no. 2 (1997): 237–67, and David Jardini, "Out of the Blue Yonder: The Rand Corporation's Diversification into Social Welfare Research, 1946–1968" (Ph.D. diss., Carnegie Mellon University, 1996). Note that Gaither persuaded the Ford Foundation to give RAND $1 million to help it transform itself from Project RAND to the RAND Corporation.

42. "Report of the Study for the Ford Foundation on Policy and Program," Ford Foundation, 1949, pp. 14–15.

43. Berelson, "The Ford Foundation Behavioral Sciences Program," pp. 4–6, 8; Berelson, "The Ford Foundation Behavioral Sciences Division Report, June 1953," pp. 12–13. The term *behavioral science* appears to have been coined in the late 1940s by a group of social scientists and biologists at the University of Chicago who were interested in social phenomena. James G. Miller, "Toward a General Theory for the Behavioral Sciences," in *The State of the Social Sciences*, ed. Leonard D. White (Chicago: University of Chicago Press, 1956).

44. Berelson, "The Ford Foundation Behavioral Sciences Program," pp. 3–4.

45. Ibid., pp. 7–9. In the last quotation, Berelson is quoting from the Study Committee report.

46. Ibid., pp. 8–9, 30.

47. Ibid., pp. 32, 34.

48. Simon, "Letter to Bernard Berelson, December 1951"; Herbert Simon, "Letter to Bernard Berelson, May 1952," 5/26/52, HSP, Box 4, ff 119. Also see Simon's marginal comments on his copy of Berelson, "The Ford Foundation Behavioral Sciences Program."

49. Herbert Simon, marginal comments on Berelson, "The Ford Foundation Behavioral Sciences Program," pp. 22, 36.

50. Simon wrote a very big "NO!" in the margins next to this proposal on his copy of the report; ibid., p. 32, 34.

51. Simon, "Letter to Bernard Berelson, December 1951," p. 1.

52. Herbert Simon, marginal comments on Berelson, "The Ford Foundation Behavioral Sciences Program," p. 36.

53. Simon, "Letter to Bernard Berelson, December 1951," p. 1.

54. Herbert Simon, marginal comments on Berelson, "The Ford Foundation Behavioral Sciences Program," pp. 23–24.

55. Simon, "Letter to Bernard Berelson, December 1951," p. 3.

56. Simon, marginal comments on Berelson, "The Ford Foundation Behavioral Sciences Division Report, June 1953," pp. 30–34.

57. Roger Geiger notes that the CASBS became not a "spearhead" but a "retreat" for social scientists, which is perhaps to give too much credit to Berelson's original plan, which Simon (quite accurately, I believe) saw as creating a place where everyone could go to study at Paul Lazarsfeld's feet. Herbert Simon, interview by Hunter Crowther-Heyck, 10/19/97. Berelson expected that "the Center will work itself out of existence after a number of years." Berelson, "The Ford Foundation Behavioral Sciences Division Report, June 1953," p. 28.

58. O. Meredith Wilson, "Letter to Herbert Simon," 3/14/69, Box 10, ff: CASBS—Stanford University—Correspondence.

59. Simon, "Letter to Bernard Berelson, December 1951," p. 3.

60. Berelson regarded the summer institutes and inventories as disappointments and the Center for Advanced Studies in the Behavioral Sciences as a success. Simon, on the other hand, believed that the summer institutes were extremely important, that his inventory, at least, was valuable, and that the CASBS was of little benefit to anyone other than Stanford, for whom it was a good recruiting tool. Roger Geiger, *Research*

and Relevant Knowledge: American Research Universities since World War II (New York: Oxford University Press, 1993), p. 102; Simon, interview.

61. The first three of these contracts are described in the first two annual reports, cited above. The Ford Foundation grant is discussed in G. L. Bach, "GSIA Annual Report for 1953–54," 7/7/54, GSIA Papers, Box 4, ff: Annual Report 1954.

62. Bach, "First Annual Report, 1949–50," p. 14.

63. Gleeson and Schlossman, "George Leland Bach," p. 17.

64. Herbert Simon, "Notes of Conversation with W. Cooper—July 10, 1951," 7/10/51, HSP, Box 61, ff: Autobiography—Source Documents—1942–82.

65. Robert A. Gordon and James E. Howell, *Higher Education for Business* (New York: Columbia University Press, 1959); Frank Pierson, *The Education of American Businessmen: A Study of University-College Programs in Business Administration* (New York: McGraw-Hill, 1959).

66. Gleeson and Schlossman, "George Leland Bach."

67. Also note that the percentage of papers in the leading social science journals citing research support from a nonuniversity source leapt from near zero in the late 1940s to more than 70 percent by the early 1960s. In the *Psychological Review* of 1949, for example, only two out of thirty-seven articles cited any form of research support, and only one of those two was from a nonuniversity source. By 1958, that number had risen to twenty out of thirty, and by 1963, to twenty-five out of thirty. (These numbers do not include a small number of papers from foreign authors, almost none of whom cite nonuniversity support.)

68. Robert Freed Bales, "Letter to Herbert Simon," 12/20/56, HSP, Box 16, ff: Correspondence—F—1956–59.

69. See the Essay on Sources for a discussion of military patronage and postwar science.

70. See Charles C. Holt and Herbert Simon, "Optimal Decision Rules for Production and Inventory Control," in *Proceedings of the Conference on Operations Research in Production and Inventory Control* (Cleveland, OH: Case Institute of Technology, 1954); Charles C. Holt, Franco Modigliani, and Herbert Simon, "A Linear Decision Rule for Production and Employment Scheduling," *Management Science* 2 (1955): 1–30; Herbert Simon, "Dynamic Programming under Uncertainty with a Quadratic Criterion Function," *Econometrica* 24 (1956): 74–81; W. W. Cooper, "Report of Progress: Project Scoop, Research Project for the Study of Intra-Firm Behavior," 6/5/52, GSIA Papers, Box: GSIA (Cyert), ff: Airforce Project Papers.

71. Herbert Simon and Harold Guetzkow, "Memorandum to G. L. Bach on Research into Behavior in Organizations—Proposed Program," 2/28/52, HSP, Box 7, ff 240.

72. Anonymous, "Summer Research Training Institutes: A New Council Program," *SSRC Items* (1954).

73. Herbert Simon, "Final Report: A Research Training Institute in Techniques for the Computer Simulation of Cognitive Processes," 8/1/63, HSP, Box 13, ff: SSRC—Summer Seminars—Mailing Lists, Final Report, Memoranda, Correspondence, 1962–63.

74. Simon, *Models of My Life*, p. 144.

75. Simon, "Notes of Conversation with W. Cooper—July 10, 1951," pp. 1–2.

76. Herbert Simon, "Letter to G. L. Bach," 7/26/51, Box 61, ff: Autobiography—Source Documents—1942–82.

77. Simon, "Notes of Conversation with W. Cooper—July 10, 1951," p. 2.

78. Ibid., pp. 4, 1.

79. Ibid., p. 3.

80. Gleeson, "The Rise of Graduate Management Education," p. 216.

81. Melvin Anshen, quoted in ibid., p. 217.

82. Melvin Anshen, quoted in ibid., p. 219.

83. Some examples, other than systems science, might be molecular biology, solar system astronomy, materials science, and solid-state physics. See Lily Kay, *The Molecular Vision of Life: Caltech, the Rockefeller Foundation, and the Rise of the New Biology* (New York: Oxford University Press, 1993); Ronald Edmund Doel, *Solar System Astronomy in America: Communities, Patronage, and Interdisciplinary Science, 1920–1960* (New York: Cambridge University Press, 1996); Stuart W. Leslie, *The Cold War and American Science: The Military-Industrial-Academic Complex at MIT and Stanford* (New York: Columbia University Press, 1993).

84. Note that although the NIH and ARPA did have explicit missions other than the advance of science and thus funded mission-oriented basic research rather than wholly "pure" science (if such an animal ever existed), their grant-making and review processes usually were structured along disciplinary lines. In addition, ARPA, the NIH, and the NSF all had explicit "capacity-building" missions, which tended to be interpreted in terms of discipline-building. As we shall see in chapter 11, NIMH in the late 1950s and early 1960s was something of an exception to this rule, in part because it had more money than it knew what to do with.

85. Kent C. Redmond and Thomas M. Smith, *From Whirlwind to MITRE: The R&D Story of the Sage Air Defense Computer* (Cambridge, MA: MIT Press, 2000). System Development Corporation, "The System Development Corporation and System Training," *American Psychologist* 12, no. 8 (1957): 524–27.

86. Simon, interview; "Transcript of Interview with Herbert Simon, April 1975," 4/9/75, HSP, Box 52, ff: Pamela McCorduck Interviews—1975, pp. 19–20.

EIGHT: Islands of Theory

1. See, for example, Talcott Parsons, *The Social System* (New York: Free Press, 1951); David Easton, *The Political System: An Inquiry into the State of Political Science* (New York: Knopf, 1953); Eliot Chapple and Carleton Coon, *Principles of Anthropology* (New York: Holt, 1947); Karl W. Deutsch, "Mechanism, Organism, and Society: Some Models in Natural and Social Science," *Philosophy of Science* 18, no. 3 (1951): 230–52; Karl W. Deutsch, *The Nerves of Government: Models of Political Communication and Control* (New York: Free Press of Glencoe, 1963); James G. Miller, "Toward a General Theory for the Behavioral Sciences," in *The State of the Social Sciences*, ed. Leonard D. White (Chicago: University of Chicago Press, 1956). For additional works in the flourishing sciences of control, see the online Essay on Sources.

2. Peter Buck, "Adjusting to Military Life: The Social Sciences Go to War, 1941–50," in *Military Enterprise and Technological Change: Perspectives on the American Experience,* ed. Merritt Roe Smith (Cambridge, MA: The MIT Press, 1985).

3. Talcott Parsons, "Letter to Dean Paul Buck," 4/3/44, Talcott Parsons Papers, Harvard University Archives, HUG (FP) 15.2, Correspondence @1930–59, ff: Buck.

4. James G. Miller, "Toward a General Theory for the Behavioral Sciences," *American Psychologist* 10, no. 9 (1955): 513–31.

5. Hunter Crowther-Heyck, "Full Employment in a Free Society: Science and Democratic Values in Paul Samuelson's *Economics*" (paper presented at the Humanities and Technology Association Annual Meeting, 1995).

6. Some exemplary works in these fields were John Von Neumann and Oskar Morgenstern, *The Theory of Games and Economic Behavior* (Princeton, NJ: Princeton University Press, 1944); Robert M. Thrall, Clyde Coombs, and Robert L. Davis, eds., *Decision Processes* (New York: Wiley, 1954); Kenneth J. Arrow, *Social Choice and Individual Values* (New Haven, CT: Yale University Press, 1951); Anthony Downs, *An Economic Theory of Democracy* (New York: Harper & Brothers, 1957); Tjalling C. Koopmans, ed., *Activity Analysis of Production and Allocation* (New York: Wiley, 1951); Abraham Wald, *Statistical Decision Functions* (New York: Wiley, 1950). For additional exemplars of the sciences of choice in the first decade after World War II, see the online Essay on Sources.

7. See Paul Samuelson, *Foundations of Economic Analysis* (Cambridge, MA: Harvard University Press, 1947), and Von Neumann and Morgenstern, *The Theory of Games and Economic Behavior.*

8. Allen Newell and Herbert Alexander Simon, *Human Problem-solving* (Englewood Cliffs, NJ: Prentice-Hall, 1972), p. 10.

9. Herbert Simon, "Some Strategic Considerations in the Construction of Social Science Models," 1951, HSP, Box 4, ff 120, pp. 2–3.

10. Ibid., p. 3.

11. See the Essay on Sources for a discussion of the historiography of political science, including the behavioral revolution.

12. David Easton, "Introduction: The Current Meaning of 'Behavioralism' in Political Science," in *The Limits of Behavioralism in Political Science,* ed. James C. Charlesworth (Philadelphia: American Academy of Political and Social Science, 1962); Herbert Simon, *Models of My Life* (New York: Basic Books, 1991), p. 169.

13. Albert Somit and Joseph Tanenhaus, *The Development of Political Science: From Burgess to Behavioralism* (Boston: Allyn and Bacon, 1967), p. 157. Taylor Cole was *APSR* editor 1949–52. He was much more sympathetic to behavioral social science, especially in its theoretical forms, than was Ogg, who had been editor since 1925. Elsbree, Cole's successor, was broadly sympathetic to behavioralism, but the mathematical orientation of many behavioralists was alien to him.

14. Simon, "Some Strategic Considerations in the Construction of Social Science Models," p. 1.

15. Herbert Simon, *Models of Man: Social and Rational. Mathematical Essays on Rational Behavior in a Social Setting* (New York: Wiley, 1957), p. 1.

16. Herbert A. Simon, "On a Class of Skew Distribution Functions," *Biometrika* 42 (December 1955). This article is reprinted in Simon, *Models of Man,* pp. 145–64, quotation from p. 145.

17. Simon, *Models of Man,* p. 90, emphasis added.

18. Another important agency in the spread of mathematical techniques through

the social sciences was the SSRC's Committee on the Mathematical Training of Social Scientists, for which Simon consulted. See Herbert Simon, "Memorandum: Mathematical Training of Social Scientists," 9/13/52, HSP, Box 4, ff 121. Also, a startling number of important works in mathematical social science in the 1950s and early 1960s were published by one company, Wiley, indicating that its editor for social science was particularly fond of mathematical approaches—and a good judge of talent.

19. Simon, *Models of Man*, p. 97.

20. Simon, "Some Strategic Considerations in the Construction of Social Science Models," p. 1.

21. Herbert Simon, "Letter to Kenneth May," 11/2/53, HSP, Box 6, ff 213, p. 2.

22. Herbert Simon, "The Classical Concepts of Mass and Force," Fall 1947, HSP, Box 2, ff 55.

23. Herbert A. Simon, "A Formal Theory of Interaction in Social Groups," *American Sociological Review* 17 (April 1952). This article is reprinted in Simon, *Models of Man*, pp. 99–114; page references are to the latter version. Note that Robert K. Merton makes a point of emphasizing the mathematical origins of the concept of function in *Social Theory and Social Structure: Toward the Codification of Theory and Research* (Glencoe, IL: Free Press, 1949), especially in chap. 1, "Manifest and Latent Functions."

24. James March, "The 1978 Nobel Prize in Economics," *Science* 202 (1978): 858–61.

25. A major source for such thinking about the importance of theory in science was Lawrence J. Henderson, "An Approximate Definition of a Fact," *University of California Studies in Philosophy* 14 (1932): 179–99. As noted in chapter 5, this notion of a conceptual scheme is very like Thomas Kuhn's concept of the paradigm. According to Henderson, before the creation of a systematic conceptual scheme, a body of knowledge is just a mass of mere data, not a science, just as a pre-paradigmatic body of knowledge is not a science for Kuhn; Thomas Kuhn, *The Structure of Scientific Revolutions* (Chicago: University of Chicago Press, 1962).

26. A particularly powerful concrete system model may become the root of a heuristic model, as when Skinnerian models of pigeon behavior inspired analogies between pigeons and humans or when Simon's models of human problem-solving behavior inspired the broader analogy between humans and computers.

27. Note that I am providing a typology here, not a description of the actual sequence of development in Darwin's thought.

28. The irony in this historical development is that the basic heuristic model for Darwin, and for most scientists in the mid-nineteenth century, was that natural systems were like economies and thus were best understood in terms of the concepts of political economy.

29. Simon termed some organizations *unitary organizations,* defining them as organizations designed to serve a single purpose. Adapting Simon's language, one could describe a concrete system model that only used one theory as a *unitary model;* such models typically involve a very high degree of abstraction.

30. The phrase "islands of theory" appears in many places in Simon's work, most particularly in Simon, "Some Strategic Considerations in the Construction of Social Science Models," and Herbert Simon and Harold Guetzkow, "General Scientific Framework (Orientation Notes)," 11/5/53, HSP, Box 4, ff 152.

31. Simon, *Models of My Life*, p. 89.

32. Herbert A. Simon, "A Formal Theory of the Employment Relation," *Econometrica* 19 (July 1951). This article is reprinted in Simon, *Models of Man,* pp. 183–95. Page citations are to the reprinted version.

33. Simon refers to his "Homans Model" as emerging directly from his work on the Controller's Study. Herbert Simon, "Annual Report of Activities, July 1, 1950–June 30, 1951," 5/3/51, HSP, Box 6, ff 224, p.2.

34. Simon, *Models of Man,* p. 166.

35. Herbert A. Simon, "A Formal Theory of the Employment Relation," in *Models of Man, Social and Rational: Mathematical Essays on Rational Behavior in a Social Setting,* ed. Herbert Simon (New York: Wiley-Interscience, 1951), p. 184.

36. Ibid.

37. The relationship to game theory is clear in Simon's analysis of this process: the possible behavior patterns chosen by the employer for the employee correspond to a set of possible strategies that a player might choose in a game.

38. Simon, "A Formal Theory of the Employment Relation," pp. 184, 194.

39. Ibid., pp. 192, 195.

40. Herbert Simon, "A Formal Theory of Interaction in Social Groups," in *Models of Man, Social and Rational: Mathematical Essays on Rational Behavior in a Social Setting,* ed. Herbert Simon (New York: Wiley, 1957), p. 99.

41. Ibid.

42. Ibid., p. 100. Since the units in which these variables can be measured are "somewhat arbitrary," Simon strives to "make us only of the *ordinal* properties of the measuring scales—the relations of greater or less" rather than using cardinal numbers.

43. Ibid., p. 101.

44. Ibid., p. 103. Here Simon cites Paul Samuelson's *Foundations of Economic Analysis* as the prime exemplar of the virtues of comparative statics.

N I N E : A New Model of Mind and Machine

1. David Mindell, *Between Human and Machine: Feedback, Control, and Computing before Cybernetics* (Baltimore: Johns Hopkins University Press, 2002); Thomas Hughes, *Networks of Power: Electrification in Western Society, 1880–1930* (Baltimore: Johns Hopkins University Press, 1983).

2. On the historiography of the systems sciences, see the Essay on Sources.

3. The movement of a large number of physicists into work on radar and electronics during the war had a powerful impact on electrical engineering, reinforcing its connections to "basic" science and abstract mathematics at the same time it grounded the physicists in technological concerns. As a result, it is often hard to tell the physicists from the electrical engineers in postwar electronics research. On this fusion of electrical engineering and physics see Michael Riordan and Lillian Hoddeson, *Crystal Fire: The Birth of the Information Age* (New York: W. W. Norton, 1997); Paul Forman, "Behind Quantum Electronics: National Security as Basis for Physical Research in the United States, 1940–60," *Historical Studies in Physical and Biological Sciences* 18, no. 1 (1987): 149–229; and Stuart W. Leslie, *The Cold War and American Science: The Military-Industrial-Academic Complex at MIT and Stanford* (New York: Columbia University Press, 1993).

4. On human-machine analogies, see Peter L. Galison, "The Ontology of the

Enemy: Norbert Wiener and the Cybernetic Vision," *Critical Inquiry* 21 (autumn 1994): 228–66; Hunter Crowther-Heyck, "George A. Miller, Language, and the Computer Metaphor of Mind," *History of Psychology* 2, no. 1 (1999): 37–64; and Tara Abraham, "(Physio)Logical Circuits: The Intellectual Origins of the Mcculloch-Pitts Neural Networks," *Journal of the History of the Behavioral Sciences* 38, no. 1 (2002): 3–25.

5. Claude Shannon's information theory gave the new sciences of machines a unifying concept and a ready unit of measurement. Warren Weaver and Claude E. Shannon, *The Mathematical Theory of Communication* (Urbana: University of Illinois Press, 1949). On the spread of the concept of information into biology, see Evelyn Fox Keller, *Refiguring Life: Metaphors of Twentieth Century Biology* (New York: Columbia University Press, 1995). On information theory in psychology, see Crowther-Heyck, "George A. Miller, Language, and the Computer Metaphor of Mind."

6. Galison, "The Ontology of the Enemy."

7. Arturo Rosenblueth, Norbert Wiener, and Julian Bigelow, "Behavior, Purpose, and Teleology," *Philosophy of Science* 10 (1943): 18–24.

8. Norbert Wiener, *Cybernetics: Or Control and Communication in the Animal and the Machine* (Cambridge, MA: Technology Press, 1948).

9. Steve J. Heims, *The Cybernetics Group* (Cambridge, MA: MIT Press, 1991).

10. Talcott Parsons, *The Social System* (New York: Free Press, 1951), passim.

11. Herbert Simon, "Letter to Howard Cirker of Dover Publications," 8/31/53, HSP, Box 5, ff 202.

12. Herbert Simon, "Letter to Ross Ashby," 6/15/53, Box 5, ff 197.

13. W. Ross Ashby, "Annual Report," Barnwood House, 1952.

14. Henderson and Cannon were renowned for their development of the concept of homeostasis in organisms and for their application of this concept to the analysis of social systems. On Henderson, see John Louis Parascandola, "Lawrence J. Henderson and the Concept of Organized Systems" (Ph.D. diss., University of Wisconsin, 1968). Cannon's ideas are best expressed in Walter B. Cannon, *The Wisdom of the Body* (New York: W. W. Norton, 1932). These ideas are discussed in chapter 5 above.

15. W. Ross Ashby, *Design for a Brain* (New York: Wiley, 1952), p. 7.

16. Ibid., p. v. Note also that Ashby takes pains to argue that the study will "make no use of the subjective elements of experience" (p. 10).

17. Ibid., p. v.

18. Ibid., pp. 7, 54.

19. All quotes from W. Ross Ashby, "Letter to Herbert Simon," 7/23/53, HSP, Box 5, ff 197.

20. Ashby, *Design for a Brain*, chaps. 9–14.

21. Ibid., p. 103.

22. John M. O'Donnell, *The Origins of Behaviorism: American Psychology, 1870–1920* (New York: New York University Press, 1985).

23. George A. Miller and Noam Chomsky, "Finite State Languages," *Information and Control* 1 (1958): 91–112.

24. Ashby, *Design for a Brain*, p. 199.

25. Ibid., p. 29.

26. Herbert Simon, *Models of My Life* (New York: Basic Books, 1991), p. 166.

27. Herbert A. Simon, "Application of Servomechanism Theory to Production

Control," *Econometrica* 20 (April 1952). This article is reprinted in *Models of Man, Social and Rational: Mathematical Essays on Rational Behavior in a Social Setting,* ed. Herbert Simon (New York: Wiley, 1957). Citations are to the reprinted version.

28. Simon, "Application of Servomechanism Theory to Production Control," p. 219. Note that this quote reveals not only the intended generality of servomechanism theory but also the closeness of the links between proto-computer science and the needs of large-scale organizations for ever more sophisticated bureaucratic systems.

29. Ibid., p. 221.

30. Ibid., p. 223.

31. Herbert Simon, "Some Strategic Considerations in the Construction of Social Science Models," 1951, HSP, Box 4, ff 120, p. 32.

32. David Truman, *The Governmental Process: Political Interests and Public Opinion* (New York: Knopf, 1951); Harold Lasswell and Abraham Kaplan, *Power and Society: A Framework for Political Inquiry* (New Haven, CT: Yale University Press, 1950). See also Hans Joachim Morgenthau, *Scientific Man vs. Power Politics* (Chicago: University of Chicago Press, 1946).

33. Kenneth J. Arrow, *Social Choice and Individual Values* (New Haven, CT: Yale University Press, 1951); Anthony Downs, *An Economic Theory of Democracy* (New York: Harper & Brothers, 1957).

34. Herbert Simon, "On the Definition of the Causal Relation," *Journal of Philosophy* 49, no. 16 (1952); Herbert A. Simon, "Causal Ordering and Identifiability," in *Studies in Econometric Method,* ed. William C. Hood and T. C. Koopmans (New York: Wiley, 1953). Both are reprinted in Herbert Simon, *Models of Man: Social and Rational. Mathematical Essays on Rational Behavior in a Social Setting* (New York: Wiley, 1957), pp. 50–61, 10–36, respectively. Citations are to the reprinted versions.

35. Herbert Simon, "Causal Ordering and Identifiability," in *Models of Man, Social and Rational: Mathematical Essays on Rational Behavior in a Social Setting,* ed. Herbert Simon (Boston: D. Reidel, 1957), p. 12.

36. Ibid.

37. Ibid., p. 22.

38. Ibid., pp. 13–22. The higher level variables affect the variables in the subsystem by determining the coefficients of the equations linking the variables.

39. Ibid., p. 22.

40. Ibid., p. 26.

41. Herbert A. Simon, "Notes on the Observation and Measurement of Political Power," *Journal of Politics* 15 (November 1953). Citations are to the draft version of fall 1952: Herbert Simon, "Draft of 'Notes on the Observation and Measurement of Political Power,'" n.d. [1952], Box 2, ff 52.

42. Simon, "Draft of "Notes on the Observation and Measurement of Political Power,'" p. 2.

43. Hugh Elsbree, "Letter to Herbert Simon," 3/25/53, Box 2, ff 52, p. 1.

44. Ibid., pp. 1–2.

45. Herbert Simon, "Letter to Dwight Waldo," 3/10/53, HSP, Box 6, ff 222. In a letter to Hugh Elsbree about this affair, Simon noted that he agreed with Waldo's assessment of him as a "dyed-in-the-wool positivist, whose tolerance for heresy is nil"; Herbert Simon, "Letter to Hugh Elsbree," 3/9/53, HSP, Box 10, ff: *American Political Science Review*—Correspondence—1950–71.

46. Simon, "Letter to Hugh Elsbree," p. 2.

47. Simon, "Draft of 'Notes on the Observation and Measurement of Political Power,'" p. 6.

48. Ibid., pp. 6–7.

49. Ibid., p. 20.

50. Simon, *Models of My Life*, pp. 37, 44, 111.

51. The key works of the Gestaltists, for Simon and Guetzkow, were Otto Selz, *Die Gezete Der Productiven Und Reproductiven Geistestätigkeit* (Bonn: Cohen, 1924); Adrian de Groot, *Het Denken Van Den Shaker* (Amsterdam: N. H. Uitg, 1946), translated as *Thought and Choice in Chess* (New York: W. W. Norton, 1965); Karl Duncker, *The Psychology of Productive Thinking* (Washington, DC: American Psychological Association, 1945); and Max Wertheimer, *Productive Thinking* (New York: Harper & Brothers, 1945). On the history of Gestalt psychology, see Mitchell G. Ash, *Gestalt Psychology in German Culture, 1890–1967: Holism and the Quest for Objectivity* (New York: Cambridge University Press, 1995).

52. On George Miller, see Crowther-Heyck, "George A. Miller, Language, and the Computer Metaphor of Mind." For Jerome Bruner's ideas about thinking, see Jerome Bruner, Jacqueline Goodnow, and George Austin, *A Study of Thinking* (New York: Wiley, 1956), and Jerome Bruner, *In Search of Mind* (New York: Harper & Row, 1983).

53. Herbert Simon and Harold Guetzkow, "Memorandum to G. L. Bach on Research into Behavior in Organizations—Proposed Program," 2/28/52, HSP, Box 7, ff 240, pp. 1–6.

54. Ibid. Bavelas's studies had been conducted under Air Force sponsorship at MIT. They were classified but well known within the small community of systems scientists; hence, here one sees one of the many ways in which it was vital to be on the inside of this emerging community. See Alex Bavelas et al., "Project Rand Research Memorandum Rm-358: The Performance of Task-Oriented Groups as Influenced by Their Communications Network," 3/20/50, HSP, Box 4, ff: RAND Corporation—Paper "The Performance of Task-Oriented Groups as Influenced by Their Communications Network"—3/20/50.

55. Bavelas et al., "Performance of Task-Oriented Groups." Also see Leon Festinger, "Letter to Herbert Simon," 3/2/53, HSP, Box 2, ff 71.

56. Festinger, "Letter to Herbert Simon." This work was published as Herbert Simon and Harold Guetzkow, "A Model of Short and Long-Run Mechanisms Involved in Pressures toward Uniformity in Groups," *Psychological Review* 62 (January 1955): 56–68, and reprinted in Simon, *Models of Man*, pp. 115–30.

57. Simon first visited the SRL in the summer of 1952 while attending a conference (hosted by RAND) on "Decision Processes." This conference is discussed in chapter 7.

58. Robert Chapman, William Biel, John Kennedy, and Allen Newell, "The Systems Research Laboratory and Its Program, Project Rand Document D-1166," 1/7/52, HSP, Box 4, ff: RAND Corp—1/7/52, p. 1.

59. On SAGE, see Paul Edwards, *The Closed World: Computers and the Politics of Discourse in Cold War America* (Cambridge, MA: MIT Press, 1996); Thomas Parke Hughes, *Rescuing Prometheus*, 1st ed. (New York: Pantheon Books, 1998); Kenneth Flamm, *Creating the Computer: Government, Industry, and High Technology* (Washington, DC: Brookings Institution, 1988); Kent C. Redmond and Thomas M.

Smith, *From Whirlwind to MITRE: The R&D Story of the SAGE Air Defense Computer* (Cambridge, MA: MIT Press, 2000).

60. Edwards, *The Closed World*, chap. 6; Crowther-Heyck, "George A. Miller, Language, and the Computer Metaphor of Mind." For descriptions of work at the HRRL and at the affiliated Human Communications Research Group (the Bavelas group) at the Air Force Cambridge Research Laboratories, see J. C. R. Licklider, George A. Miller, and Jerome Wiesner, "Psychological Research Program for the Air Force Human Resources Research Laboratories Quarterly Progress Report," 12/16/52, HSP, Box 6, ff 213, and William Huggins, "Letter to Herbert Simon," 1/19/51, HSP, Box 5, ff 206.

61. The Air Force's own Maxwell Air Force Base was another major center for such research, but it was more isolated from the mainstream of psychological research. Herbert Simon, "Letter to Albert Biderman," 6/24/52, HSP, Box 5, ff 198. On Lincoln Labs, MIT, and RAND, see the Essay on Sources.

62. Licklider, Miller, and Wiesner, "Psychological Research Program"; Huggins, "Letter to Herbert Simon."

63. Many at HRRL, for example, had trained at Harvard's Psycho-Acoustics Laboratory under S. S. Stevens; Edwards, *The Closed World*, chap. 6.

64. Huggins, "Letter to Herbert Simon"; Licklider, Miller, and Wiesner, "Psychological Research Program."

65. Chapman et al., "The Systems Research Laboratory and Its Program," p. 1.

66. Allen Newell and Joseph Kruskal, "Organization Theory in Miniature," 5/18/51, HSP, Box 4, ff 147.

67. Chapman et al., "The Systems Research Laboratory and Its Program," pp. 8, 10.

68. Ibid., p. 20.

69. Ibid., pp. 4, 7, 13.

70. Ibid., p. 20.

71. These later tests are described in Robert Chapman et al., "The Systems Research Laboratory's Air Defense Experiments: Rand Paper No. P-1202," 10/23/57, HSP, Box 4, ff: H. S. RAND Corp—The Systems Research Laboratory's Air Defense Experiments.

72. Simon, *Models of My Life*, pp. 168, 111.

73. Ibid., p. 198. Simon titles chapter 13 of his autobiography "Climbing the Mountain: Artificial Intelligence Achieved."

74. Herbert A. Simon, "A Behavioral Model of Rational Choice," *Quarterly Journal of Economics* 69 (1955); Herbert Simon, "Rational Choice and the Structure of the Environment," *Psychological Review* 63 (1956): 129–138. These articles are reprinted in Simon, *Models of Man*, pp. 241–60, 261–73, respectively. Citations are to the reprinted versions.

75. Herbert Simon, "A Behavioral Model of Rational Choice," in *Models of Man*, ed. Herbert Simon (New York: Wiley, 1957), p. 241.

76. Simon, *Models of Man*, p. 202.

77. Ibid., p. 198, emphasis in original. One should remember that, though this is the first published use of the term *bounded rationality* by Simon, the concept had played a prominent role in his work since the first version of *Administrative Behavior*, written in 1942.

78. Simon, "A Behavioral Model of Rational Choice," pp. 241, 242.

79. Ibid., p. 245.

80. Ibid., p. 246.

81. Herbert Simon, "Letter to Ward Edwards," 8/16/54, HSP, Box 5, ff 203, p. 1.

82. Simon, "A Behavioral Model of Rational Choice," p. 243.

83. Simon, *Models of Man*, p. 199.

84. Simon, "A Behavioral Model of Rational Choice," p. 256.

85. There is an interesting parallel here to the trajectory of L. J. Henderson's interests, for he moved from studying the homeostatic mechanisms of the organism to studying the "fitness of the environment" as well. Lawrence J. Henderson, *The Fitness of the Environment* (New York: Macmillan, 1913).

86. Herbert Simon, "Rational Choice and the Structure of the Environment," in *Models of Man*, ed. Herbert Simon (New York: Wiley, 1957), pp. 262, 271.

87. Simon, "Letter to Ward Edwards," p. 1.

88. Simon, "Rational Choice and the Structure of the Environment," pp. 270–71.

89. Simon, *Models of Man*, p. 200.

90. Newell came to CIT in the fall of 1954 to "get his union card" (his Ph.D.) under Simon. Newell continued to be a paid member of the RAND staff through the 1950s. Newell began graduate training in mathematics but left Princeton to work on applied mathematics at RAND. He was twenty-seven in 1954.

91. Herbert Simon, "Dynamic Programming under Uncertainty with a Quadratic Criterion Function," *Econometrica* 24 (1956): 74–81.

92. Herbert Simon, "The Theory of Departmentalization: Ford Working Paper #8," 3/8/54, HSP, Box 7, ff 240; Herbert Simon, "Functional Analysis and Organization Theory: I," 4/1/54, HSP, Box 7, ff 239; Herbert Simon, "Functional Analysis and Organization Theory: II," 4/30/54, HSP, Box 7, ff 239.

93. James G. March, Herbert Simon, and Harold Guetzkow, *Organizations* (New York: Wiley, 1958). Although March is listed as the first author, and he undoubtedly made many important contributions to the book, the major ideas of *Organizations* are based on Simon's work, as seen in *Administrative Behavior, Public Administration*, and the papers listed in the previous note.

94. Simon, "Functional Analysis and Organization Theory: I," p. 1.

95. Ibid.

96. Ibid., p. 2.

97. Ibid., pp. 9, 10.

98. Ibid., pp. 9, 10, 12, 13.

99. Ibid., pp. 12, 14.

100. Ibid., p. 6.

101. Herbert Simon, "Concepts and Propositions for Possible Application to Social Behavior Laboratory," undated, presumably 1953, HSP, Box 4, ff 116, pp. 1–2.

102. Simon, "Functional Analysis and Organization Theory: I," p. 9. In his view, such activities were primarily the tasks of the executive in the business firm, rather than the middle-manager, though every individual necessarily had to be capable of such inductive, adaptive, problem-solving behavior else he or she could not survive.

103. Herbert A. Simon, *Administrative Behavior*, 2nd ed. (New York: Macmillan, 1961), pp. 149, 57.

104. The appendix was deleted by the editor. Herbert Simon, interview by Hunter Crowther-Heyck, 10/19/97.

TEN: The Program *Is* the Theory

1. On instruments in the history of science, see the Essay on Sources.

2. George Lakoff and Mark Johnson, *Metaphors We Live By* (Chicago: University of Chicago Press, 1980). On metaphors in the history of science, see the Essay on Sources.

3. The "trading zone" is a concept developed by Peter Galison to describe the interdisciplinary meeting ground where new concepts—and even new languages—are created through common experiences in a shared social-technical environment. Galison is primarily interested in the instruments and practices associated with them as shapers of the subculture of a trading zone; I would add that certain root metaphors play equally significant—and strikingly parallel—roles in the construction of such subcultures. Peter Louis Galison, *Image and Logic: A Material Culture of Microphysics* (Chicago: University of Chicago Press, 1997).

4. Herbert Simon, *Models of My Life* (New York: Basic Books, 1991), p. 168.

5. Herbert A. Simon, "Letter to Bernard Berelson, March 1957," 3/25/57, HSP, Box 10, ff: Ford Foundation—Correspondence—1957–59, p. 3.

6. K. Anders Ericsson and Herbert Simon, "Verbal Reports as Data," *Psychological Review* 87, no. 3 (1980): 215–51.

7. Herbert Simon, "Allen Newell (1927–1992)," *Annals of the History of Computing* 20 (1998): 63–76, p. 67.

8. Herbert Simon and Harold Guetzkow, "Memorandum to G. L. Bach on Research into Behavior in Organizations—Proposed Program," 2/28/52, HSP, Box 7, ff 240.

9. Ibid., pp. 6–9; Harold Guetzkow, "Organization Behavior Laboratory Research Working Paper No. 3, Proposal for Second Experimental Study," March 1953, HSP, Box 4, ff 116, p. 3.

10. Alan Turing, "On Computable Numbers, with an Application to the Entscheidungsproblem," *Proceedings of the London Mathematical Society, Series 2*, 42 (1936): 230–65; Alan Turing, "Computing Machinery and Intelligence," in *Computers and Thought*, ed. Edward Feigenbaum and Julian Feldman (1950; reprint, New York: McGraw-Hill, 1963). On Turing, see Alan Hodges, *Alan Turing: The Enigma of Intelligence* (New York: Simon & Schuster, 1983).

11. Simon, *Models of My Life*, p. 201.

12. Claude E. Shannon, "A Symbolic Analysis of Relay and Switching Circuits" (M.S. thesis, MIT, 1938); Warren McCullough and Walter Pitts, "A Logical Calculus of the Ideas Immanent in Nervous Activity," *Bulletin of Mathematical Biophysics* 5 (1943): 115–37.

13. In this they were like W. Ross Ashby's homeostat, which had required a human to intervene whenever it needed to change its "way of behaving."

14. An analogy might be to a switchboard; every new call that comes in requires the switchboard operator to make a new connection, plugging a wire into a new socket.

15. Martin Campbell-Kelly and William Aspray, *Computer: A History of the Infor-*

mation Machine, 1st ed. (New York: Basic Books, 1996); William Aspray, *John Von Neumann and the Origins of Modern Computing* (Cambridge, MA: MIT Press, 1990); Paul E. Ceruzzi, *A History of Modern Computing,* 2nd ed. (Cambridge, MA: MIT Press, 2003).

16. The "First Draft of a Report on EDVAC" has been called the "founding document of modern computing." Though it was never published, it was circulated widely. Ceruzzi, *A History of Modern Computing,* p. 21.

17. Von Neumann's idea of the program as a sequence of operations to be applied in a given situation bears a strong kinship to his idea of a strategy in a game, an idea advanced but a year earlier in John Von Neumann and Oskar Morgenstern, *The Theory of Games and Economic Behavior* (Princeton, NJ: Princeton University Press, 1944). Many of those who pursued game theory in the late 1940s and 1950s, as a result, came to talk of both humans and machines as pursuing determinate "courses of action," "plans," or "strategies." By the late 1950s, all these sibling concepts had begun to be folded into the general concept of the program.

18. To those who objected that the computer is still "just" a machine because it is "merely" carrying out a program it did not create, Simon replied that the same was true of humans, for do we not carry out the program inscribed within our DNA?

19. Herbert A. Simon and Allen Newell, "What Have Computers to Do with Management?," 5/21/59, HSP, Box 28, ff: RAND Corp.—What Have Computers to do With Management?—1959, p. 6.

20. JOHNNIAC was named after John von Neumann.

21. Simon, *Models of My Life,* p. 189.

22. Ibid., pp. 203–4; Herbert A. Simon, "Some Notes on the Early History of LT," 7/9/57, HSP, Box 29, ff: Paper "Some Notes on the Early History of LT."

23. Simon, *Models of My Life,* pp. 206–7. Richard Feynman tells a similar story regarding the organization of human "computers" who had to check the calculations of the IBM machines at Los Alamos. Lilian Hoddeson, ed., *Critical Assembly: A Technical History of Los Alamos During the Oppenheimer Years, 1943–45* (New York: Cambridge University Press, 1993).

24. Allen Newell and Herbert Simon, "Heuristic Problem Solving," *Journal of the Operations Research Society of America* 6 (1958): 1–10. This paper is reprinted in *Models of Bounded Rationality, Volume I: Economic Analysis and Public Policy,* ed. Herbert Simon (Cambridge, MA: MIT Press, 1982), pp. 380–89, 381. A further example of the intimacy of the connection between organization theory and problem-solving programming is that Simon developed the kernel of his second heuristic problem-solving program, the *General Problem Solver,* during the Summer Institute on Organizations held at Carnegie Tech (and sponsored by the Ford Foundation via the SSRC). See Herbert A. Simon, "Logic Problem Program," 7/6/57, HSP, Box 61, ff: Autobiography—Source Documents—1942–82.

25. Simon, *Models of My Life,* p. 206.

26. Newell and Simon, "Heuristic Problem Solving," p. 385.

27. The fact that a program is a "virtual machine" is not always recognized: programs initially were understood by the courts to be like texts rather than virtual machines and so were held to be covered by copyright rather than patent law. Robert X. Cringely, *Accidental Empires: How the Boys of Silicon Valley Make Their Millions,*

Battle Foreign Competition, and Still Can't Get a Date (New York: Harper Business, 1993).

28. Interestingly, it actually worked "backwards," through substitution, taking the theorem to be proven as its input and testing it against a set of givens (axioms and previously proven theorems). The Logic Theorist is described in Allen Newell and Herbert Simon, "The Logic Theory Machine: A Complex Information Processing System," *IRE Transactions on Information Theory* 1 (1956): 61–79, and Allen Newell and Herbert Simon, "Current Developments in Complex Information Processing, Rand Paper P-850," 5/1/56, HSP, Box 4, ff: RAND Corporation—Paper 5/1/56.

29. Newell and Simon, "Heuristic Problem Solving."

30. Simon, "Allen Newell (1927–1992)," p. 68.

31. Newell and Simon, "Heuristic Problem Solving," p. 382.

32. Ibid., pp. 386–87.

33. Ibid., p. 387.

34. Allen Newell, J. C. Shaw, and Herbert Simon, "Elements of a Theory of Human Problem Solving," *Psychological Review* 65, no. 3 (1958): 151–66, p. 151.

35. Turing, "On Computable Numbers, with an Application to the Entscheidungsproblem"; Philip Mirowski, "What Were Von Neumann and Morgenstern Trying to Accomplish?," in *Toward a History of Game Theory, History of Political Economy* 24, supp., ed. E. Roy Weintraub (Durham, NC: Duke University Press, 1992).

36. As David Mindell has shown in his superb book, *Between Human and Machine,* these mathematicians did not invent the concepts of feedback or of the servomechanism, nor were they the first to try to describe feedback and control systems in the language of communications engineering. They did, however, attempt to develop new mathematical formalisms to describe such systems. Simon and Newell drew on "classical servo theory, circa 1940," plus Ashby's biological reformulations of servo theory, for their understanding of servomechanisms and feedback, but they drew on Turing, Shannon, Wiener, von Neumann, and other logicians (Alonzo Church, George Polya, Emil Post) in their attempts to develop a new formal language for describing path-dependent systems. David Mindell, *Between Human and Machine: Feedback, Control, and Computing before Cybernetics* (Baltimore: Johns Hopkins University Press, 2002).

37. George A. Miller and Fred Frick, "Statistical Behavioristics and Sequences of Responses," *Psychological Review* 56 (1949): 311–25; Hunter Crowther-Heyck, "George A. Miller, Language, and the Computer Metaphor of Mind," *History of Psychology* 2, no. 1 (1999): 37–64; Abraham Wald, *Statistical Decision Functions* (New York: Wiley, 1950); M. M. Flood, "Report of a Seminar on Organization Science, Rm-709," 10/29/51, HSP, Box 28, ff: The RAND Corporation—"Report of a Seminar on Organization Science"—1951; Robert J. Leonard, "Creating a Context for Game Theory," in Weintraub, *Toward a History of Game Theory,* p. 71.

38. Mindell, *Between Human and Machine;* Weintraub, *Toward a History of Game Theory;* Thomas P. Hughes and Agatha C. Hughes, eds., *Systems, Experts, and Computers* (Cambridge, MA: MIT Press, 2000).

39. Later, Simon and Newell would emphasize the concept of the "production system" as the vital core of programming. A production system specifies a set of conditions and an action or actions to be taken when the conditions are satisfied.

ELEVEN: The Cognitive Revolution

1. Hunter Crowther-Heyck, "George A. Miller, Language, and the Computer Metaphor of Mind," *History of Psychology* 2, no. 1 (1999): 37–64.

2. On behaviorism and the history of psychology more generally, see the Essay on Sources.

3. John B. Watson, "Psychology as the Behaviorist Views It," *Psychological Review* 20 (1913): 158–77.

4. "Wundtian" after Wilhelm Wundt, one of the founders of experimental psychology. Ibid., pp. 159–60, 171–73, 175.

5. Ibid., p. 167.

6. Edwin G. Boring, *A History of Experimental Psychology*, 1st ed. (New York: D. Appleton-Century, 1929).

7. Clark Hull, *The Principles of Behavior: An Introduction to Behavior Theory* (New York: D. Appleton-Century, 1943), p. 19.

8. Edward C. Tolman, *Purposive Behavior in Animals and Men* (New York: Century, 1932), chap. 1, "Behavior, a Molar Phenomenon"; Hull, *The Principles of Behavior: An Introduction to Behavior Theory*, chap. 1.

9. On positivism and behaviorism, see Laurence D. Smith, *Behaviorism and Logical Positivism: A Reassessment of the Alliance* (Stanford, CA: Stanford University Press, 1986); B. F. Skinner, "Review of *Behaviorism and Logical Positivism*," *Journal of the History of the Behavioral Sciences* 23 (1987): 206–10. Note that in his review of Smith, Skinner states that he "followed a strictly Machian line" whereas Tolman and Hull had both been strongly influenced by the Vienna Circle.

10. Clark Hull, for example, often describes the brain as being like a telephone switchboard, but he is careful to present this metaphor as being merely a pedagogical aid, not an explanatory model. Hull, *The Principles of Behavior: An Introduction to Behavior Theory*, p. 40. The logical positivists were not the first to object to the use of metaphorical language in science, of course. Watson, in "Psychology as the Behaviorist Views It," for example, rails against reasoning by analogy in psychology. The logical positivists, however, raised such opposition to metaphorical language to a key position in their philosophy. Interestingly, one common feature of articles by cognitive psychologists in the late 1950s and early 1960s—the early days of the "revolution"—is a defense of reasoning by analogy, which reveals that though operationalism and logical positivism were kin, they were not twins. See, for example Alphonse Chapanis, "Men, Machines, and Models," *American Psychologist* 16, no. 2 (1961): 113–31.

11. Rudolf Carnap's logical positivism had its adherents among the behaviorists, but perhaps the most important statement of scientific philosophy for psychologists in the late 1920s to 1930s was Percy Bridgman, *The Logic of Modern Physics* (New York: Macmillan, 1927). For Bridgman's influence on Hull, see Hull, *The Principles of Behavior*, pp. 30–31. For Bridgman's influence on S. S. Stevens, see Stevens, "The Operational Basis of Psychology," *American Journal of Psychology* 43 (1935): 323–30, and "The Operational Definition of Psychological Concepts," *Psychological Review* 42 (1935): 517–42.

12. Hull, *The Principles of Behavior*, pp. 17–31; Clark Hull, *A Mathematico-Deductive Theory of Rote Learning* (New Haven, CT: Yale University Press, 1940).

13. Hull, *The Principles of Behavior,* p. 27.

14. Nadine M. Weidman, "Mental Testing and Machine Intelligence: The Lashley-Hull Debate," *Journal of the History of the Behavioral Sciences* 30 (1994): 162–80.

15. Karl Lashley, for example, was hostile to theory in general. One of his favorite tactics was to set forth several theories only to dismantle them, without offering one of his own (ibid., pp. 163–64).

16. Stevens, "The Operational Basis of Psychology"; Stevens, "The Operational Definition of Psychological Concepts."

17. Karl Lashley, "The Behavioristic Interpretation of Consciousness," *Psychological Review* 30, no. 1 (1923): 237–72, 329–53.

18. B. F. Skinner, *The Behavior of Organisms* (New York: Appleton-Century Crofts, 1938); B. F. Skinner, "The Concept of the Reflex in the Description of Behavior," *Journal of General Physiology* 5 (1931): 427–58.

19. George A. Miller, interview by Hunter Crowther-Heyck, 3/1/93.

20. Mitchell G. Ash, *Gestalt Psychology in German Culture, 1890–1967: Holism and the Quest for Objectivity* (New York: Cambridge University Press, 1995).

21. Wolfgang Kohler, "Gestalt Psychology Today," *American Psychologist* 14, no. 12 (1959): 727–34.

22. George A. Miller, "George A. Miller," in *A History of Psychology in Autobiography,* ed. Gardner Lindzey (Stanford, CA: Stanford University Press, 1989), p. 393.

23. B. F. Skinner, *Verbal Behavior* (New York: Appleton-Century-Crofts, 1957).

24. Miller, interview.

25. Claude E. Shannon, "A Mathematical Theory of Communication, Part I," *Bell System Technical Journal* (July 1948); Claude E. Shannon, "A Mathematical Theory of Communication, Part II," *Bell System Technical Journal* (October 1948). These two articles are reprinted together in Warren Weaver and Claude E. Shannon, *The Mathematical Theory of Communication* (Urbana: University of Illinois Press, 1949).

26. Hull, *The Principles of Behavior,* p. 40. Indeed, Hull, among others, often referred to the brain as being similar to a telephone switchboard.

27. O. H. Mowrer discusses psychology's aversion to the study of language in "The Psychologist Looks at Language," *American Psychologist* 9, no. 11 (1954): 660–94.

28. Watson, "Psychology as the Behaviorist Views It," p. 175.

29. Wendell Garner, "The Contributions of Information Theory to Psychology," in *The Making of Cognitive Science,* ed. William Hirst (New York: Cambridge University Press, 1988); George A. Miller, "What Is Information Measurement?," *American Psychologist* 8 (1953): 3–11. For a description of the research projects in the early 1950s of several soon-to-be leaders of cognitive psychology, including Miller, William McDill, J. C. R. Licklider, and Ulric Neisser, see J. C. R. Licklider, George A. Miller, and Jerome Wiesner, "Psychological Research Program for the Air Force Human Resources Research Laboratories Quarterly Progress Report," 12/16/52, HSP, Box 6, ff 213.

30. George A. Miller, "The Magical Number Seven, Plus or Minus Two," *Psychological Review* 63 (1956): 81–97, p. 82.

31. Ibid., p. 86.

32. Ibid., p. 92.

33. Ibid., p. 93.

34. George A. Miller, Eugene Galanter, and Karl Pribram, *Plans and the Structure of Behavior* (New York: Henry Holt, 1960), p. 132.

35. Miller, "The Magical Number Seven, Plus or Minus Two," p. 93.

36. Miller, interview.

37. Ibid.

38. Miller, "George A. Miller."

39. Noam Chomsky, *Syntactic Structures* (The Hague: Mouton, 1957); Noam Chomsky, "Three Models for the Description of Language," *IRE Transactions on Information Theory* 1 (1956): 113–24.

40. George A. Miller and Noam Chomsky, "Finite State Languages," *Information and Control* 1 (1958): 91–112; Chomsky, "Three Models for the Description of Language"; Chomsky, *Syntactic Structures;* George A. Miller, "Some Psychological Studies of Grammar," *American Psychologist* 17, no. 11 (1962): 748–62; Bernard Baars, "Interview with George A. Miller," in *The Cognitive Revolution in Psychology,* ed. Bernard Baars (New York: Guilford Press, 1986), pp. 338–41.

41. A Turing Machine is a machine capable of simulating the behavior of any other machine. Turing proved in his famed 1936 paper "On Computable Numbers" that it was possible to conceive of such a machine and that it need only be capable of performing a very few basic operations. Turing is also famous for his later proposal of the "Turing Test" for artificial intelligence: assume there is a teletype terminal connected to something in another room, perhaps to a computer, perhaps to a person. A computer capable of deceiving a human on the other end of the wire into thinking that it was human must itself be intelligent, argued Turing, for the only evidence of intelligence we ever really have, even regarding other people, is their behavior. Thus the effective simulation of intelligence *is* intelligence, to Turing. Alan Turing, "On Computable Numbers, with an Application to the Entscheidungsproblem," *Proceedings of the London Mathematical Society, Series 2* 42 (1936): 230–65; Alan Turing, "Computing Machinery and Intelligence," in *Computers and Thought,* ed. Edward Feigenbaum and Julian Feldman (New York: McGraw-Hill, 1963).

42. "Miller recalls that Newell told him that Chomsky 'was developing exactly the same kind of ideas for language that he and Herb Simon were developing for theorem proving.'" Paul Edwards, *The Closed World: Computers and the Politics of Discourse in Cold War America* (Cambridge, MA: MIT Press, 1996), p. 229.

43. Applying a little math to the numbers Arthur Norberg cites, one arrives at approximately 250 computers in existence in the world in 1955. Arthur Norberg, Judy O'Neill, and Kerry Freedman, *Transforming Computer Technology: Information Processing for the Pentagon, 1962–1986* (Baltimore: Johns Hopkins University Press, 1996), p. 75.

44. Allen Newell, J. C. Shaw, and Herbert Simon, "Elements of a Theory of Human Problem Solving," *Psychological Review* 65, no. 3 (1958): 151–66; reprinted in *Models of Thought,* ed. Herbert Simon (New Haven, CT: Yale University Press, 1989).

45. Miller, Galanter, and Pribram, *Plans and the Structure of Behavior;* Crowther-Heyck, "George A. Miller, Language, and the Computer Metaphor of Mind."

46. Simon, Newell, and Shaw, "Elements of a Theory," pp. 6–7.

47. Ibid., pp. 7–8.

48. Ibid., p. 8.

49. Ibid., p. 10.

50. Ibid., p. 13.

51. Ibid., pp. 14–15.

52. Kurt Danziger, *Constructing the Subject: Historical Origins of Psychological Research* (New York: Cambridge University Press, 1990).

53. Edwards, *The Closed World;* Norberg, O'Neill, and Freedman, *Transforming Computer Technology.* For a specific case in point, see Herbert A. Simon, "Letter to S. L. Seaton of the U.S. Continental Army Command," 3/1/61, HSP, Box 17, ff: Correspondence—S—1959–65.

54. Norberg, O'Neill, and Freedman, *Transforming Computer Technology.*

55. J. C. R. Licklider, "Man-Computer Symbiosis," *IRE Transactions on Human Factors in Engineering* HFE-1, no. 1 (1960): 4–11.

56. Norberg, O'Neill, and Freedman, *Transforming Computer Technology,* pp. 286–87 (centers of excellence) and 102 (support of programming languages and time-sharing at CMU).

57. As noted in chapter 2, many of the leaders of this younger generation cut their teeth on work for the National Resources Planning Board in the late 1930s and 1940s.

58. Talcott Parsons et al., "Letter to President Conant," 10/16/45, Paul Buck Papers, Harvard University Archives, Box: "Social Sciences-Z," Correspondence, 1945–46, UA III, 5.55.26, Dean (FAS), pp. 1–2.

59. The powerful pull of the anticipated NSF is all the more remarkable in the case of Parsons and the Harvard Department of Social Relations, for Parsons was actually between generations. He was committed to making social science more rigorous and sought closer connections to federal patrons, but he was not mathematically sophisticated nor was he employed in military research during the war.

60. Jessica Wang, "Liberals, the Progressive Left, and the Political Economy of Postwar American Science: The National Science Foundation Debate Revisited," *Historical Studies in the Physical and Biological Sciences* 26, no. 1 (1995): 139–66.

61. The social sciences, by one calculation, had received 31 percent of federal funds for research before the war, but only 6 percent of a vastly expanded research budget after the war. These figures were drawn from John Riley's report to the SSRC board of directors on the state of the social sciences. The years of comparison were 1938 and 1948. Social Science Research Council, "Minutes of Meeting, Board of Directors, September 1949," 9/11–14/49, SSRC Papers, Rockefeller Archives Center (RAC), Tarrytown, NY, Box 358, ff 2100, AC 1, Series 9, p. 11.

62. Roger Geiger, "American Foundations and Academic Social Science, 1945–1960," *Minerva* 26 (1988): 315–41; Roger Geiger, *Research and Relevant Knowledge: American Research Universities since World War II* (New York: Oxford University Press, 1993). In the latter, see chap. 4, "Private Foundations and Research Universities, 1945–60," esp. pp. 94–110. Geiger argues quite specifically that the social sciences were excluded from federal support during the 1950s, but this is because he focuses on the National Science Foundation. The same is also true of Samuel Klausner in "The Bid to Nationalize American Social Science," in Samuel Klausner and Victor Lidz, eds., *The Nationalization of the Social Sciences* (Philadelphia: University of Pennsylvania Press, 1986).

63. Henry Riecken, "Underdogging: The Early Career of the Social Sciences in the

NSF," in *The Nationalization of the Social Sciences*, ed. David Klausner and Victor Lidz (Philadelphia: University of Pennsylvania Press, 1986). Also see the Appendix.

64. Herbert Simon, "Social and Behavioral Science Programs in the National Science Foundation: Final Report," National Academy of Sciences, Committee on the Social Sciences in the National Science Foundation, 1976.

65. Gerald N. Grob, *From Asylum to Community: Mental Health Policy in Modern America* (Princeton, NJ: Princeton University Press, 1991), pp. 60–67.

66. Robert Felix, "Mental Disorders as Public Health Problems," *American Journal of Psychiatry* 106 (1949): 401–6.

67. "Training Grant Program Evaluation," National Institute of Mental Health (U.S.). Division of Extramural Research Programs, Program Analysis and Evaluation Section, 1958.

68. Ibid., p. 15.

69. NIMH Program Analysis and Evaluation Section, *Behavioral Science and Mental Health*, National Institute of Mental Health, Division of Extramural Research Programs, 1970.

70. Grob, *From Asylum to Community: Mental Health Policy in Modern America*, p. 67.

71. See the Appendix for a discussion of funding for the postwar social and behavioral sciences.

72. Edwards, *The Closed World*, chaps. 6–8, passim.

73. National Academy of Sciences Behavioral and Social Science Survey Committee, *The Behavioral and Social Sciences: Outlook and Needs* (Englewood Cliffs, NJ: Prentice-Hall, 1969); Kenneth Clark and George A. Miller, eds., *Psychology* (Englewood Cliffs, NJ: Prentice-Hall, 1970), p. 107.

74. Herbert A. Simon, "Letter to Bernard Berelson, March 1957," 3/25/57, HSP, Box 10, ff: Ford Foundation—Correspondence—1957–59, p. 3.

75. Thomas H. Leahey, "The Mythical Revolutions of American Psychology," in *Evolving Perspectives on the History of Psychology*, ed. Wade Pickren and Donald Dewsbury (Washington, DC: American Psychological Association, 2002); John Mills, *Control: A History of Behavioral Psychology* (New York: New York University Press, 1998).

76. Norton Wise and Crosbie Smith, "Work and Waste: Political Economy and Natural Philosophy in Nineteenth-Century Britain (I) and (II)," *History of Science* 27 (1989): 263–301, 391–449; Norton Wise and Crosbie Smith, "Work and Waste: Political Economy and Natural Philosophy in Nineteenth-Century Britain (III)," *History of Science* 28 (1990): 221–61.

77. Simon's 1980 article on "Verbal Reports as Data" became his most frequently cited article in psychology, leading to a book on the subject as well. K. Anders Ericsson and Herbert Simon, "Verbal Reports as Data," *Psychological Review* 87, no. 3 (1980): 215–51; K. Anders Ericsson and Herbert Alexander Simon, *Protocol Analysis: Verbal Reports as Data* (Cambridge, MA: MIT Press, 1984).

T W E L V E : *Homo Adaptivus*, the Finite Problem Solver

1. Bernard Berelson, "Letter to Herbert Simon," 3/18/57, HSP, Box 10, ff: Ford Foundation—Correspondence—1957–59; Richard Sheldon, "Letter to Herbert

Simon," 4/2/57, HSP, Box 10, ff: Ford Foundation—Correspondence—1957–59; Herbert Simon, "Letter to Richard Sheldon," 4/11/57, HSP, Box 10, ff: Ford Foundation—Correspondence—1957–59.

2. Computer science was such a new area of research that even RAND, home to perhaps one-eighth of all computer programmers in the nation in the 1950s, did not create a computer science division until the early 1960s. Previously, computer work was carried out primarily by the staff of its Systems Research Laboratory and by members of its applied mathematics group. David Hounshell, "The Cold War, Rand, and the Generation of Knowledge, 1946–1962," *Historical Studies in the Physical and Biological Sciences* 27, no. 2 (1997): 237–67; David Jardini, "Out of the Blue Yonder: The Rand Corporation's Diversification into Social Welfare Research, 1946–1968" (Ph.D. diss., Carnegie Mellon University, 1996).

3. Herbert Simon, "Draft Proposal for Support of Research on Cognitive Processes at the Carnegie Institute of Technology," 2/18/60, HSP, Box 8, ff 282; Carnegie Institute of Technology, "Draft Press Release Re: Carnegie Corporation Grant to GSIA," 6/13/60, HSP, Box 8, ff 282.

4. Jerome Bruner, quoted in Ellen Condliffe Lagemann, *The Politics of Knowledge: The Carnegie Corporation, Philanthropy, and Public Policy* (Middletown, CT: Wesleyan University Press, 1989), p. 210.

5. Simon, "Draft Proposal for Support of Research on Cognitive Processes," pp. 1–2.

6. Ibid., p. 3.

7. Ibid., pp. 3–4.

8. Ibid., p. 4.

9. Lagemann, *The Politics of Knowledge,* pp. 10–11.

10. Carnegie Tech's computing center was funded by the Information Processing Techniques Office (IPTO) of the Department of Defense's Advanced Research Projects Agency (ARPA).

11. Lagemann, *The Politics of Knowledge,* pp. 10–11.

12. Carnegie Institute of Technology (CIT), "Description of a New Doctoral Program in Systems and Communications Sciences at Carnegie Institute of Technology," 1961, HSP, Box 40, ff: Systems and Communications Sciences Proposal—Correspondence, Budget, Memoranda, etc. 1961–64.

13. Lee and Hal Leavitt Gregg, "Memo to Behavioral Science Faculty," 4/19/60, HSP, Box 2, ff 85.

14. CIT, "Draft Press Release Re: Carnegie Corporation Grant to GSIA"; Herbert A. Simon, "Letter to Jack (Warner, President of Cit)," 11/4/61, Box 61, ff: Autobiography—Source Documents—1942–82. The letter to Gilmer (11/2/61) is attached to the letter to Warner.

15. Bert F. Green, interview by Hunter Crowther-Heyck, 4/28/99. One should note that at CIT, as at many technical schools, department chairs actually have some authority as well as responsibility.

16. "Co-evolved" is more precise. The program in Systems and Communication Sciences continued to exist after the new Department of Computer Science was created in 1965, though it eventually dissolved into the department.

17. Allen Newell, "Notes on Preliminary Discussion of Funding of the Systems and Communication Sciences," 4/9/62, HSP, Box 40, ff: Systems and Communications Sciences Program—Correspondence, Budgets, Memos, Proposals, 1961–64, p. 1.

18. Herbert A. Simon, "Application for Research Grant to NIH," 11/1/62, HSP, Box 31, ff: NIH Grant 1, Detailed Budget Statements 1963–64.

19. Newell, "Notes on Preliminary Discussion of Funding of the Systems and Communication Sciences," pp. 1–2.

20. Simon also notes that this lack of limitations is unusual, for "computer simulation, while it does not compete in cost with atom smashing or moon shooting, is not inexpensive." Herbert Simon, *Models of Thought*, vol. 1 (New Haven, CT: Yale University Press, 1979), p. xv.

21. Simon, "Application for Research Grant to NIH," pp. 8, 20.

22. Ibid., p. 10.

23. Ibid., p. 16.

24. Allen Newell, J. C. Shaw, and Herbert Simon, "Elements of a Theory of Human Problem Solving," *Psychological Review* 65, no. 3 (1958): 151–66, p. 156.

25. For a description of the ways in which verbal reports of thinking processes play a role in Simon's science, see K. Anders Ericsson and Herbert Alexander Simon, *Protocol Analysis: Verbal Reports as Data* (Cambridge, MA: MIT Press, 1984). Ericsson and Simon's shorter defense of verbal reports is Simon's most frequently cited recent work: K. Anders Ericsson and Herbert Simon, "Verbal Reports as Data," *Psychological Review* 87, no. 3 (1980): 215–51.

26. Allen Newell and Herbert Alexander Simon, *Human Problem Solving* (Englewood Cliffs, NJ: Prentice-Hall, 1972), p. 5.

27. Ibid., p. 9.

28. Ibid., p. 10.

29. Ibid., pp. 10, 11.

30. Ibid., p. 11.

31. Ibid., p. 12.

32. Ibid., p. 13.

33. Ibid., pp. 20–21.

34. Ibid., pp. 22–23.

35. Ibid., pp. 29–30.

36. Ibid., p. 808.

37. Ibid. This is also why we often find an EM useful, since accessing information in an EM in ready view can be considerably quicker than accessing information in the LTM.

38. On the "von Neumann" architecture, see Herman Goldstine, *The Computer from Pascal to Von Neumann* (Princeton, NJ: Princeton University Press, 1972); Paul Ceruzzi, *A History of Modern Computing*; William Aspray and Arthur Burks, eds., *Papers of John Von Neumann on Computing and Computer Theory* (Cambridge, MA: MIT Press, 1987).

39. Newell and Simon, *Human Problem Solving*, p. 870.

40. For the quintessential statement of Simon's views on computer science as empirical science, see Allen Newell and Herbert Simon, "Computer Science as Empirical Enquiry: Symbols and Search," in *The Philosophy of Artificial Intelligence*, ed. Margaret Boden (New York: Oxford University Press, 1990).

41. Newell and Simon, *Human Problem Solving*, pp. 3–4.

42. Ulric Neisser, "The Imitation of Man by Machine," *Science* 139 (1963): 193–97, p. 195.

43. Herbert Simon, "Motivational and Emotional Controls of Cognition," in

Models of Thought, ed. Herbert Simon (New Haven, CT: Yale University Press, 1979). References are to the reprinted version.

44. Ibid., pp. 31–33.

45. Ibid., pp. 31–35.

46. Ibid., pp. 34–35.

47. Ibid., p. 35. Ayer's key work is A. J. Ayer, *Language, Truth, and Logic* (London: V. Gollancz, 1936). Simon routinely assigned Ayer's book to his graduate students.

48. Herbert A. Simon, "Letter to George Miller," 4/30/57, HSP, Box 16, ff: Correspondence—M—1955–58, p. 1.

49. John Searle, *The Rediscovery of the Mind* (Cambridge, MA: MIT Press, 1992); John Searle, "Minds, Brains, and Programs," in *The Philosophy of Artificial Intelligence,* ed. Margaret Boden (New York: Oxford University Press, 1990).

50. Herbert Alexander Simon, *The Sciences of the Artificial* (Cambridge, MA: MIT Press, 1969), pp. 15–20.

51. Joseph Weizenbaum, *Computer Power and Human Reason: From Judgment to Calculation* (San Francisco: W. H. Freeman, 1976); Hubert Dreyfus, *What Computers Can't Do: The Limits of Artificial Intelligence,* rev. ed. (New York: Harper & Row, 1979).

52. Daniel Dennett, quoted in Dreyfus, *What Computers Can't Do,* p. 2.

53. Ibid., pp. 3, 62.

54. This letter is reprinted, almost complete, in Herbert Simon, *Models of My Life* (New York: Basic Books, 1991), pp. 273–75, quotation from p. 274.

55. Herbert A. Simon and Allen Newell, "What Have Computers to Do with Management?" 5/21/59, HSP, Box 28, ff: RAND Corp.—What Have Computers to do With Management?—1959, pp. 3–7.

56. Herbert A. Simon, Donald Smithburg, and Victor Thompson, *Public Administration* (New Brunswick, NJ: Transaction, 1950), pp. 95–102.

57. Herbert Simon, "Administrative Behavior: A Study of Decision-Making Processes in Administrative Organization," 1945, Box 20, ff: Administrative Behavior—Preliminary ed.—1945, p. 43.

58. Herbert Simon, Allen Newell, and J. C. Shaw, "Elements of a Theory of Human Problem Solving," in *Models of Thought,* ed. Herbert Simon (New Haven, CT: Yale University Press, 1989), p. 6.

THIRTEEN: Scientist of the Artificial

1. Herbert Simon, *The Sciences of the Artificial,* 3rd ed. (Cambridge, MA: MIT Press, 1996).

2. Simon was the first social scientist invited to give these prestigious lectures. The other speakers had been Niels Bohr, Otto Struve, André Lwoff, and Isidor I. Rabi.

3. The most noticeable change over the editions has been the steadily greater importance given to the topic of complexity.

4. Simon, *The Sciences of the Artificial,* pp. ix, x.

5. Ibid., p. xi.

6. Ibid., p. 7. Simon later would pursue this sociobiological connection in his well-known article "A Mechanism for Social Selection and Successful Altruism," *Science* 250 (1990): 1665–68. In it, Simon suggests that if the capacity for social learning is heredittable, then it is quite likely that nature will select in favor of the altruistic.

7. Simon, *The Sciences of the Artificial,* p. 8.

8. Ibid., p. 9.

9. Ibid., p. 15.

10. Ibid.

11. Ibid., pp. 15, 16–17.

12. Ibid., p. 18.

13. Ibid., pp. 20, 21.

14. Ibid., p. 22.

15. Ibid., p. 25.

16. Ibid.

17. Ibid., pp. 1, 2.

18. Ibid., p. 53.

19. Ibid., p. 55.

20. Ibid., p. 58.

21. Ibid., pp. 79–80.

22. Indeed, it informs the BASS report's recommendation that the federal government fund the creation of a graduate school of applied behavioral science. National Academy of Sciences Behavioral and Social Science Survey Committee, *The Behavioral and Social Sciences: Outlook and Needs* (Englewood Cliffs, NJ: Prentice-Hall, 1969).

23. Simon, *The Sciences of the Artificial*, p. 81.

24. Ibid., p. 82.

25. Ibid., pp. 82–83.

26. George Miller, "Psychology as a Means of Promoting Human Welfare," in *Psychology: The Science of Mental Life*, ed. Robert Buckhout (New York: Harper & Row, 1973).

27. Simon, *The Sciences of the Artificial*, p. 75.

28. Ibid.

29. Herbert Simon, "Letter to E. D. Smith," 5/29/51, HSP, Box 6, ff 220, p. 1.

30. Stephen Waring, *Taylorism Transformed: Scientific Management Theory since 1945* (Chapel Hill: University of North Carolina Press, 1991); Dwight Waldo, "Political Science: Tradition, Discipline, Profession, Science, Enterprise," in *Handbook of Political Science*, ed. Fred Greenstein and Nelson Polsby (Reading, MA: Addison-Wesley, 1975), pp. 1–130; Dwight Waldo, "Development of Theory of Democratic Administration," *American Political Science Review* 46, no. 1 (1952): 81–103.

31. On the Marshall Plan, see Michael J. Hogan, *The Marshall Plan: America, Britain, and the Reconstruction of Western Europe, 1947–1952* (Cambridge: Cambridge University Press, 1987); Anthony Carew, *Labour under the Marshall Plan: The Politics of Productivity and the Marketing of Management Science* (Manchester: Manchester University Press, 1987); Herbert Simon, "Birth of an Organization: The Economic Cooperation Administration," *Public Administration Review* 13 (1953): 227–36.

32. The report was published as "Report of the Public Members of the Governor's Milk Control Inquiry Committee," Commonwealth of Pennsylvania, 1965. See, in particular, Herbert Simon, J. E. Holtzinger, and F. K. Miller, "Appendix: Economics of Milk Production and Distribution in Pennsylvania," in the report.

33. Herbert Simon, "Letter to William B. Travis," 4/15/70, HSP, Box 4, ff 159, p. 1.

34. Herbert A. Simon, "Letter to Kathie and David," 5/23/65, HSP, Box 61, ff: Materials for Autobiography—1982.

35. Herbert Simon, "Letter to the Editor of the Tartan, 9/28/70," 9/28/70, HSP, Box 61, ff: Autobiography—Source Documents, 1969–88.

36. Herbert Simon, "Letter to the Editor of the Tartan, 10/9/70," HSP, Box 61, ff: Autobiography—Source Documents, 1969–88.

37. Herbert Simon, "Simon Says," HSP, Box 61, ff: Materials for Autobiography, p. 2.

38. Ibid., pp. 8–9.

39. Ibid., p. 12.

40. Ibid., pp. 12–13. Simon's very critical (and very accurate) views of Toffler are also expressed in a memo to E. E. David, the President's Science Advisor: "Letter to E. E. David, 9/18/70," 9/18/70, HSP, Box 8, ff 260. In another communication to Dr. David, Simon writes that "I think the current attack on science and technology is real, widespread, and vigorous" and that scientists need to be able to discuss "the role of science in directing change, supplying reason to our choices, and in providing feasible alternatives for choices that may solve our problems"; "Letter to E. E. David, 9/23/70," 9/23/70, HSP, Box 8, ff 260. Toffler has since embraced technological change, especially computer-mediated communications, with a vengeance. See Alvin Toffler, *Third Wave* (New York: Morrow, 1980), and Esther Dyson, George Gilder, George Keyworth, and Alvin Toffler, "Cyberspace and the American Dream: A Magna Carta for the Information Age," release 1.2 (August 1994): www.pff.org/publications/ecommerce/fi1.2magnacarta.html.

41. Simon, "Simon Says," pp. 13–14.

42. Ibid., p. 15.

43. Simon, *The Sciences of the Artificial*, p. ix.

44. Herbert Simon, "The Architecture of Complexity: Some Common Properties of Complex Systems," 4/23/62, HSP, Box 2, ff 72, pp. 10–11. Simon corresponded with Warren Weaver in the mid-1950s about this issue, and the watch example is derived from Weaver's analysis; Warren Weaver, "Letter to Herbert Simon, 5/18/62," 5/18/62, HSP, Box 2, ff 72. The majority of Simon's sources for this argument are attempts to apply information theory to biology: Herbert Simon, "Some Observations on Complex Systems," 2/14/56, HSP, Box 4, ff 147.

45. Herbert Simon, *Models of Discovery: And Other Topics in the Methods of Science* (Boston: D. Reidel, 1977), pp. 260–61.

46. Simon also addresses these issues in Nicholas Rescher and Herbert Simon, "Cause and Counterfactual," *Philosophy of Science* 33 (1966): 323–40.

47. Simon, "The Architecture of Complexity," pp. 30, 17–18, 26.

48. Ibid., pp. 36–37, 31.

FOURTEEN: The Expert Problem-Solver

1. Herbert Simon to Hunter Crowther-Heyck, "Comments on Your Dissertation," 2000, possession of author.

2. Herbert Simon, "Computers—Non-Numerical Computation," *Proceedings of the National Academy of Sciences of the United States of America* 77, no. 11 (1980): 6264–68.

3. Allen Newell, J. C. Shaw, and Herbert Simon, "Chess-Playing Programs and the

Problem of Complexity," *IBM Journal of Research and Development* 2, no. 4 (1958): 320–35, p. 320.

4. Robert McFadden, "Computer in the News: Kasparov's Inscrutable Conqueror," *New York Times,* May 12, 1997, accessed at: www.nytimes.com/library/cyber/week/ 051297blue.html; Bruce Weber, "IBM Chess Machine Beats Humanity's Champ," *New York Times,* May 12, 1997, accessed at: www.nytimes.com/library/cyber/week/ 051297weber.html. Significantly, Deep Blue "did not play chess like a computer" but like a person—a person whose past Kasparov did not know, whose body he could not see, and whose personality he therefore could not understand.

5. Newell, Shaw, and Simon, "Chess-Playing Programs and the Problem of Complexity"; Herbert Simon, "Progress Report: Research on Decision-Making under Uncertainty under ONR-Cowles Commission Contract," 4/11/53, HSP, Box 6, ff 226.

6. Herbert A. Simon, "Some Notes on the Early History of LT," 7/9/57, HSP, Box 29, ff: Paper "Some Notes on the Early History of LT."

7. Newell, Shaw, and Simon, "Chess-Playing Programs and the Problem of Complexity," p. 326.

8. Ibid.

9. Feigenbaum went on to have an illustrious career as a computer scientist. He is known chiefly for his pioneering work in the design of expert systems.

10. Herbert Simon and Edward Feigenbaum, "An Information-Processing Theory of Some Effects of Similarity, Familiarization, and Meaningfulness in Verbal Learning," *Journal of Verbal Learning and Verbal Behavior* 3 (1964): 385–96; Herbert Simon and Kenneth Kotovsky, "Human Acquisition of Concepts for Sequential Patterns," *Psychological Review* 70 (1963): 534–46.

11. Simon and Feigenbaum, "An Information-Processing Theory"; Simon and Kotovsky, "Human Acquisition of Concepts for Sequential Patterns"; Howard B. Richman, James J. Staszewski, and Herbert A. Simon, "Simulation of Expert Memory Using EPAM IV," *Psychological Review* 102, no. 2 (1995): 305–30.

12. Richman, Staszewski, and Simon, "Simulation of Expert Memory Using EPAM IV," pp. 305–7.

13. Ibid., p. 307.

14. Hunter Crowther-Heyck, "Mind and Network" (paper presented to the History of Science Society Annual Meeting, 2001).

15. Richman, Staszewski, and Simon, "Simulation of Expert Memory Using EPAM IV."

16. Ibid.

17. Ibid.

18. Simon to Crowther-Heyck; Herbert A. Simon and Fernand Gobet, "Expertise Effects in Memory Recall: Comment on Vicente and Wang (1998)," *Psychological Review* 107, no. 3 (2000): 593–600.

19. "But the findings refute only the incautious prediction of the experimenters [including Simon], who did not use MAPP [a program that simulates rule induction] to check their guess (illustrating the frequent superiority in reasoning from precise models rather than words)"; Simon and Gobet, "Expertise Effects in Memory Recall," p. 596.

20. Herbert Simon, *The Sciences of the Artificial,* 3rd ed. (Cambridge, MA: MIT Press, 1996), pp. 15, 20.

21. David Klahr and Herbert Simon, "Studies of Scientific Discovery: Complementary Approaches and Convergent Findings," *Psychological Bulletin* 125, no. 5 (1999): 524–43.

22. Ibid., p. 524.

23. Ibid., pp. 531–32.

24. Ibid., p. 535.

25. Herbert A. Simon, "Letter to William Cooper (January)," 1/17/46, HSP, Box 1, ff 26. Note that Simon distinguishes not only between induction and deduction but also between deduction from a priori principles and deduction from broad empirical generalizations.

26. Klahr and Simon, "Studies of Scientific Discovery," p. 529.

27. Herbert Simon, "Discovery, Invention, and Development: Human Creative Thinking," *Proceedings of the National Academy of Sciences of the United States of America* 80, no. 14 (1983): 4570–71. For a more complete description of BACON, see Pat Langley and Herbert A. Simon, "Applications of Machine Learning and Rule Induction," *Communications of the ACM* 1995. BACON simulates certain aspects of Kepler's work quite well: given his original data, it arrives at Kepler's original, erroneous equations for circular motion of the planets. Given his later data, and his later confidence in their accuracy, BACON generates the correct equation for the third law, just as Kepler did. The apparent success of BACON and its descendants has not endeared Simon to scholars who emphasize the social context of science, particularly advocates of the "strong programme" in the sociology of science, such as Harry Collins. Simon, for his part, thought the strong programme was so patently foolish that it was not worth his time to rebut it, though he acknowledged (in principle, though perhaps not in practice) the salience of the "weak programme." For more on these debates, see the exchanges in the 1989 and 1990 volumes of *Social Studies of Science.*

28. Simon and Gobet, "Expertise Effects in Memory Recall," pp. 595, 598.

29. Klahr and Simon, "Studies of Scientific Discovery," p. 539.

30. Ibid.

31. This final insight is particularly applicable to debates about why some ideas or technologies flourished despite being equal, or perhaps even inferior, to their competitors at some point in their evolution. Humans, unlike nonsentient creatures, are able to look ahead and to choose among several different paths to their goals. For example, much has been made of the superior abilities of analog computers versus their digital competitors during the 1950s and early 1960s, making the decision of the computer industry and its governmental patrons to emphasize the development of digital technologies appear so foolish as to be almost sinister. Yet it was clear to almost everyone in the industry that the superior technology, in the end, would be a digital machine, because of its enormous potential for flexibility. Analog computing, to use Simon's phrasing, might lead to a local maximum, but digital computing would lead to a more global maximum, and virtually everyone involved knew it.

32. It was a letter to his daughter Kathie and her husband regarding the student protests of the 1960s that Simon reprinted almost verbatim in his autobiography.

33. David Klahr, interview by Hunter Crowther-Heyck, 7/15/2003.

34. Ibid.

CONCLUSION: A Model Scientist

1. In this context, "adaptive" means capable of responding to environmental stimuli, with the implicit goal of maintaining the internal equilibrium of the system.

2. E. J. Dijksterhuis, *The Mechanization of the World Picture* (Oxford: Clarendon Press, 1961).

3. Lewis Mumford, *Technics and Civilization* (New York,: Harcourt, 1934); Carolyn Merchant, *The Death of Nature: Women, Ecology, and the Scientific Revolution* (New York: Harper & Row, 1989); Steven Shapin, *The Scientific Revolution* (Chicago: University of Chicago Press, 1996); Peter Robert Dear, *Revolutionizing the Sciences: European Knowledge and Its Ambitions, 1500–1700* (Princeton, NJ: Princeton University Press, 2001); John Henry, *The Scientific Revolution and the Origins of Modern Science,* 2nd ed. (New York: Palgrave, 2001); A. Rupert Hall, *The Scientific Revolution, 1500–1800; The Formation of the Modern Scientific Attitude,* 2nd ed. (London: Longmans, 1962).

4. J. David Bolter, *Turing's Man: Western Culture in the Computer Age* (Chapel Hill: University of North Carolina Press, 1984); Otto Mayr, *Authority, Liberty, & Automatic Machinery in Early Modern Europe* (Baltimore: Johns Hopkins University Press, 1986); Crosbie Smith and Norton Wise, "Work and Waste: Political Economy and Natural Philosophy in Nineteenth Century Britain (I) and (II)," *History of Science* 27 (1989): 263–301, 391–449; Crosbie Smith and Norton Wise, "Work and Waste: Political Economy and Natural Philosophy in Nineteenth Century Britain (III)," *History of Science* 28 (1990): 221–61.

5. Anson Rabinbach, *The Human Motor: Energy, Fatigue, and the Origins of Modernity* (Berkeley: University of California Press, 1990); Laura Otis, "The Metaphoric Circuit," *Journal of the History of Ideas* 63, no. 1 (2002): 105–28; Laura Otis, *Networking: Communicating with Bodies and Machines in the Nineteenth Century* (Ann Arbor: University of Michigan Press, 2002).

6. Paul Edwards, *The Closed World: Computers and the Politics of Discourse in Cold War America* (Cambridge, MA: MIT Press, 1996); N. Katherine Hayles, *How We Became Posthuman: Virtual Bodies in Cybernetics, Literature, and Informatics* (Chicago: University of Chicago Press, 1999); Manuel Castells, *The Rise of the Network Society* (Malden, MA: Blackwell Publishers, 1996); Donna Haraway, "A Manifesto for Cyborgs," in *Simians, Cyborgs, and Women: The Reinvention of Nature,* ed. Donna Haraway (London: Free Association, 1991).

7. Simon Schaffer, "Babbage's Intelligence: Calculating Engines and the Factory System," *Critical Inquiry* 21 (autumn 1994): 203–27.

8. Emily Martin, *The Woman in the Body: A Cultural Analysis of Reproduction: With a New Introduction* (Boston: Beacon Press, 2001); Emily Martin, *Flexible Bodies: Tracking Immunity in American Culture from the Days of Polio to the Age of Aids* (Boston: Beacon, 1994).

9. Herbert A. Simon, "Memorandum to 'Various and Sundry of My Friends,'" n.d. [late 1957], Box 36, ff: ORSA—Correspondence—Memo 1957–61.

10. On the organizational synthesis, see the Essay on Sources.

11. George Lakoff and Mark Johnson, *Metaphors We Live By* (Chicago: University of Chicago Press, 1980).

12. One index of the widespread acceptance of this link between computers and

bureaucracies is the oft-noted fact that computers in popular films and novels of the 1960s, 1970s, and early 1980s almost invariably were associated with totalitarian bureaucracies: witness HAL in *2001*, Colossus in *Colossus: The Forbin Project*, Skynet in *The Terminator*, and the unnamed computing systems in *Dr. Strangelove* and Apple's famed *1984* advertisement. As with all potent metaphors, however, the analogy between organism, organization, and programmed computer contained within it the seeds of its own transformation. As computers changed, becoming networked communications devices as well as information processors, so too did their metaphorical implications: computers in the films and fiction of the mid 1980s and 1990s could be many things, from robotic agents of a totalitarian bureaucracy as in *Terminator 2* and *The Matrix* to lovable (in a saccharine way) little devices with all the personality and individuality of a "real" person, as in *Short Circuit*, *The Bicentennial Man*, and *AI*.

13. The point here is less complicated than it sounds. Simply put, one never can point at a bureaucracy; one only can point at a set of people who behave as if they were part of something we have called a bureaucracy. The typical way of demonstrating that such a set of entities is in fact an integrated system, such as a bureaucracy, is to measure energy or information flows among the components. Following this reasoning, many biologists at midcentury concluded that an organism was defined by its communication lines, just as political scientists concluded that a nation was defined by its patterns of communication. See Karl W. Deutsch, "Mechanism, Organism, and Society: Some Models in Natural and Social Science," *Philosophy of Science* 18, no. 3 (1951): 230–52; Karl W. Deutsch, *The Nerves of Government: Models of Political Communication and Control* (New York: Free Press of Glencoe, 1963).

14. John Gunnell, *The Descent of Political Theory: The Genealogy of an American Vocation* (Chicago: University of Chicago Press, 1993); Vincent Ostrom, *The Intellectual Crisis in American Public Administration*, 2nd ed. (Tuscaloosa: University of Alabama Press, 1989); Alvin W. Gouldner, *The Coming Crisis of Western Sociology* (New York: Basic Books, 1970); Seymour Bernard Sarason, *Psychology Misdirected* (New York: Free Press, 1981).

15. Daniel Bell and Irving Kristol, *The Crisis in Economic Theory* (New York: Basic Books, 1981).

16. For Simon's recent views on the state of economics, see Herbert Simon, "Organizations and Markets," *Journal of Economic Perspectives* 5, no. 2 (1991): 25–44; Herbert Simon, "The State of Economic Science," in *The State of Economic Science: Views of Six Nobel Laureates*, ed. Werner Sichel (Kalamazoo, MI: W. E. Upjohn Institute for Employment Research, 1989); Herbert Alexander Simon and Claudio Demattáe, *An Empirically Based Microeconomics* (Cambridge: Cambridge University Press, 1997).

17. Robert M. Collins, *More: The Politics of Economic Growth in Postwar America* (New York: Oxford University Press, 2000); Michael Bernstein, *A Perilous Progress: Economists and Public Purpose in Twentieth-Century America* (Princeton, NJ: Princeton University Press, 2001); Herman E. Daly, John B. Cobb, and Clifford W. Cobb, *For the Common Good: Redirecting the Economy toward Community, the Environment, and a Sustainable Future*, 2nd ed. (Boston: Beacon, 1994).

18. Herbert Simon, "The Architecture of Complexity," *Proceedings of the American Philosophical Society* 106 (1962): 467–82.

19. Two classic statements of the perils of unintended consequences are Jeffrey L.

Pressman, Aaron B. Wildavsky, and Oakland Project, *Implementation: How Great Expectations in Washington Are Dashed in Oakland; or, Why It's Amazing That Federal Programs Work at All, This Being a Saga of the Economic Development Administration as Told by Two Sympathetic Observers Who Seek to Build Morals on a Foundation of Ruined Hopes* (Berkeley: University of California Press, 1973), and Charles A. Murray, *Losing Ground: American Social Policy, 1950–1980* (New York: Basic Books, 1984). *Losing Ground* has been described as the "bible" of the first Reagan administration.

Essay on Sources

A WORD ON SOURCES AND NOTES

In the interest of readability, notes in the main text generally have been used to reference direct quotations and similarly specific debts. The works that have shaped my analysis on a broad level are discussed in the following essay. A complete bibliography, organized by topic, is available at http://faculty-staff.ou.edu/c/hunter.a.crowther-heyck-1/.

ARCHIVAL SOURCES AND INTERVIEWS

The collections of the Carnegie Mellon University Archives primarily consist of the Herbert Simon Papers, which run to over 100 linear feet, the Allen Newell Papers, which are even more extensive, the papers of the various university presidents (I found Doherty's to be most useful), and the papers of the various administrative units of the university, such as the GSIA. The latter are less extensive than one might hope. The Simon papers contain many thousands of Simon's letters, as well as drafts, memos, and so forth. The archived correspondence, it should be noted, is purely professional. The CMU Archives is in the process of digitizing many documents from the Simon and Newell papers, so many of the most important documents are accessible via the Web.

The University of Chicago Archives is a great place to do research on the social sciences in the early twentieth century. The Charles Merriam Papers are particularly valuable because he corresponded with almost everyone in the social sciences, and he kept just about everything. The Henry Schultz Papers provide insight into the state of economics in the 1920s and 1930s. The Rashevsky papers and the Papers of the Committee on Mathematical Biology, the Department of Economics, and the Unity of Science Movement are not extensive, but the documents in them do provide an interesting window on mathematical social science in the 1930s and 1940s.

The Talcott Parsons Papers, the Paul Buck Papers, and the Papers of the Department of Social Relations / Sociology at the Harvard University Archives are extremely valuable for anyone interested in the history of sociology and in the struggle (on the part of sociologists) to find a way to integrate economics with the other social sciences.

The Rockefeller Archive Center in Tarrytown, New York, is probably the best run archive I have seen: when I noted one day that one of the boxes I had ordered had not come up with the other six, the archivist on duty leapt to his phone and cried "Retrieval on 3! Retrieval on 3!" About five minutes later one of the staff came up the stairs at a run, carrying the box in question. (I did not have the heart to tell them I did not need it until that afternoon.) The SSRC papers and the Rockefeller Foundation Papers are extensive, well indexed, and indispensable for research on the social sciences from the 1920s through the 1960s.

I had the good fortune to be able to interview Herbert Simon in 1999. The interview was quite revealing, both regarding Simon's ideas and his personality. At first, Simon frequently had to ask me to fill in certain names and dates. (Simon: "It was about this time that I read, oh, what was his name, he dedicated his book to the lab rat?" HCH: "That would be Tolman.") This went on for about a half an hour, leaving me to wonder whether Simon's memory was beginning to go. Then it dawned on me that these fill-in-the-blanks questions were getting harder and harder. I was being tested. Finally, I filled in one last blank: Simon: "I had just read that French philosopher, you know, the one who was so influential for Proust." HCH: "Henri Bergson?"

After that, Simon's discourse suddenly became so flawless as to seem rehearsed. The next ninety minutes were a truly remarkable performance, one that taught me a great deal about how he saw science and the world—and that revealed how powerful and persuasive a conversationalist Simon could be. I was glad that I came to the interview after having done much of my research already, for it was hard to talk to Simon for very long without starting to see the world from his perspective. Simon later favored me with twenty-five pages (single-spaced) of comments on my dissertation, which, contrary to my fears, were written in exactly the right spirit: he corrected me on factual errors but cheerfully acknowledged my right to disagree with him on matters of interpretation. (Then he made the case for his interpretations, with vigor.)

I also was able to interview Bert Green, David Klahr, and George Miller, all of whom worked with Simon and knew him well. Specific information from each of these interviews is cited in the notes, but I should mention here that all three often spoke of Simon in a curious tone of respect, even awe, mixed with bemusement, as if they were discussing a precocious but headstrong youth. I am not quite sure what this says about Simon, other than that he impressed some very talented people as being in another league—and sometimes, as being from another world.

SIMON

There is a paucity of historical work on Herbert Simon, which is, of course, one reason why I wrote this book. Fortunately, the studies that do exist are quite good, though most are narrowly conceived. All, except his autobiography, focus on one slice of Simon's intellectual life. His autobiography, *Models of My Life* (New York: Basic Books, 1991), has been discussed at several points in the text, so here I will restrict myself to noting that *Models of My Life* is quite well written, though curiously cool and detached. It reveals very little about Simon's emotional life, although the strength of his bond with Allen Newell does come through. Also, because *Models of My Life* is persuasive and engaging, just as Simon himself was, it is important to remember that it was crafted as a kind of case study in support of his later views of scientific discovery.

For overviews of Simon's contributions to management theory, see Hunter Crowther-Heyck, "Herbert A. Simon," in the *Biographical Dictionary of Management*, ed. Morgen Witzel (Bristol, UK: Thoemmes, 2003), and Stephen Waring, *Taylorism Transformed: Scientific Management Theory since 1945* (Chapel Hill: University of

North Carolina Press, 1991). Waring's analysis is astute, but he, like many other critics of Simon, sees him as much more interested in bolstering centralized authority than I do. As with many a New Deal liberal, Simon's politics truly were as democratic as they were managerial.

Other important works on Simon focus on the connections between his work in economics and operations research, writ broadly. These include Mie Augier, "Models of Herbert A. Simon," *Perspectives on Science* 8, no. 4 (2000): 407–43, and Esther-Mirjam Sent, "Herbert A. Simon as a Cyborg Scientist," *Perspectives on Science* 8, no. 4 (2000): 380–406. These articles take opposite sides of the change vs. continuity argument regarding Simon's career, a debate in which I come down squarely in the middle. I see great continuity in Simon's basic goals, intellectual habits, and philosophical outlook on science and society, but I see marked changes in the ways that the above were expressed in his science, largely as a result of his embrace of computer simulation as his primary research technique.

Simon is only one of many characters in Philip Mirowski's, *Machine Dreams: Economics Becomes a Cyborg Science* (New York: Cambridge University Press, 2002), but Mirowski's analysis of Simon is well worth reading. Mirowski invariably is insightful and provocative, and he is right on target in his argument that Simon and a number of other important pioneers of mathematical economics were very strongly influenced by operations research, game theory, cybernetics, and early AI. I would add that a fascination with communication, both as a mode of control and as a means of representing the world, also characterized many in this group. (Mirowski probably would agree, though he does not emphasize it as strongly as I might.) Also, Mirowski's analysis, as scholarly as it is, has something of the tone of an exposé, as if the connection to operations research, in itself, should cast doubt on modern economics. There are many reasons to be skeptical of postwar economics, but its connection to operations research does not seem to me to be high on the list.

THE SOCIAL SCIENCES

The place to begin any study of the social sciences in the United States is Dorothy Ross, *The Origins of American Social Science* (Cambridge: Cambridge University Press, 1991). Ross's book is magisterial, a comprehensive, detailed exploration of social science in America through the early twentieth century. Its greatest strength lies in its analysis of Progressive-era social science—the period when the social sciences began to embrace a new professional identity based upon a new vision of science. Her argument centers on the notion of "American exceptionalism"—the idea, pervasive in American intellectual life, that the country has a special place in history. In Ross's view, until the late nineteenth century America's special place was understood to be outside history: it was exempt from the degradation and decline of European society because of its republican institutions and virtuous citizenry. During the Progressive era, however, America was understood to have reentered the stream of history—at the leading edge of a universal process of social evolution, of course.

Ross is absolutely correct that a change in historical consciousness was a vital part of the turn to scientism in American social science, but she may overemphasize the contribution of the idea of American exceptionalism to this changed consciousness. Exceptionalism was—and still is—an important part of American thought, but it seems to have been more of a *context* for than a *cause* of change in the social sciences. One of the novel goals of social science in the Progressive era, on the other hand, was to create a more truly universal social science, one that could be applied to Europe or to "primitive societies" as readily as to the United States. The way to harmonize this search for universal laws with American uniqueness was to believe in universal laws of social development, with America being the most developed nation.

Other works on the social sciences that have shaped my perspective in significant ways are Mary Jo Deegan, *Jane Addams and the Men of the Chicago School, 1892–1918* (New Brunswick, NJ: Transaction Books, 1988); Mary Furner, *Advocacy and Objectivity: A Crisis in the Professionalization of American Social Science, 1865–1905* (Lexington: University Press of Kentucky, 1975); Thomas Haskell, *The Emergence of Professional Social Science: The ASSA and the Nineteenth-Century Crisis of Authority* (Urbana: University of Illinois Press, 1977); Barry D. Karl, *Charles E. Merriam and the Study of Politics* (Chicago: University of Chicago Press, 1974); Theodore Porter, *Trust in Numbers: The Pursuit of Objectivity in Science and Public Life* (Princeton, NJ: Princeton University Press, 1995); Robert N. Proctor, *Value-Free Science? Purity and Power in Modern Knowledge* (Cambridge, MA: Harvard University Press, 1991); Edward A. Purcell Jr., *The Crisis of Democratic Theory: Scientific Naturalism and the Problem of Value* (Lexington, KY: University Press of Kentucky, 1973); and Morton White, *Social Thought in America: The Revolt against Formalism* (New York: Viking, 1949).

I have drawn many things from the works above, but probably the most important is the idea of instrumental knowledge as the core of the Progressive era transformation of the social sciences (see chapter 2). All observers (present and past) agree that the period witnessed a turn toward a harder scientism in social science, but opinions vary markedly as to what, exactly, the essential features of this shift were. Dorothy Ross, as noted above, argues that the important change lay in historical consciousness, seeing scientism primarily as a rejection of historicism. Others, however, see scientism as a rejection of idealism (Fritz Ringer), formalism (Morton White), or advocacy (Mary Furner). Still others see it as the embrace of a novel kind of naturalism (Edward Purcell) or the rise of the concept of interdependence (Thomas Haskell).

Probably the most commonly expressed view, especially in histories written during the 1960s and 1970s, is that the turn toward scientism was a rejection of social science's reform-driven past. Mary Furner and Thomas Haskell give the most sophisticated versions of this view, with the many practitioner-written histories of the various social science disciplines generally depicting this shift in starker form. Other work has made it clear, however, that social scientists did not give up their ambitions to reform society when they embraced the research ethos. Barry Karl's superb biography of Merriam makes this point very effectively.

Each of these accounts of Progressive era social science captures an aspect of the changes afoot, but none seems sufficiently comprehensive to me. I think that one can come to a more complete view if one distinguishes between two different levels of thought: one level concerned with changes in specific ideas about the social world— where the problems of change, interdependence, and subjectivity were paramount— and another level concerned with changes in ideas about the nature and *purpose* of knowledge. Instrumentalism was a change on this second level, a change that shaped how social scientists responded to the specific challenges of the day.

Beyond Ross, Furner, Porter, Proctor, and Haskell, the histories of the social sciences almost always focus on a single discipline. Most of these histories are chronicles of a particular discipline's Progressive-era birth; only a few carry the analysis past World War II. The various crises of the 1970s produced a number of practitioner-written disciplinary histories, some of which are quite good, but almost all of which have a large axe to grind. There is a clear need for works on the postwar social and behavioral sciences, especially ones that trace changes across fields.

For political science, I recommend John Gunnell, *The Descent of Political Theory: The Genealogy of an American Vocation* (Chicago: University of Chicago Press, 1993); and Raymond Seidelman, *Disenchanted Realists: Political Science and the American Crisis, 1884–1984* (Albany: State University of NY Press, 1985). For sociology, see Robert Bannister, *Sociology and Scientism: The American Quest for Objectivity, 1880–1940* (Chapel Hill: University of North Carolina Press, 1987); Martin Bulmer, *The Chicago School of Sociology: Institutionalization, Diversity, and the Rise of Sociological Research* (Chicago: University of Chicago Press, 1984); and Charles Camic, "Introduction: Talcott Parsons before the Structure of Social Action," in *Talcott Parsons: The Early Essays,* ed. Charles Camic (Chicago: University of Chicago Press, 1991).

For psychology, see Mitchell Ash, *Gestalt Psychology in German Culture, 1890–1967: Holism and the Quest for Objectivity* (New York: Cambridge University Press, 1995). Ash's book is an exemplary study in the history of intellectual culture. Any scholar who wants to see how broad cultural movements are manifested in specific institutions, ideas, and individuals would do well to read it.

On the history of psychology in America, the places to start are Kurt Danziger, *Constructing the Subject: Historical Origins of Psychological Research* (New York: Cambridge University Press, 1990); Kurt Danziger, *Naming the Mind: How Psychology Found Its Language* (Thousand Oaks, CA: Sage, 1997); Ellen Herman, *The Romance of American Psychology: Political Culture in the Age of Experts* (Berkeley: University of California Press, 1995); Jill Morawski, "Organizing Knowledge and Behavior at Yale's Institute of Human Relations," *Isis* 77 (1986): 219–42; John O'Donnell, *The Origins of Behaviorism: American Psychology, 1870–1920* (New York: NYU Press, 1985); and Laurence Smith, *Behaviorism and Logical Positivism: A Reassessment of the Alliance* (Stanford, CA: Stanford University Press, 1986).

Of the above, I found Jill Morawski's article particularly illuminating, for there were many parallels between Clark Hull's goals for Yale's Institute of Human Relations and Simon's goals for the GSIA. In addition, Smith's work on behaviorism and logical positivism is quite interesting, for it argues that the behaviorists of the 1920s and 1930s, such as Tolman, Skinner, and Hull, were not influenced directly by the Vienna Circle of logical positivists. While Smith's argument appears to be correct as regards direct influences, Smith's own analysis shows that many of the behaviorists already held ideas very similar to those of the logical positivists before they encountered the works of the Vienna Circle. Thus, even if they were not influenced by the positivists, they saw them as natural allies. In addition, several leading behaviorists, such as Clark Hull and S. S. Stevens, were committed operationalists influenced strongly by Percy Bridgman.

A powerful but almost wholly unexplored philosophical influence on the human sciences was Alfred North Whitehead, whose *Science in the Modern World* (New York: Macmillan, 1925) provided a vision of science that was both positivist and organicist. There is a great book to be written on Whitehead and the human sciences; I hope someone writes it soon.

For economics, the works I have found most useful are Michael Bernstein, *A Perilous Progress: Economists and Public Purpose in Twentieth-Century America* (Princeton, NJ: Princeton University Press, 2001); Mirowski, *Machine Dreams;* Philip Mirowski, *More Heat Than Light: Economics as Social Physics, Physics as Nature's Economics* (New York: Cambridge University Press, 1989); Mary Morgan, *The History of Econometric Ideas* (Cambridge: Cambridge University Press, 1990); E. Roy Weintraub, *How Economics Became a Mathematical Science* (Durham, NC: Duke University Press, 2002); and E. Roy Weintraub, ed., *Toward a History of Game Theory,* supplement to the *History of Political Economy,* vol. 24 (Durham, NC: Duke University Press, 1992).

As the works above indicate, the historiography of modern economics focuses on mathematization. Mirowski's *More Heat Than Light* is both a fine history of the introduction of ideas, methods, and equations from nineteenth-century field physics into economics (the story in a nutshell: economists come to equate utility with energy) and an indictment of neoclassical economists for not understanding their thermodynamics. Both Nicholas Georgescu-Roegen and Herman Daly have made similar critiques of modern neoclassical economics, though they still find value in the analogy between energy and utility: they want economists to understand the analogy between energy and utility more fully, while Mirowski wants economists to circumscribe its use. On this point, see Herman Daly and John Cobb Jr., *For the Common Good: Redirecting the Economy toward Community, the Environment, and a Sustainable Future* (Boston: Beacon, 1989).

On the same topic of mathematization, the Weintraub-edited collection on the history of game theory is very good, with many of the pieces illuminating the remarkable linkages, both intellectual and institutional, between operations research, game theory, decision theory, mathematical economics, and applied mathematics. Weintraub's *How Economics Became a Mathematical Science* and Mary Morgan's *History of Econometric Ideas* also tell the story of mathematization, focusing more on the internal development of ideas and techniques. Those interested in the statistical turn in economics should read Gerd Gigerenzer et al., *The Empire of Chance: How the Science of Probability Changed Science and Everyday Life* (Cambridge: Cambridge University Press, 1989).

In addition to the works above, my thinking on the history of the social sciences also has been influenced by recent historical work on gender. The works that have played the biggest role in shaping my understanding of gender in human science are Ann Taylor Allen, "Feminism, Social Science, and the Meanings of Modernity: The Debate on the Origin of the Family in Europe and the United States, 1860–1914," *American Historical Review* 104, no. 4 (1999): 1085–13; JoAnne Brown, *The Definition of a Profession: The Authority of Metaphor in the History of Intelligence Testing, 1890–1930* (Princeton, NJ: Princeton University Press, 1992); Deegan, *Jane Addams and the Men of the Chicago School;* Donna Haraway, *Primate Visions: Gender, Race, and Nature in the World of Modern Science* (New York: Routledge, 1989); Donna Haraway, *Simians, Cyborgs, and Women: The Reinvention of Nature* (New York: Routledge, 1991); Elizabeth Lunbeck, *The Psychiatric Persuasion: Knowledge, Gender, and Power in Modern America* (Princeton, NJ: Princeton University Press, 1994); and Eileen Yeo, *The Contest for Social Science: Relations and Representations of Gender and Class* (London: Rivers Oram Press, 1996).

As Mary Jo Deegan's work shows, social scientists in the Progressive era struggled to define their fields as true sciences, which involved distancing them from the female-dominated field of social work. In a similar vein, JoAnne Brown reveals that even occupations that were largely female attempted to describe their work in masculine, scientific terms. (Occupations in which women were prominent, however, were not successful in claiming professional status, no matter how masculine their rhetoric.) Haraway, Yeo, and Lunbeck all show how gendered preconceptions of rationality and normalcy shaped social scientists' ideas about both women and society at fundamental levels (and often in frightening ways).

In some ways, Simon is a difficult subject to study in relation to gender. Gender simply was not a relevant category of analysis for him, and he never made public pronouncements on the proper roles of the different sexes. I think he would have been stunned if someone came to him with evidence that his organizational or cognitive theories had to be altered to describe women as well as men. He thought his work was much more universal than that.

His work, however, clearly was gendered. Simon's vision of human science as an objective, experimental science clearly fit in well with an ideal of virile but controlled masculinity. In this ideal, the male was passionate and driven yet thoroughly subject to the discipline of reason, thanks to his fierce will. Simon's goals for his science meshed almost perfectly with this gender ideal, which was widely held in the 1920s–1950s. His goals also meshed well with a number of other, quite different ideals, however, and the connection of his scientific ideals to gender do not appear to have been conscious, making him quite different from Karl Lashley, say, or Harlow Shapley.

THE SYSTEMS SCIENCES: OPERATIONS RESEARCH, CYBERNETICS, AND OTHERS

Historians have begun to call operations research, cybernetics, decision theory, game theory, systems analysis, artificial intelligence and related fields the "systems sciences," and a number of recent studies have explored the intersections among these fields in the 1940s–1960s. As I argue, members of these fields shared a common interest in mathematization, modeling, and simulation; assumed that the world was made up of systems and subsystems; described such systems in terms of behaviors and functions; and understood humans to be adaptive, problem-solving organisms operating in a complex world.

My understanding of these fields and their intersection is probably closest to that of Paul Edwards, whose book *The Closed World: Computers and the Politics of Discourse in Cold War America* (Cambridge: MIT Press, 1996) is simply terrific. In this ambitious study of the relationship between the "discourse" of computing and Cold War geopolitical strategy, Edwards makes a persuasive argument that the systems analysts of RAND, certain technophiliac leaders of the military, and many pioneers in computer development all shared a "closed world" discourse. In this discourse, the world was pictured as a closed system and seen as a battlefield; the contest between East and West was understood as a zero-sum game; and the key to victory was thought to be central direction of the global battlefield via computerized communications, command and control technologies, such as SAGE and strategic defense initiative (SDI) or "star wars." Hence the military's interest in and support of both communications theory and computer science. In Edwards's view, the computer metaphor of mind played an important role in a redefinition of humans and their relationship to machines that, in turn, was a key part of this "closed world" discourse.

The best work on the systems sciences before the war is David Mindell, *Between Human and Machine: Feedback, Control, and Computing before Cybernetics* (Baltimore: Johns Hopkins University Press, 2002). Mindell shows that Norbert Weiner and the postwar systems scientists did not invent the concepts of feedback or of the servo-mechanism, nor were they the first to try to describe feedback and control systems in the language of communications engineering. Mindell's book is excellent, but he focuses his analysis on engineering and physical science, leaving out the equally important traditions of "systems thinking" in biology (especially physiology) and the social sciences.

There have been several recent studies of operations research's origins, all of which emphasize the context of World War II as vital for its development: Michael Fortun and Silvan Schweber, "Scientists and the Legacy of World War II: The Case of Operations Research (OR)," *Social Studies of Science* 23 (1993): 595–642; Andy Pickering, "Cyborg History and the World War II Regime," *Perspectives on Science* 3, no. 1 (1995): 1–48; and Erik Rau, "Combat Scientists: The Emergence of Operations Research in the United States during World War II" (Ph.D. diss., University of Pennsylvania, 1999). Stephen Waring explores how OR was transformed during the cold war in "Cold Calculus: The Cold War and Operations Research," *Radical History Review* 63 (1995): 28–51. Andrew Abbott also analyzes the postwar transformation of OR as one case study in *The System of Professions: An Essay on the Division of Expert Labor* (Chicago: University of Chicago Press, 1988).

Other scholars have focused on the universalizing ambitions of cybernetics and information theory: see Tara Abraham, "(Physio)Logical Circuits: The Intellectual Origins of the McCulloch-Pitts Neural Networks," *Journal of the History of the Behavioral Sciences* 38, no. 1 (2002): 3–25; Geoff Bowker, "How to Be Universal: Some Cybernetic Strategies, 1943–70," *Social Studies of Science* 23 (1993): 107–27; Peter Galison, "The Ontology of the Enemy: Norbert Wiener and the Cybernetic Vision," *Critical Inquiry* 21 (autumn 1994): 228–66; Thomas P. Hughes and Agatha C. Hughes, eds., *Systems, Experts, and Computers* (Cambridge, MA: MIT Press, 2000); N. Katherine Hayles, *How We Became Posthuman: Virtual Bodies in Cybernetics, Literature, and Informatics* (Chicago: University of Chicago Press, 1999); Steve J. Heims, *The Cybernetics Group* (Cambridge, MA: MIT Press, 1991); Lily Kay, "Cybernetics, Information, Life: The Emergence of Scriptural Representations of Heredity," *Configurations* 5, no. 1 (1997): 23–91; and Evelyn Fox Keller, *Refiguring Life: Metaphors of Twentieth Century Biology* (New York: Columbia University Press, 1995).

Two studies that emphasize the importance of managing large-scale projects for the postwar development of operations research and the systems sciences are Thomas P. Hughes, *Rescuing Prometheus* (New York: Pantheon Books, 1998), and Steven Johnson, "Three Approaches to Big Technology: Operations Research, Systems Engineering, and Project Management," *Technology and Culture* 38, no. 4 (1997): 891–919. Hughes emphasizes the common elements involved in large-scale technical projects, such as SAGE and the "Big Dig" in Boston, describing these projects as efforts in the construction of complex social-technical systems. Johnson, on the other hand, emphasizes the distinctiveness of the different approaches to managing such large projects. As with all debates between "lumpers" and "splitters," both arguments have merit. Like Simon, however, I am at heart a lumper in search of broad patterns, not a splitter seeking that which makes every actor and every action unique, so Hughes's approach is more congenial to me.

PATRONAGE

An enormous amount has been written in the past two decades on the subject of patronage in science. This is all to the good so long as we remember that following the money can tell us some, but not all, things. The work that has influenced my understanding of patronage relationships most strongly is Mario Biagioli, *Galileo, Courtier* (Chicago: University of Chicago Press, 1993). Biagioli may slight the more purely intellectual roots of Galileo's novel views, but his exploration of the relationships involved in court patronage is fascinating. In particular, his emphasis on patronage as a two-way relationship and his observations on the role of brokers in such relationships are eye-opening.

As regards patronage and science in the postwar era, the works I have depended on most often are Ronald Doel, *Solar System Astronomy in America: Communities, Patronage, and Interdisciplinary Science, 1920–1960* (New York: Cambridge University Press, 1996); Paul Forman, "Behind Quantum Electronics: National Security as Basis for Physical Research in the United States, 1940–60," *Historical Studies in the Physical and Biological Sciences* 18, no. 1 (1987): 149–229; Peter Galison and Bruce Hevly, eds., *Big Science: The Growth of Large-Scale Research* (Stanford, CA: Stanford University Press, 1992); and Stuart W. Leslie, *The Cold War and American Science: The Military-Industrial-Academic Complex at MIT and Stanford* (New York: Columbia University Press, 1993).

All of these works show how military goals shaped research priorities in the cold war era, with Leslie's book going a step further to reveal the influence of military social values on academic researchers. Doel's book is particularly illuminating because of the way he links the patronage system of postwar astronomy and the use of certain powerful, expensive instruments (radio telescopes) to the construction of a consciously interdisciplinary field of research—a story that has many parallels to that of the systems sciences in the same period.

INSTRUMENTS, MODELS, AND METAPHORS IN SCIENCE

It may appear odd to group instruments together with models and metaphors, but all are "tools to think with," to adapt a phrase. Instruments, models, and metaphors are ways of representing the world and thereby structure the discourses, perceptions, and social interactions of research communities. Instruments do so in a very material fashion while models and metaphors are more ethereal, but the functions of all three are often quite similar, especially when one is dealing with very sophisticated, expensive pieces of equipment, be they mental or physical.

The essential sources for the study of instruments are: Lorraine Daston and Peter Galison, "The Image of Objectivity," *Representations* 40 (fall 1992): 81–128; Peter Galison, *Image and Logic: A Material Culture of Microphysics* (Chicago: University of Chicago Press, 1997); Thomas Hankins and Robert J. Silverman, *Instruments and the*

Imagination (Princeton, NJ: Princeton University Press, 1995); Steven Shapin and Simon Schaffer, *Leviathan and the Air Pump: Hobbes, Boyle, and the Experimental Life* (Princeton, NJ: Princeton University Press, 1985); Albert Van Helden and Thomas L. Hankins, eds., *Instruments*, *Osiris* 9 (1994); Mary Winkler and Albert Van Helden, "Representing the Heavens," *Isis* 83 (1992): 195–217; and M. Norton Wise, "Materialized Epistemology," *Studies in the History and Philosophy of Modern Physics* 30B (1999): 547–53.

On metaphors and models in general, see George Lakoff and Mark Johnson, *Metaphors We Live By* (Chicago: University of Chicago Press, 1980). For specific studies of metaphor in science, see Geoffrey Cantor, "Weighing Light: The Role of Metaphor in 18th Century Optical Discourse," in *Figural and Literal: Problems of Language in the History of Science and Philosophy, 1630–1800*, ed. Andrew Benjamin (Manchester, UK: Manchester University Press, 1986): 124–46; Hunter Crowther-Heyck, "George A. Miller, Language, and the Computer Metaphor of Mind," *History of Psychology* 2, no. 1 (1999): 37–64; Laura Otis, "The Metaphoric Circuit," *Journal of the History of Ideas* 63, no. 1 (2002): 105–28; Mary Morgan and Margaret Morrison, eds., *Models as Mediators: Perspectives on Natural and Social Science* (Cambridge: Cambridge University Press, 2000); Anson Rabinbach, *The Human Motor: Energy, Fatigue, and the Origins of Modernity* (Berkeley: University of California Press, 1990); and Crosbie Smith and M. Norton Wise, "Work and Waste: Political Economy and Natural Philosophy in Nineteenth-Century Britain," published in three parts: *History of Science* 27 (1989): 263–301, 391–449, and *History of Science* 28 (1990): 221–61.

THE BIG PICTURE

My perspective on the history of science has been shaped most strongly by the works above and by what has been called the "organizational synthesis" in modern American history. The organizational synthesis focuses on the development of large-scale organizations, both public and private, and on the rise of the professions, which was closely tied to the emergence of the large-scale organization. The organizational synthesis at times suffers from an overly structural-functionalist approach, and so sometimes downplays both contingency and human agency (not to mention race and gender), but it seems to me to be absolutely right in the importance it attaches to the rise of large-scale organizations and the attendant bureaucratization of life. Surprisingly, there have been few attempts to introduce this perspective into the study of American intellectual culture. I hope the present study does its part to show the virtues of placing certain intellectual developments against the background of the organizational revolution.

The works associated with the organizational synthesis that have shaped my thinking most strongly are Andrew Abbott, *The System of Professions: An Essay on the Division of Expert Labor* (Chicago: University of Chicago Press, 1988); Guy Alchon, *The Invisible Hand of Planning: Capitalism, Social Science, and the State in the 1920s* (Princeton, NJ: Princeton University Press, 1985); Brian Balogh, "Reorganizing the Organizational Synthesis: Federal-Professional Relations in Modern America," *Studies in American Political Development* 5 (spring 1991): 119–72; James Beniger, *The Control Revolution: Technological and Economic Origins of the Information Society* (Cambridge, MA: Harvard University Press, 1986); Alfred Chandler, *The Visible Hand: The*

Managerial Revolution in American Business (Cambridge, MA: Belknap Press of Harvard University Press, 1977); Louis Galambos, "Technology, Political Economy, and Professionalization: Central Themes of the Organizational Synthesis," *Business History Review* 57, no. 4 (1983): 471–93; Stephen Skowronek, *Building a New American State: The Expansion of National Administrative Capacities, 1877–1920* (Cambridge: Cambridge University Press, 1982); Paul Starr, *The Social Transformation of American Medicine* (New York: Basic Books, 1982); and Robert Wiebe, *The Search for Order, 1877–1920* (New York: Hill and Wang, 1967).

Of these works, Andrew Abbott's deserves special mention. Abbott argues that the possession of a body of abstract knowledge is the key to a profession's claim to authority—or jurisdiction—over some social function. Abbott's account is in the functionalist tradition of sociology and so shares many of its strengths and weaknesses. He has a marvelous sense of the interconnections among the professions and of the ways each profession is a subsystem of the larger system of professions, but the functions over which the various professions compete for jurisdiction often seem to exist *a priori*. Still, Abbott is much more aware of the contingencies of history than his functionalist forebears, and his accounts of the rise of psychiatry and the demise of operations research are fascinating.

Index